Fluid Mechanics

Volume 4

Take anything in the universe, put it in a box, and heat it up. Regardless of what you start with, the motion of the substance will be described by the equations of fluid mechanics. This remarkable universality is the reason why fluid mechanics is important.

The key equation of fluid mechanics is the Navier-Stokes equation. This textbook starts with the basics of fluid flows, building to the Navier-Stokes equation while explaining the physics behind the various terms and exploring the astonishingly rich landscape of solutions. The book then progresses to more advanced topics, including waves, fluid instabilities, and turbulence, before concluding by turning inwards and describing the atomic constituents of fluids. It introduces ideas of kinetic theory, including the Boltzmann equation, to explain why the collective motion of 10^{23} atoms is, under the right circumstances, always governed by the laws of fluid mechanics.

David Tong is a Professor of Theoretical Physics at the University of Cambridge and a Fellow of Trinity College. He is known for his contributions to quantum field theory and its application to diverse areas of physics, including particle physics, condensed matter, cosmology, and quantum gravity. His lecture notes on theoretical physics have gained a global following due to their clear explanations and easy-going, accessible style.

Lectures on Theoretical Physics

Fluid Mechanics

Lectures on Theoretical Physics, Volume 4

DAVID TONG

University of Cambridge

CAMBRIDGE
UNIVERSITY PRESS

Shaftesbury Road, Cambridge CB2 8EA, United Kingdom

One Liberty Plaza, 20th Floor, New York, NY 10006, USA

477 Williamstown Road, Port Melbourne, VIC 3207, Australia

314–321, 3rd Floor, Plot 3, Splendor Forum, Jasola District Centre, New Delhi – 110025, India

103 Penang Road, #05–06/07, Visioncrest Commercial, Singapore 238467

Cambridge University Press is part of Cambridge University Press & Assessment, a department of the University of Cambridge.

We share the University's mission to contribute to society through the pursuit of education, learning and research at the highest international levels of excellence.

www.cambridge.org
Information on this title: www.cambridge.org/highereducation/isbn/9781009594738

DOI: 10.1017/9781009594714

First published 2025

Printed in the United Kingdom by CPI Group Ltd, Croydon CR0 4YY

A catalogue record for this publication is available from the British Library

A Cataloging-in-Publication data record for this book is available from the Library of Congress

ISBN 978-1-009-59473-8 Hardback
ISBN 978-1-009-59469-1 Paperback

For my sister, Katie.

Contents

Preface

Take anything in the universe, throw a bunch of it in a box, and turn up the heat. Regardless of what you start with, the motion of this substance will be governed by the equations of fluid dynamics.

This is a remarkable statement. There are lots of different things in the universe and we go to great lengths to understand their properties. Yet if you heat them, nearly all their differences disappear. When things get hot, everything looks the same.

Here are some examples. Take any element in the periodic table and heat it until it melts, so that it is either a liquid or a gas. The motion of every element is then governed by the same set of equations. The only reminder of what you started with is to be found in a handful of parameters of these equations which describe, among other things, the density and viscosity of the fluid. These will differ from element to element. But the basic set of equations are the same, regardless of whether you're working with an alkaline earth metal or an inert gas.

This same story holds if we turn our attention to more exotic substances. For example, inside every proton and neutron sit three quarks. They have been trapped there since the Big Bang, held in place by the grip of the strong nuclear force. Earlier this century experimenters succeeded in colliding nuclei together with energies that were so high that the protons and neutrons themselves melted, freeing their imprisoned quarks and forming a novel state of matter known as the quark-gluon plasma. This plasma lasts for only a fraction of a second before it cools and once again forms protons and neutrons. But during that fraction of a second it moves. And the movement is described by the laws of fluid mechanics.

Here is an even more extreme example. Take spacetime itself. When densities get too large, spacetime collapses to form a black hole and, due to the work of Hawking, we know that these black holes are hot objects. Said differently, a black hole can be viewed as a way to heat spacetime. If you

1

look at the equations that govern the event horizon of a black hole, you will once again find the laws of fluid mechanics.

All of which is to say that there is a wonderful universality to the laws that govern fluids. In certain circumstances, these laws describe literally everything. And this makes them interesting.

Moreover, the equations of fluid dynamics are not restricted to academic pursuits like quarks and black holes. They have enormous practical application. They explain, for example, why planes fly. (As we will recount later in this book, one of the more embarrassing episodes in the history of theoretical physics occurred in 1903 when the Wright brothers took to the air before physicists were able to adequately explain either lift or drag!) Fluid mechanics explains how oil flows through pipes and how the motion of the atmosphere manifests itself in the climate, and how many decades of focussing on the former has resulted in an urgent and desperate need to better understand the latter.

The purpose of this book is to describe the basics of fluid mechanics, also referred to (a little archaically) as hydrodynamics. Our focus will not be on quarks and black holes, but nor will it be on any particular application of fluid mechanics. Instead our goal is simply to understand the different things that fluids can do. Fluids are everywhere and they have a tendency to move. We will construct the equations that govern this motion and explore their properties. Prominent among these is the Navier–Stokes equation, together with one or two further equations that describe the conservation of mass and, in some situations, the flow of heat.

One of the themes of fluid mechanics is that this simple set of equations gives rise to a wonderful diversity of different behaviour. As this book progresses, we will find ourselves falling into a routine. Like Monet and his haystacks, we will return to these same equations over and over again, not because we did anything wrong the first time but because there is always something new to see. Attacking the same set of equations, but with a slight change to the boundary conditions, or a novel approximation scheme, will often yield something new and surprising. One of the delights of the subject lies in finding such riches sitting inside such simplicity.

In the later chapters of this book, we change tack. Instead of looking more closely at the solutions to the equations of fluid mechanics, we instead ask: Why are these the equations that govern everything and anything, provided that it's suitably hot? Where does this universality come from? To answer this, we look inwards at the atomic constituents of fluids.

At the microscopic level, fluids are ridiculously complicated objects, consisting of, say, 10^{23} atoms, each following its own path while acting through various forces upon the atoms around it. But much of this motion is fleeting and we lose little if we ignore it. Instead, we care only about patterns in the collective motion of the atoms that survive over long time scales. It turns out that these long-lived modes are all related to familiar conservation laws – conservation of mass, momentum, and energy – and these conservation laws are universal and obeyed by all substances. This, ultimately, is why all fluids look the same: the equations of fluid dynamics are essentially the equations that govern how conserved quantities evolve in time. By focussing on these conservation laws, we will be able to start with the ridiculously complicated equations that govern all 10^{23} atoms and see how they reduce, in a certain limit, to the Navier–Stokes and other equations of fluid mechanics.

How to Read This Book

The first two chapters in this book contain an introduction to the basic equations of fluid mechanics. We start, in Chapter 1, by ignoring viscosity, which has the advantage that our equations are significantly simpler and the disadvantage that we've thrown away a whole lot of interesting physics. This is remedied in Chapter 2 where we introduce viscosity and meet the Navier–Stokes equation for the first time.

The next three chapters are concerned with waves, instabilities, and turbulence, respectively. It makes sense to learn about the basics of waves before looking at instabilities, but otherwise these sections can be read more or less independently. The chapter on turbulence contains less in the way of solutions for the obvious reason that none are known, and focusses more on the statistical formalism to describe the turbulent nature of fluids.

Chapters 6 and 7 cover kinetic theory and diffusion. They are something of a departure from the main narrative of the book, turning inwards to look at the microscopic constituents of the fluid. In particular, in the chapter on kinetic theory we will show how to derive the Navier–Stokes equation from first principles. Thus, from a logical perspective, Chapter 6 should come before Chapters 1 and 2 and you should certainly feel free to read them in this order. From a pedagogical perspective, my guess is that the traditional ordering offered by the natural numbers will prove more illuminating.

Either way, you should be comfortable with the basics of Newtonian mechanics before embarking on this book. The Navier–Stokes equation is just $F = ma$, admittedly with a complicated F and a slightly subtle a. The relevant prerequisites can be found in the first six chapters of Volume 1 in this series on Classical Mechanics. In addition, it will be useful to have some familiarity with Volume 2 on Electromagnetism, although more for the mathematical methods used as much as the underlying physics since both subjects deal with fields, rather than particles. Finally, the derivation of the Boltzmann equation in Chapter 6 will use the Hamiltonian formulation of classical mechanics.

Just One Book Among Many

This is volume 4 in a series of $N \gg 1$ books covering theoretical physics. Most of these books have intertwined narratives and complicated codependencies. This book, however, is something of an exception and stands on its own more than the others. In particular, the laws of fluid mechanics will not play much of a role in later volumes.

This is by no means a reflection on the importance of fluid mechanics, only a statement about my own prejudice, which means that this series of books is very much focussed on delving deeper into the fundamental laws of physics rather than doing anything useful with them. If you choose not to study the equations of fluid mechanics then it won't be much of a hindrance in learning, say, quantum field theory. It's just that you will be a less-rounded human being.

The contents of this book will have greatest overlap with the forthcoming book on Statistical Physics. This should be no surprise. The purpose of statistical mechanics is to describe physical systems when things get hot and, as we described above, the motion of hot things is governed by fluid mechanics. There are a number of places in the present book, especially in Chapters 6 and 7, where the ideas resonate with results that we will see in the Statistical Physics book.

That said, it's perfectly possible to study much of fluid mechanics without worrying about statistical physics, and vice versa. It's worth explaining why, because this will serve to highlight the logical connection between these topics.

If you isolate a hot system, and leave it alone for long enough, then it will relax down to a state of *equilibrium*. Equilibrium is defined to be a state in which, at least on the coarse-grained level, things don't change. That's not to say that nothing is moving: if you have a hot system and you look closely enough, then everything is flying around on the atomic level. But if you focus only on macroscopic variables, like temperature, pressure, and density, then, in equilibrium, all is calm. The macroscopic, equilibrium properties of a system are described by the subject of *thermodynamics*.

Fluid mechanics and statistical physics are two different ways to generalise the ideas of thermodynamics. Fluid mechanics is what you get if you take thermodynamics and splash it. The macroscopic variables now become local variables, meaning that they can vary in both time and space. Meanwhile, the subject of statistical mechanics is concerned with how we derive the equilibrium properties from first principles, by thinking about the underlying atomic constituents.

Both of these are interesting, but logically separate, questions that you can ask starting from the idea of equilibrium. The link between them is to be found in the subject of kinetic theory, described in Chapter 6 of this volume. This takes the microscopic perspective and looks at how systems approach equilibrium, and what "splashing" means from the atomic viewpoint.

This summarises the main relations between this book and others. However, no subject in physics is completely isolated, and there will be a number of minor, but entertaining, connections that we meet along the way, from the role of topology in quantum mechanics and fluids, to the mystery of singularities in gravity and fluids.

Problems

I have not included any exercises in this series of books. However, the Faculty of Mathematics at the University of Cambridge has long had a policy of making all problems (and, indeed, exam questions) publicly available. Problem sheets, aligned with material covered in the first half of this book, can be downloaded from:

www.damtp.cam.ac.uk/user/tong/books/fluids.html

I will also include an errata on this page.

Conventions

Pick up any research paper or textbook on fluid mechanics, and you will find pressure denoted as a little p. Not so here. Throughout this series of books, pressure is denoted as a capital P.

You might think this is relatively unimportant. But conventions matter and the choice of P rings alarm bells for any expert in fluid mechanics and screams that I'm not part of their club. It's not quite as bad as writing papers in Microsoft Word rather than LaTeX but it's not far off.

My reasons for failing to conform are twofold. First, I am apparently alone in the world in thinking that little p looks way too much like the density ρ for them to happily cohabit in the same equation. Second, and more importantly, in all other areas of physics, p is obviously momentum. That doesn't cause too many problems in much of fluid mechanics where particles rarely get a mention, but it becomes an issue when we talk about kinetic theory in Chapter 6 and also when we discuss Statistical Physics in a later volume. So you will just have to suck up capital P.

Acknowledgements

The contents of the book largely follow a series of lecture courses that are taught in the mathematics degree at the University of Cambridge. These courses have been honed over many years by many of my colleagues, among them Natasha Berloff, Colm-Cille Caulfield, Stephen Cowley, Stuart Dalziel, Julia Gog, Ray Goldstein, Peter Haynes, John Hinch, Rich Kerswell, Eric Lauga, Paul Linden, John Lister, Michael McIntyre, Keith Moffat, Jerome Neufeld, Nigel Peake, Tim Pedley, Mike Proctor, John Taylor, and Grae Worster, In addition to the textbooks listed in the bibliography, I found the publicly available lecture notes of Stephen Cowley, Julia Gog, Peter Haynes, John Hinch, Michael McIntyre, and John Taylor extremely useful. Chapters 6 and 7 grew out of a graduate course that I taught at Cambridge.

My thanks to Matt Davison, Daniele Dorigoni, Sean Hartnoll, Andreas Karch, John Lister, John McGreevy, and Jorge Santos for helpful discussions on some of the topics contained here, and to Mihalis Dafermos for an enjoyable summary of the current status of the singularities in classical physics. I'm grateful to Rubaiyat Khondaker, Eric Lauga, and Matt Smith for giving feedback and for their superhuman typo-spotting abilities. Special thanks also go to Colm-Cille Caulfield for an extraordinarily careful reading of the manuscript, and for taking the time to educate me on many subtleties of fluid flow.

The front cover depicts an example of the Rayleigh–Taylor instability (described in Chapter 4 of this book). The beautiful photograph was taken by Megan Davies Wykes and appeared in a paper she wrote with Stuart Dalziel, "Efficient mixing in stratified flows: experimental study of Rayleigh–Taylor unstable interface within an otherwise stable stratification", published in the Journal of Fluid Mechanics, 756 (2014). The blurry photo on the back cover shows the great fluid dynamicist G.I. Taylor testing his own design for a parachute. The photograph is part of the G.I. Taylor archive housed in the Wren Library, Trinity College Cambridge.

Finally, to Alex. Thank you for the many years of love and support. I couldn't have done this without you.

1 Inviscid Flows

In this chapter, we lay down the basic principles and equations of fluid mechanics and start to explore some properties of fluid flows.

Unusually, and a little defensively, the title of this chapter highlights what we won't talk about, rather than what we will. Fluids have a property known as *viscosity*. This is an internal friction force acting within the fluid as different layers rub together. It is crucially important in many applications.

In spite of its importance, we will start our journey into the world of fluids by ignoring viscosity altogether. Such flows are called *inviscid*. This will allow us to build intuition for the equations of fluid mechanics without the complications that viscosity brings. Moreover, the flows that we find in this chapter will not be wasted work. As we will see later in Chapter 2, they give a good approximation to viscous flows in certain regimes where the more general equations reduce to those studied here.

1.1 The Basics

When we were kids, we are told that there are three phases of matter: solid, liquid, and gas. As we grow older, we learn that this is a hopelessly naive view of the world. Nonetheless, it is the one that we will adopt for the purposes of this book which is concerned only with the latter two. Liquids and gases are both examples of fluids. Roughly speaking, a fluid is a substance that flows when pushed. More rigorously, fluids are objects that are well described by the equations in this book.

The subject of fluid mechanics starts with a lie. (Applied mathematicians prefer the term "approximation".) The lie, sometimes dubbed the *continuum hypothesis*, is that fluids are indivisible continuous objects. The fluid can be then described by two continuous fields,

- the density $\rho(\mathbf{x}, t)$,
- the velocity $\mathbf{u}(\mathbf{x}, t)$.

Of course, we know that in reality fluids are made of molecules and the continuum hypothesis must break down on atomic scales. But we also know from experience that if we look on suitably large scales, where we are coarse graining over many many molecules, then the continuum description is remarkably good.

It is appropriate to start this book by stressing that we are dealing with an approximation. It will not be our last. The study of fluids is all about the art of approximation. The equations of fluid mechanics, simple as they are, cannot be solved in full generality and we will make progress only by simplifying. The skill is in learning what to keep and what to ignore. And we start by ignoring the existence of atoms. We will ultimately make up for this deficiency in Chapter 6, where we revisit the basic equations of fluid mechanics from an atomic perspective.

It's not just the discreteness of matter that is swept under the rug in the continuum description. We also ignore the vast majority of the motion of the constituent atoms and molecules that make up the fluid. At room temperature, these constituents are flying around at speeds of $100 \mathrm{~m\,s}^{-1}$ or so. (This is certainly true of gases. For liquids, the molecules are more closely bound to their neighbours and we have to think more carefully about what the velocity of a single molecule really means.) But most of this underlying atomic motion is neglected in our coarse-grained description. Instead, the velocity field $\mathbf{u}(\mathbf{x}, t)$ describes the average, macroscopic motion of the fluid. In particular, there is a state of the fluid in which $\mathbf{u}(\mathbf{x}, t) = 0$ and we pretend that the fluid is completely still, even though the underlying particles are still flying around, just with no direction preferred over any other.

(As an aside: the internal motion of the constituents doesn't show up in the velocity field $\mathbf{u}(\mathbf{x}, t)$, but it does manifest itself in the temperature of the fluid which is another field $T(\mathbf{x}, t)$. We will pay more attention to the role of temperature as the book progresses, but for now it is something that can be put on the back burner.)

It is also worth elaborating on how to think about the position $\mathbf{x}$ that appears in the argument of the fields $\rho(\mathbf{x}, t)$ and $\mathbf{u}(\mathbf{x}, t)$. This is some fixed position in space. This means, in particular, that $\mathbf{u}(\mathbf{x}, t)$ is the velocity that would be measured by some fixed array of sensors embedded in the fluid, as opposed to sensors that drift along with the fluid. The use of fields $\rho(\mathbf{x}, t)$ and $\mathbf{u}(\mathbf{x}, t)$ is called the *Eulerian* description.

There is a different viewpoint, in which we think of individual "parcels of fluid", each initially sitting at some position $\mathbf{x}$ and then following the flow by travelling at speed $\mathbf{u}(\mathbf{x}, t)$. It's not so easy to define what we mean by these "parcels of fluid" given that the underlying atoms are, as we described above, wandering off in all sorts of directions, often at high speed, with only the most scant regard for the velocity field $\mathbf{u}(\mathbf{x}, t)$. But the concept of a fluid parcel that keeps its identity as the fluid moves is an extremely useful pretence. We will sometimes talk about a "particle" of fluid and we have in mind these parcels rather than the underlying atoms. The perspective in which we follow the trajectories of these parcels, and study the forces that act on them as if they were particles in classical mechanics, is called the *Lagrangian* description.

Throughout this book, all our equations will be written in the Eulerian description, using the velocity field $\mathbf{u}(\mathbf{x}, t)$, but some intuition will come from a more Lagrangian way of thinking. Moreover, we will certainly have a need to understand the trajectories of particles that are embedded within the fluid. To this end, we kick off with some simple observations.

1.1.1 Pathlines and Streamlines

There are a number of ways to visualise the flow $\mathbf{u}(\mathbf{x}, t)$ of a fluid. Here are the two most useful:

- A *pathline* is the trajectory followed by a particle embedded within the fluid.

- A *streamline* is a tangent to $\mathbf{u}(\mathbf{x}, t)$ at every point $\mathbf{x}$ for fixed time t. In general, the tangents to a vector field $\mathbf{F}(\mathbf{x})$ are said to be integral curves for $\mathbf{F}$. So the streamlines are integral curves for the velocity field at a fixed time.

If the flow is *steady*, meaning that $\partial \mathbf{u}/\partial t = 0$, then the pathlines and streamlines coincide. But, for time-dependent flows, they differ. To see this, let's drape some equations around the definitions above.

First consider the pathline. A particle within the fluid will follow some trajectory $\mathbf{x}(t)$. At any time t, the velocity of this particle is given by the velocity field $\mathbf{u}$ evaluated at the position of the particle, meaning

$$\frac{d\mathbf{x}}{dt}(t) = \mathbf{u}(\mathbf{x}(t), t) . \tag{1.1}$$

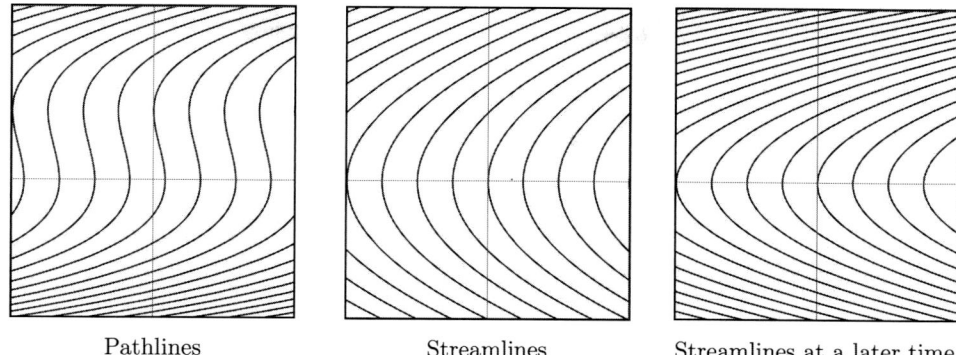

| Pathlines | Streamlines | Streamlines at a later time |

Fig. 1.1 The pathlines for particles in the flow $\mathbf{u} = (yt, 1)$ are shown on the left. These are a history of the flow. The middle and right-hand figures show streamlines, with the right-hand figure at a later time.

Given some initial starting point $\mathbf{x}(t = 0) = \mathbf{x}_0$, we can solve this equation to find the pathline.

In contrast, a streamline is a trajectory $\mathbf{x}(s)$ such that the tangents of $\mathbf{x}(s)$ coincide with the velocity field at a *fixed* time t,

$$\frac{d\mathbf{x}}{ds}(s) = \mathbf{u}(\mathbf{x}(s), t) \ . \tag{1.2}$$

In words, the streamline is a snapshot of the flow at some fixed time, while the pathline tells us about the actual history of the particle.

An Example

Consider the two-dimensional flow given by

$$\mathbf{u}(\mathbf{x}, t) = \begin{pmatrix} \alpha yt \\ \beta \end{pmatrix} \tag{1.3}$$

for some fixed coefficients α and β. The pathline obeys

$$\frac{d\mathbf{x}}{dt} = \begin{pmatrix} \dot{x} \\ \dot{y} \end{pmatrix} = \begin{pmatrix} \alpha yt \\ \beta \end{pmatrix} \ . \tag{1.4}$$

The y-component is solved by $y = y_0 + \beta t$, while the equation for the x-component becomes $\dot{x} = \alpha yt = \alpha(y_0 t + \beta t^2)$, which gives $x = x_0 + \frac{1}{2}\alpha y_0 t^2 + \frac{1}{3}\alpha\beta t^3$. To get the pathline, we eliminate t to get the family of curves in the (x, y)-plane

$$x = x_0 + \frac{\alpha}{2\beta^2} y_0 (y - y_0)^2 + \frac{\alpha}{3\beta^2} (y - y_0)^3 \ . \tag{1.5}$$

These are shown in the left-hand plot of Figure 1.1 for various values of the starting point (x_0, y_0).

In contrast, to find the streamlines we instead solve

$$\frac{d\mathbf{x}}{ds} = \begin{pmatrix} x' \\ y' \end{pmatrix} = \begin{pmatrix} \alpha y t \\ \beta \end{pmatrix} \tag{1.6}$$

where the prime means d/ds. These now have the solutions $y = y_0 + \beta s$ and $x = x_0 + \alpha y_0 t s + \frac{1}{2}\alpha\beta t s^2$, where t is now some fixed parameter. These are shown in the middle and right-hand plots of Figure 1.1 for $t > 0$. Note that the pathlines and streamlines are not similar in this example: the former is a cubic curve, the latter a parabola. (Or, in the special case of $t = 0$, straight lines.) Moreover, the streamlines are time-dependent: the right-hand figure is a snapshot of the flow at a later time than the middle figure.

1.1.2 The Material Time Derivative

As we stressed above, the density $\rho(\mathbf{x}, t)$ and velocity field $\mathbf{u}(\mathbf{x}, t)$ are measured in the Eulerian sense at some fixed point $\mathbf{x}$. But this leaves us with the question: How do we see things change in time if we're drifting along with the fluid?

Specifically, suppose that there is some field $\phi(\mathbf{x}, t)$ that we would like to measure. This might be the density of the fluid itself, or something else. The explicit time dependence in $\phi(\mathbf{x}, t)$ tells us how this quantity changes with time if we're sitting at some fixed position $\mathbf{x}$. But if we're drifting with the fluid, then we follow a pathline $\mathbf{x}(t)$ defined by (1.1). The value of the field along this trajectory is given by $\phi(\mathbf{x}(t), t)$ and the total time derivative is

$$\frac{d}{dt}\phi(\mathbf{x}(t), t) = \frac{\partial\phi}{\partial t} + \dot{\mathbf{x}} \cdot \nabla\phi = \frac{\partial\phi}{\partial t} + \mathbf{u} \cdot \nabla\phi \ . \tag{1.7}$$

The additional $\mathbf{u} \cdot \nabla\phi$ term captures the change in ϕ because of the way we're swept along by the fluid. The transport of some object as it's carried along by a fluid is known as *advection* and, correspondingly, $\mathbf{u} \cdot \nabla\phi$ is called the *advective* rate of change. This idea of a total time derivative will be important, so much so that we introduce some new notation for it (even though we already have perfectly good notation in $d\phi/dt$!). We write

$$\frac{D\phi}{Dt} = \frac{\partial\phi}{\partial t} + \mathbf{u} \cdot \nabla\phi \tag{1.8}$$

and call this the *material derivative*. It can be thought of as a bridge between the Eulerian description in terms of a fixed point $\mathbf{x}$ and the Lagrangian description which moves with the fluid.

1.1.3 Conservation of Mass

Our first equation of fluid mechanics is the simplest: it captures the fact that mass is conserved. Moreover, like all conservation laws in physics, mass is conserved *locally*. This means that if the mass of the fluid decreases at some point in space then it must have moved to a neighbouring point.

This fact is captured by the conservation equation, relating the density ρ and the velocity $\mathbf{u}$

$$\frac{\partial \rho}{\partial t} + \nabla \cdot (\rho \mathbf{u}) = 0 \ . \tag{1.9}$$

Equations of this kind are commonplace in physics because they appear whenever we have a conservation law. An identical equation appeared in the book on Electromagnetism where, in that context, ρ is the electric charge density and $\mathbf{J} = \rho \mathbf{u}$ is the electric current density. For us, ρ is the mass density and $\rho \mathbf{u}$ is the *mass flux density*.

To see why (1.9) captures the conservation of mass, consider the mass M of fluid in some fixed region V

$$M = \int_V \rho \ dV \ . \tag{1.10}$$

The change of this mass is given by

$$\frac{dM}{dt} = \int_V \frac{\partial \rho}{\partial t} \ dV = -\int_V \nabla \cdot (\rho \mathbf{u}) \ dV = -\int_S \rho \mathbf{u} \cdot d\mathbf{S} \tag{1.11}$$

where we have used the divergence theorem to convert the volume integral into a surface integral over $S = \partial V$, the boundary of the region V. This tells us that if there is no net flow of mass flux through the boundary S then the total mass M inside the region V remains constant. In other words, mass is conserved.

We can also write the mass conservation equation (1.9) using our new material derivative notation. It becomes

$$\frac{D\rho}{Dt} + \rho \nabla \cdot \mathbf{u} = 0 \ . \tag{1.12}$$

Incompressible Flows

Throughout much of this book we will make one further approximation: we will assume that fluids are *incompressible* which, in the simplest case, means that $\rho(\mathbf{x}, t)$ is a constant. Taking $\dot{\rho} = \nabla \rho = 0$, the continuity equation (1.9) becomes simply

$$\nabla \cdot \mathbf{u} = 0 \ . \tag{1.13}$$

In the language of vector calculus, we say that the fluid flow is solenoidal or divergence-free. Much of this book will be devoted to finding the wonderfully diverse solutions to equation (1.13).

In fact, the requirement that $\dot{\rho} = \nabla \rho = 0$ can be loosened. We see from (1.12) that we only really require $D\rho/Dt = 0$ for the incompressible condition (1.13) to be enforced. This means that any individual parcel of fluid should not change its density as it's swept along, but different parts of the larger fluid may have different densities. Such a situation is said to be *stratified* and arises, for example, in the ocean, where the water is more dense at the bottom than the top yet, to a good approximation, the flow can be viewed as incompressible in the sense that $D\rho/Dt = 0$. We'll meet stratified flows in Chapter 3 when we discuss some aspects of waves.

The assumption that fluid flow is incompressible is not totally innocent. In fact, the phenomenon of fluids compressing and expanding as their density changes is so common that we give it a special name. This name is "sound"! It turns out that the assumption of incompressibility is good when the speed of the fluid $|\mathbf{u}|$ is much less than the speed of sound. For air at atmospheric pressure, the speed of sound is $340 \ \mathrm{m\,s}^{-1}$; for water at room (or ocean) temperature it is around $1500 \ \mathrm{m\,s}^{-1}$.

For much of this book, we will restrict ourselves to flows that are much slower than the sound speed and assume that $\nabla \cdot \mathbf{u} = 0$. But, in Section 3.4, we will discuss the propagation of sound waves and then we will be forced to look more closely at the equations that govern compressible fluids.

1.1.4 The Stream Function

For incompressible flows, satisfying $\nabla \cdot \mathbf{u} = 0$, we can write the velocity field as

$$\mathbf{u} = \nabla \times \mathbf{A} \ . \tag{1.14}$$

For many fluid flows, this isn't particularly helpful since we have just swapped one vector field $\mathbf{u}$ for another $\mathbf{A}$. However, when the flow is two-dimensional (in some sense) this provides a useful simplification because it means that we get to exchange the vector field $\mathbf{u}$ for a scalar field ψ called the *stream function*.

For example, suppose that the flow is independent of the z-direction, so that the velocity field takes the form

$$\mathbf{u} = (u_1(x, y, t), u_2(x, y, t), 0) . \tag{1.15}$$

Then the vector potential $\mathbf{A}$ can be written as

$$\mathbf{A} = (0, 0, \psi(x, y, t)) \quad \Longrightarrow \quad u_1 = \frac{\partial \psi}{\partial y} \quad \text{and} \quad u_2 = -\frac{\partial \psi}{\partial x} \tag{1.16}$$

and the degrees of freedom are captured by the stream function $\psi(x, y, t)$. It has the nice property that lines of constant ψ are streamlines of the flow. To see this, note that lines of constant ψ have a normal $\mathbf{n}$ given by

$$\mathbf{n} = \nabla \psi = \left(\frac{\partial \psi}{\partial x}, \frac{\partial \psi}{\partial y}, 0 \right) \tag{1.17}$$

and so $\mathbf{u} \cdot \mathbf{n} = 0$. This is telling us that vectors that are normal to lines of constant ψ are also normal to streamlines. But in 2d, if you're normal to two different curves, then those curves must be the same. (In 2d, the enemy of an enemy is necessarily a friend.) So lines of constant ψ are streamlines.

We can also use a stream function in cylindrical polar coordinates, with $\mathbf{A} = (0, 0, \psi(r, \theta, t))$. In this case, the resulting flow is

$$\mathbf{u} = \frac{1}{r} \frac{\partial \psi}{\partial \theta} \hat{\mathbf{r}} - \frac{\partial \psi}{\partial r} \hat{\boldsymbol{\theta}} . \tag{1.18}$$

We'll make good use of the stream function in a number of places throughout this book, starting when we discuss potential flows in Section 1.4.

There is another, closely related object that is adapted to axisymmetric flows. In cylindrical polar coordinates, we consider vector potentials of the form $\mathbf{A} = \frac{1}{r}\Psi(r, z, t)\hat{\mathbf{e}}_\phi$ with Ψ the *Stokes stream function*. It gets a capital Ψ to reflect the fact that it has a different dimension from the usual stream function ψ. We won't have call to use Ψ in this book.

1.2 The Euler Equation

We have already met the mass conservation equation (1.9)

$$\frac{\partial \rho}{\partial t} + \nabla \cdot (\rho \mathbf{u}) = 0 . \tag{1.19}$$

We will assume that the flow is incompressible, so that this becomes

$$\nabla \cdot \mathbf{u} = 0 . \tag{1.20}$$

But we need one more equation to describe the motion of fluids. This second equation comes from what fluid dynamicists sometimes call "momentum balance". It is what everyone else calls "$F = ma$".

Consider some fixed region in space that we call V. The momentum in this volume is $\int_V \rho \mathbf{u} \, dV$ and Newton's second law tells us that the rate of change of momentum is equal to the force. The novelty here is that the momentum inside V might change simply because the fluid leaves or enters the region V. To write down the equation of motion, we need to take this into account.

Claim: The momentum flux across the boundary in some time δt is

$$\text{momentum flux} = \delta t \times \int_S (\rho \mathbf{u}) \, \mathbf{u} \cdot d\mathbf{S} \tag{1.21}$$

with $S = \partial V$ the boundary of the volume V.

Proof: Consider some small area of fluid δS lying on the surface S and watch it evolve. In some small time interval δt it sweeps out a volume $\delta(\text{Vol}) = \mathbf{u} \cdot \mathbf{n} \, \delta t \, \delta S$ where $\mathbf{n}$ is the normal to the surface as shown in the figure. As usual we write the vector area element

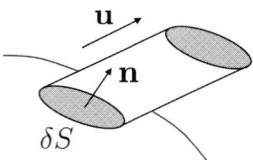

$\delta\mathbf{S} = \mathbf{n} \, \delta S$, so we have $\delta(\text{Vol}) = \mathbf{u} \cdot \delta\mathbf{S} \, \delta t$. This means that the momentum departing through the surface is $\rho \mathbf{u} \, \delta(\text{Vol})$ which gives the claimed result. $\square$

Including this extra term for the momentum that leaks through the sides, the "$F = ma$" equation of motion for the fluid is

$$\frac{d}{dt} \int_V \rho \mathbf{u} \, dV = - \int_S (\rho \mathbf{u}) \, \mathbf{u} \cdot d\mathbf{S} + \text{force} . \tag{1.22}$$

The additional term from the leaking momentum flux (1.21) sits on the right-hand side with the minus sign there, as always, to signify loss.

We'll get to the force shortly. Before we do, we can use the divergence theorem to convert the surface integral over momentum flux into a volume integral. Taking this term over to the other side, and resorting to index notation for the vectors $\mathbf{u}$, we have

$$\int_V \rho \frac{\partial u_i}{\partial t} \, dV + \int_V \rho \frac{\partial}{\partial x^j} (u_i u_j) \, dV = \text{force} . \tag{1.23}$$

Here we've used the fact that we're working with an incompressible flow, both in treating ρ as constant (or, more generally, satisfying $D\rho/Dt = 0$)

and also in the derivative in the second term which reads $\partial_j(u_i u_j) = u_i \partial_j u_j + u_j \partial_j u_i$. But the first of these vanishes because incompressible flows obey $\nabla \cdot \mathbf{u} = 0$. We're left with

$$\int_V \rho \left(\frac{\partial \mathbf{u}}{\partial t} + (\mathbf{u} \cdot \nabla)\mathbf{u} \right) dV = \int_V \rho \frac{D\mathbf{u}}{Dt} \, dV = \text{force} . \qquad (1.24)$$

In other words, the "ma" part of our equation involves the material derivative of the velocity. In hindsight, this is not surprising. The material derivative is the rate of change when you follow a parcel of fluid through the flow. This is the appropriate meaning of "rate of change" in Newton's second law.

1.2.1 Under Pressure

Next we come to the question of the forces that the fluid experiences. As we've already mentioned, we'll postpone any discussion of friction forces to Chapter 2. The fluid may be exposed to some external force, with gravity the most obvious, and we'll come to these shortly. But the most important force comes from within: this is *pressure*.

From a microscopic perspective, the pressure in a fluid comes from the motion of the underlying atoms or molecules. We'll explore this more in Chapter 6. For now, we shy away from the fundamentals and focus on the macroscopic. Here pressure manifests it- 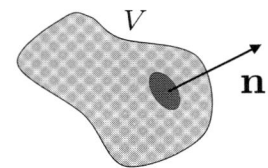 self as a force acting on the surface of any fluid element.

The *pressure* is defined as a normal force per unit area. Consider a small parcel of fluid, contained within a fixed volume V. The pressure $P(\mathbf{x}, t)$ acts on the surface $S = \partial V$ of this volume. It's an isotropic force, meaning that it is the same in all directions. The force exerted by the fluid outside V on some small region δS on the surface is

$$\mathbf{F}_{\text{pressure}} = -P\mathbf{n}\,\delta S \qquad (1.25)$$

where $\mathbf{n}$ is the outward-pointing normal as shown in the figure. The pressure $P(\mathbf{x}, t)$ should be viewed as a dynamical field that must be solved, subject to certain boundary conditions, at the same time as the velocity field $\mathbf{u}(\mathbf{x}, t)$.

Including the pressure, the equation of motion for the fluid (1.24) becomes

$$\int_V \rho \frac{D\mathbf{u}}{Dt} \, dV = -\int_S P \, d\mathbf{S} + \text{other forces} . \qquad (1.26)$$

We will assume that these other forces act on the volume of the fluid rather than the surface (this is true for external forces like gravity) and so take the integral form

$$\text{other forces} = \int_V \mathbf{f} \, dV \ . \tag{1.27}$$

Here $\mathbf{f}$ is the force per unit volume, or the *force density*.

The pressure acts on the surface of the volume V but we can massage it into a volume-type force through use of the divergence theorem. This gives

$$\int_V \rho \frac{D\mathbf{u}}{Dt} \, dV = \int_V (-\nabla P + \mathbf{f}) \, dV \ . \tag{1.28}$$

The final step is to recall that this whole derivation holds for an arbitrary volume V within the fluid. Since it holds for all such V, the two integrands must be equal everywhere. So we're left with the differential equation of motion for the fluid

$$\rho \frac{D\mathbf{u}}{Dt} = -\nabla P + \mathbf{f} \ . \tag{1.29}$$

This is the *Euler equation*. Finding solutions to this simple equation will occupy us for the rest of this section although we will, ultimately, replace it in Chapter 2 by the Navier–Stokes equation which includes the effects of viscosity. Fluids that obey the Euler equation are said to be *ideal*.

Importantly, the Euler equation is non-linear in the velocity field, although this is somewhat hidden in the notation above since the non-linearity sits in the material derivative: $D\mathbf{u}/Dt = \partial \mathbf{u}/\partial t + (\mathbf{u} \cdot \nabla)\mathbf{u}$. The non-linearity is such that the Euler equation is invariant under Galilean transformations

$$\mathbf{x} \to \mathbf{x} - \mathbf{v}t \quad \text{and} \quad \mathbf{u}(\mathbf{x}, t) \to \mathbf{u}(\mathbf{x} - \mathbf{v}t, t) + \mathbf{v} \tag{1.30}$$

for any constant $\mathbf{v}$. Neither the $\partial \mathbf{u}/\partial t$ term nor the non-linear $\mathbf{u} \cdot \mathbf{u}$ term are individually invariant under this transformation, but they cancel between themselves to ensure that the material derivative $D\mathbf{u}/Dt$ is invariant.

Note that a constant pressure P throughout the fluid does nothing. This is because the pressure is isotropic: if one piece of fluid pushes on a neighbour, then the neighbour pushes back with equal force. Interesting dynamics only arises when we have pressure differences across the fluid, as captured by ∇P.

The equations of fluid mechanics have four dynamical variables: three velocities $\mathbf{u}(\mathbf{x}, t)$ and the pressure $P(\mathbf{x}, t)$. And, happily, we have four equations governing these variables: the three components of the Euler equation (1.29), together with the incompressibility condition $\nabla \cdot \mathbf{u} = 0$.

However, the pressure $P(\mathbf{x}, t)$ is a slightly strange dynamical variable. We don't have any equation of the form $dP/dt = $ something. Instead, you should think of the pressure $P(\mathbf{x}, t)$ as more akin to a Lagrange multiplier whose role is to adjust to ensure that it's always possible to obey the incompressibility constraint $\nabla \cdot \mathbf{u} = 0$. (We described Lagrange multipliers in the book on Classical Mechanics.) Indeed, if we take the divergence of the Euler equation with $\mathbf{f} = 0$ then, invoking incompressibility, we see that the pressure obeys the Poisson equation

$$\nabla^2 P = -\rho \frac{\partial u_j}{\partial x^i} \frac{\partial u_i}{\partial x^j} . \qquad (1.31)$$

This tells us that the pressure doesn't really enjoy a life of its own. Instead, its dynamics is dictated by what the velocity field is doing. Moreover, the incompressibility condition means that there's no time delay needed for information in the pressure field to propagate from one part of the fluid to another. This is in sharp contrast to what happens in, for example, relativistic field theories like electromagnetism.

Looking Forwards: the Equation of State

If you know one thing about gases, then it will be the ideal gas law. This relates the pressure P, volume V, and temperature T of a gas by

$$PV = Nk_BT \qquad (1.32)$$

where N is the number of molecules in the gas and k_B is a universal constant of nature called Boltzmann's constant that converts energy to temperature. (For what it's worth, $k_B \approx 1.4 \times 10^{-23}$ J K^{-1}.) For our purposes, it's more useful to think of the ideal gas law in terms of the density $\rho = Nm/V$ rather than volume, where m is the mass of the constituent molecule. We then have

$$P = \frac{k_B \rho T}{m} . \qquad (1.33)$$

The ideal gas law is an example of an *equation of state*. It holds for strictly non-interacting gases. If we take into account interactions, either in gases or in liquids, it will be replaced by some other equation of state that again relates pressure P, density ρ, and temperature T. We will derive the ideal gas law in Chapter 6 when we think about fluids from the perspective of the underlying atoms. A much fuller discussion can be found in the forthcoming book on Statistical Physics.

When we first meet the ideal gas law, we think of P, ρ, and T as numbers that characterise the whole system. But it also holds if they are promoted to the kind of local fields $P(\mathbf{x}, t)$, $\rho(\mathbf{x}, t)$, and $T(\mathbf{x}, t)$ that we work with in this book.

For any flow that has ρ constant, the equation of state tells us that the temperature $T(\mathbf{x}, t)$ simply tracks the pressure $P(\mathbf{x}, t)$. From a microscopic perspective, that makes sense: both the temperature and the pressure arise from the motion of the underlying molecules, and as these molecules speed up both the pressure and temperature increase. That's the reason why we don't need to consider $T(\mathbf{x}, t)$ separately when we have fluid flows at constant density.

Things are more interesting if we have compressible fluids in which $\rho(\mathbf{x}, t)$ is another dynamical variable. In this case the mass conservation equation (1.9) and the Euler equation aren't enough information to tell us what happens and we need another equation. It turns out that in this situation the right way forward is to use the equation of state to replace $\rho(\mathbf{x}, t)$ with the temperature field $T(\mathbf{x}, t)$ and then to write down a separate equation for how heat flows in the system. (Roughly speaking, it is a version of the heat equation, with the material derivative replacing the usual time derivative.) We'll explain this further in Section 3.4 when we discuss sound waves and we will be forced to think more carefully about the thermodynamics of fluids. We will then present a more complete discussion in Chapter 6 when we discuss kinetic theory.

There are also many more subtle situations that arise in fluid mechanics that are a sort of halfway-house between the case with constant ρ and the case where $\rho(\mathbf{x}, t)$ is a fully fledged dynamical field. These are the stratified flows that we mentioned previously where ρ varies in space, but the flow is still solenoidal, meaning $\nabla \cdot \mathbf{u} = 0$. We will touch upon some properties of stratified flows later in the book.

1.2.2 The Euler Equation Is Just Momentum Conservation

Suppose that there is no external force on our fluid, so $\mathbf{f} = 0$. Then the Euler equation can be written in the characteristic form of a conservation law

$$\rho \frac{\partial \mathbf{u}}{\partial t} + \rho (\mathbf{u} \cdot \nabla)\mathbf{u} + \nabla P = 0$$

$$\implies \quad \frac{\partial (\rho u_i)}{\partial t} + \frac{\partial}{\partial x^j}\left(\rho u_i u_j + P \delta_{ij}\right) = 0 \tag{1.34}$$

where we've used the assumption that the flow is incompressible, both in taking ρ inside the derivatives and in $\partial_j u_j = 0$.

It's clear what is conserved here: it is simply the momentum in each of the three directions: $\int_V \rho u_i \, dV$. Associated to each conserved quantity is a current. The novelty here is that, because the conserved quantity is itself a vector, the associated current is a tensor Π_{ij} and the equation of momentum conservation (1.34) takes the form

$$\frac{\partial(\rho u_i)}{\partial t} + \frac{\partial \Pi_{ij}}{\partial x^j} = 0 \ . \tag{1.35}$$

Here the momentum current Π_{ij} tells us how the momentum in the ith direction is transported in the jth direction. From (1.34), it takes the form

$$\Pi_{ij} = \rho u_i u_j + P \delta_{ij} \ . \tag{1.36}$$

The first, advective contribution describes the momentum due to the motion of the fluid. The pressure contribution to momentum is perhaps more surprising. It is a hint, even at this macroscopic level, that pressure is associated to something moving around. From a microscopic perspective, this something is, of course, the constituent atoms of molecules of the fluid that we have declared irrelevant for fluid mechanics.

There is a simple way of seeing why it's not unreasonable that pressure is related to momentum. Take a container with some fluid inside and make a little hole in it. The pressure inside the box will force the fluid out of the hole. The rate at which momentum escapes from the box is equal to the pressure. (Or, more strictly, the pressure difference on either side of the hole.)

1.2.3 Archimedes' Principle

Before exploring the full content of the Euler equation, we can extract some familiar and long-known results. Take a fluid that sits in a gravitational field. (Which, let's face it, most do.) This means that we have an external force density

$$\mathbf{f} = \rho \mathbf{g} \tag{1.37}$$

where $\mathbf{g} = -g\hat{\mathbf{z}}$ is the gravitational acceleration and points downwards.

We can now look for the trivial solution to the Euler equation (1.29) in which the fluid is at rest, so $\mathbf{u} = 0$. We see that the fluid must have a pressure gradient to counteract the gravitational field

$$\nabla P = \rho \mathbf{g} \quad \implies \quad P = P_0 - \rho g z \ . \tag{1.38}$$

This is known as *hydrostatic pressure*. (If you're worried about the possibility that the pressure becomes negative in (1.38), then think of the surface of the fluid as sitting at $z = 0$, so that pressure only increases as we move down to $z < 0$.) It is the pressure gradient that pushes against the weight of the fluid above. You can think of the pressure as adjusting to ensure that the density of the fluid remains constant, despite the external force of gravity.

Suppose that we have some object fully immersed in a fluid, as shown in the figure to the right. Then we can ask: What is the force that the fluid exerts on the body? If the volume of the object is V, then the force can be computed as the integral of the pressure over the surface $S = \partial V$

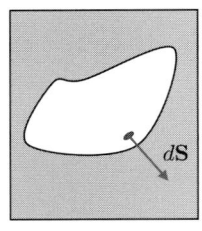

$$\mathbf{F} = -\int_S P(z) \, d\mathbf{S} \ . \tag{1.39}$$

Here the minus sign is because $d\mathbf{S}$ is taken to have outward-pointing normal as shown in the figure. The pressure of the fluid over the surface of the object is the same whether or not the object itself is there. This means that we can use the divergence theorem, together with our expression for the hydrostatic pressure (1.38), to write this as an integral over the would-be pressure inside the volume V of displaced fluid

$$\mathbf{F} = -\int_V \nabla P \, dV = -\int_V \rho \mathbf{g} \, dV \ . \tag{1.40}$$

This is telling us that the force exerted by the fluid on the object is equal to the weight of the displaced fluid. Eureka! This, of course, is Archimedes' principle. In equilibrium, the force $\mathbf{F}$ must balance the weight of the object itself.

Objects that are less dense than the fluid will float, partially submerged so that the weight of the displaced fluid is equal to the weight of the object. Objects that are denser than the fluid will sink. The discussion above hasn't brought anything new to Archimedes' famous idea. It's really just the old argument wrapped in the language of vector calculus.

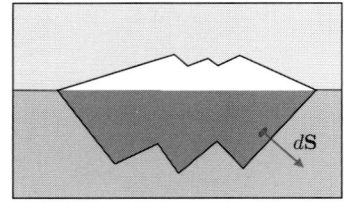

These results also give us a reason to ignore gravity for much of this book. In the presence of a gravitational field, the pressure simply adapts as in

(1.38) to cancel it. We can then write the pressure as

$$P(\mathbf{x}, t) = P_0 - \rho g z + P'(\mathbf{x}, t) \tag{1.41}$$

and Euler's equation again becomes

$$\rho \frac{D\mathbf{u}}{Dt} = -\nabla P' \tag{1.42}$$

allowing us to work with the adjusted pressure P' and ignore gravity.

1.2.4 Energy Conservation and Bernoulli's Principle

In classical mechanics, it's often useful to identify conserved quantities. The same is true in fluid mechanics, and there is a way to rewrite the Euler equation that highlights one such conserved quantity. We start with the vector identity

$$\mathbf{u} \times (\nabla \times \mathbf{u}) = \frac{1}{2}\nabla(\mathbf{u} \cdot \mathbf{u}) - (\mathbf{u} \cdot \nabla)\mathbf{u} \ . \tag{1.43}$$

We use this to substitute for the non-linear $(\mathbf{u} \cdot \nabla)\mathbf{u}$ term in the Euler equation to get

$$\rho \left(\frac{\partial \mathbf{u}}{\partial t} + \frac{1}{2}\nabla|\mathbf{u}|^2 - \mathbf{u} \times (\nabla \times \mathbf{u}) \right) = -\nabla P + \mathbf{f} \ . \tag{1.44}$$

So far this doesn't look any more useful. But now we dot with $\mathbf{u}$ to make the curly term disappear. We have

$$\rho \mathbf{u} \cdot \frac{\partial \mathbf{u}}{\partial t} + \mathbf{u} \cdot \nabla \left(\frac{1}{2}\rho|\mathbf{u}|^2 + P \right) = \mathbf{u} \cdot \mathbf{f} \ . \tag{1.45}$$

At this stage, we make one further assumption: we take the force to be conservative, meaning that we can write it in terms of a potential energy $\Phi(\mathbf{x}, t)$,

$$\mathbf{f} = -\nabla \Phi \ . \tag{1.46}$$

For example, the gravitational force can be written in this way. We then have

$$\frac{1}{2}\rho \frac{\partial |\mathbf{u}|^2}{\partial t} + \mathbf{u} \cdot \nabla \left(\frac{1}{2}\rho|\mathbf{u}|^2 + P + \Phi \right) = 0 \ . \tag{1.47}$$

This is again of the form of a conservation equation. To see this, we again pull the $\mathbf{u}$ inside the ∇ using the fact that the fluid is incompressible so $\nabla \cdot \mathbf{u} = 0$. (This is the same step that we did for the momentum conservation equation in (1.34).) We get the final form

$$\frac{1}{2}\rho \frac{\partial |\mathbf{u}|^2}{\partial t} + \nabla \cdot (\mathbf{u}H) = 0 \tag{1.48}$$

where

$$H = \frac{1}{2}\rho|\mathbf{u}|^2 + P + \Phi \ . \tag{1.49}$$

There's no mystery in what is being conserved here: the time derivative is acting on $\frac{1}{2}\rho|\mathbf{u}|^2$ which we recognise as the kinetic energy density of the fluid. The equation (1.48) is simply capturing energy conservation of the continuous fluid, with $\mathbf{u}H$ the energy flux.

The term $\nabla \cdot (P\mathbf{u})$ on the right-hand side of (1.48) deserves a special mention since it again hints at the "pressure is a Lagrange multiplier" slogan. This is the kind of term in an equation of motion that would usually arise from thinking of P as a Lagrange multiplier for $\nabla \cdot \mathbf{u} = 0$. In Section 1.6, we'll see that things are a little more subtle than that, but the basic slogan still holds.

For a steady fluid, satisfying $\partial \mathbf{u}/\partial t = 0$, we have

$$\mathbf{u} \cdot \nabla H = 0 \ . \tag{1.50}$$

This is *Bernoulli's theorem*. It states that the quantity H is constant along streamlines. The fluid flows quickly in places where the pressure is low, and more slowly when the pressure builds. However, given that pressure in incompressible fluids has no real life of its own, it's perhaps better to think of things the other way round, with the pressure adapting to the flow to enforce $\nabla \cdot \mathbf{u} = 0$.

In special cases, there are further refinements of Bernoulli's theorem. In particular, something nice happens when dealing with *irrotational* flows, obeying $\nabla \times \mathbf{u} = 0$. In this case, the third term in (1.44) vanishes automatically, and we're left with

$$\rho \frac{\partial \mathbf{u}}{\partial t} + \nabla H = 0 \ . \tag{1.51}$$

This means that for a steady, irrotational flow, we have $\nabla H = 0$ and so H is actually constant everywhere, not just along streamlines. We will make use of the time-dependent formula (1.51) in Section 3.1 when discussing surface waves.

An Example: Drinking from a Firehose

Consider water flowing down a pipe which, at some point, narrows as shown in Figure 1.2. This might, for example, be the nozzle on a firehose. We'll

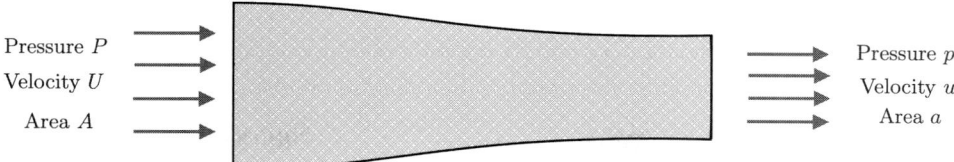

Fig. 1.2 As a pipe narrows, the velocity must increase and Bernoulli's theorem tells us that, for steady flows, the pressure decreases.

take the narrowing to be gradual so that the streamlines are smooth and follow the pipe.

Initially, the pipe has area A and the fluid has speed U. By the end the area has reduced to $a < A$ and the speed has changed to u. For incompressible flow, the speed is dictated by the conservation of mass which tells us that the volume of fluid passing through any given slice of the pipe must remain the same, so

$$UA = ua \ . \tag{1.52}$$

This immediately tells us that the speed of the flow in the narrow section is faster than in the initial, wider section: $u = UA/a$. Meanwhile, Bernoulli's theorem tells us that

$$\frac{\rho}{2}U^2 + P = \frac{\rho}{2}u^2 + p \tag{1.53}$$

where P and p are the initial and final pressure respectively, and we are ignoring any external forces. Rearranging, we have

$$p = P + \frac{\rho}{2}U^2\left(1 - \frac{A^2}{a^2}\right) \ . \tag{1.54}$$

We see that because $A > a$, the pressure actually decreases as the pipe narrows. This makes sense: the decrease in pressure in the narrow section means that there is a pressure difference and this drives the fluid to accelerate from speed U to speed u, ensuring that the flow remains solenoidal with $\nabla \cdot \mathbf{u} = 0$.

More Qualitative Applications

There are other situations where Bernoulli's principle gives us some useful intuition. For example, it's possible to levitate a ping pong ball on a fast jet of air. You can achieve this by blowing through a straw or by using a hairdryer. The question is: Why is the ball stable? Why doesn't it fall off to one side? In this situation, the air flow is turbulent and it's not entirely clear that Bernoulli's principle, which requires a steady flow, can be invoked. Nonetheless, it does provide an answer. Suppose that the ball did move slightly off to one side and out of the main flow. Then the air will be moving faster in the middle of the flow, resulting in a lower pressure and the ball gets pushed back into the middle.

The most famous application of Bernoulli's principle is to explain the lift experienced by an aerofoil. The air travels faster over the top of the wing than the bottom and the pressure difference results in a net upwards force. But this begs the question: Why does the air travel faster over the top of the wing? One popular explanation (and one that I was told in school) is that the flow must reach the trailing edge of the wing at the same time, regardless of whether it goes up or down. But that doesn't sound right! There's no principle in physics that says you must reach your goal at the same time regardless of the path you take. (If there were, we wouldn't need maps.) We will revisit this later when we study flows around objects in some detail.

1.3 Vorticity

To characterise the shape of a velocity field $\mathbf{u}$, we look at its derivatives. In general there are nine such derivatives, $\partial_i u_j$, with $i, j = 1, 2, 3$. But, for incompressible flows, we know that one linear combination vanishes: $\nabla \cdot \mathbf{u} = 0$. The remaining derivatives can be decomposed as a symmetric and anti-symmetric tensor. The symmetric one is known as the *rate of strain tensor*,

$$E_{ij} = \frac{1}{2}\left(\frac{\partial u_i}{\partial x^j} + \frac{\partial u_j}{\partial x^i}\right) .$$ (1.55)

The anti-symmetric tensor is

$$\Omega_{ij} = \frac{1}{2}\left(\frac{\partial u_i}{\partial x^j} - \frac{\partial u_j}{\partial x^i}\right) .$$ (1.56)

It contains the same information as the vector field, $\omega_i = -\epsilon_{ijk}\Omega_{jk}$, which is more familiarly written as

$$\boldsymbol{\omega} = \nabla \times \mathbf{u} .$$ (1.57)

This is the *vorticity*. It tells us how the fluid swirls at each point in space. The integral curves associated to $\boldsymbol{\omega}$ (i.e. the lines that are tangent to $\boldsymbol{\omega}$ at each point $\mathbf{x}$) are called *vortex lines*. Because $\boldsymbol{\omega} = \nabla \times \mathbf{u}$, the vortex lines are perpendicular to streamlines.

In the book on Classical Mechanics, we described how rigid bodies move. A rigid object rotating with angular velocity $\boldsymbol{\Omega}$ about the origin has velocity $\mathbf{u} = \boldsymbol{\Omega} \times \mathbf{x}$. (We called the angular velocity $\boldsymbol{\omega}$ in the book on Classical Mechanics but here we're using $\boldsymbol{\Omega}$ to avoid confusion with the vorticity.) For such rigid body motion, the vorticity would be

$$\boldsymbol{\omega} = \nabla \times \mathbf{u} = \nabla \times (\boldsymbol{\Omega} \times \mathbf{x}) = 2\boldsymbol{\Omega} \ . \tag{1.58}$$

So, for a rigid body, the vorticity is twice the angular velocity. Of course, fluids are most certainly not rigid bodies and so the vorticity can be a much more interesting function of space and time.

Examples of Flows

To get a feel for what the vorticity $\boldsymbol{\omega}$ and rate of strain E are telling us, we can look at a couple of examples.

First consider the 2d flow

$$\mathbf{u} = \alpha(-x, y, 0) \tag{1.59}$$

with α a constant. This is plotted on the left of Figure 1.3. The velocity field has $\nabla \cdot \mathbf{u} = 0$ and also $\boldsymbol{\omega} = 0$, while the rate of strain tensor is

$$E = \alpha \begin{pmatrix} -1 & 0 & 0 \\ 0 & +1 & 0 \\ 0 & 0 & 0 \end{pmatrix} \ . \tag{1.60}$$

From the figure, you can see that the fluid is squeezed in one direction (the x-direction in this case) and stretched in the other (the y-direction). This is the characteristic feature of flows with a rate of strain. To see why, note that the rate of strain tensor is symmetric and so can always be diagonalised so that it takes the form

$$E = \begin{pmatrix} E_1 & 0 & 0 \\ 0 & E_2 & 0 \\ 0 & 0 & E_3 \end{pmatrix} \ . \tag{1.61}$$

But, for incompressible flows with $\nabla \cdot \mathbf{u} = 0$, we must have $E_1 + E_2 + E_3 = 0$. So one eigenvalue is necessarily positive and another necessarily negative. These are the directions in which the flow is, respectively, stretched and squeezed.

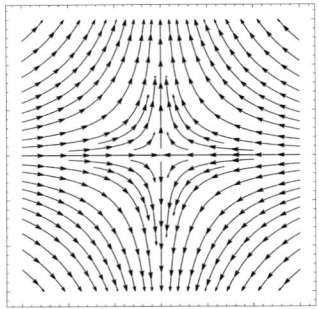 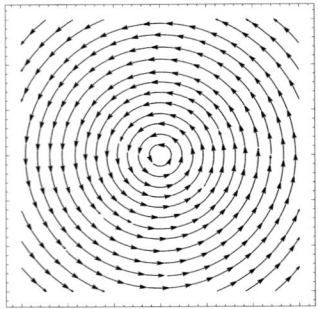

Fig. 1.3 On the left, a flow with strain and no vorticity. On the right, a flow with vorticity and no strain.

Next consider the flow

$$\mathbf{u} = \alpha(-y, x, 0) \ . \tag{1.62}$$

This has $\nabla \cdot \mathbf{u} = E = 0$ and a constant vorticity everywhere in the fluid, $\boldsymbol{\omega} = (0, 0, 2\alpha)$. It is depicted on the right of Figure 1.3. Unsurprisingly, it exhibits a rotation.

However, one should be wary of simply eyeballing a flow to decide on its vorticity. To illustrate this, consider the example

$$\mathbf{u} = f(r)(-y, x, 0) \ . \tag{1.63}$$

where $f(r)$ is any function of $r^2 = x^2 + y^2$. (Note that we're keeping the flow essentially two-dimensional.) This is a generalisation of our previous flow (1.62) and the streamlines look identical for any choice of $f(r)$. The vorticity is $\boldsymbol{\omega} = (0, 0, \omega(r))$, with

$$\omega = \frac{1}{r} \frac{d}{dr}(r^2 f) \ . \tag{1.64}$$

Now the vorticity $\omega(r)$ varies in the radial direction. This means that if we take the specific choice of $f = 1/r^2$, then the vorticity vanishes, with $\boldsymbol{\omega} = 0$ for all $r \neq 0$, even though the flow is clearly rotating around the origin. This is because a non-zero vorticity $\boldsymbol{\omega}(\mathbf{x}) \neq 0$ at some point $\mathbf{x}$ means that the fluid is rotating locally around $\mathbf{x}$, not just around the origin.

To build a more physical understanding for what vorticity means, suppose that we drop some propellers in the fluid, like those plastic windspinners that you can buy at the seaside. If you drop them in the fluid, they will move around the origin with the flow. But if the fluid has a vorticity then their orientation will also rotate as they move, as shown on the left-hand side of

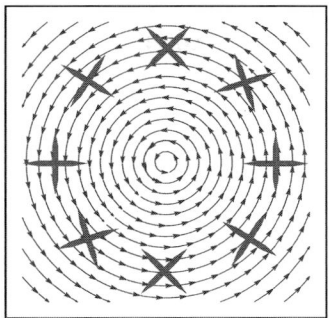 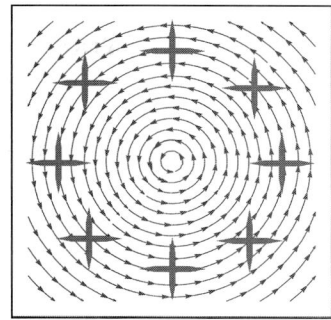

Fig. 1.4 On the left, a flow with vorticity. On the right, a flow that rotates around the origin but with vanishing vorticity (except at the origin!).

Figure 1.4. If the fluid has no vorticity, as is the case for $f = 1/r^2$, then they will remain in the same orientation as they move around, as shown in the right-hand figure.

In fact, things are a little more subtle than this. The flow that seemingly has vanishing vorticity, $\mathbf{u} = (-y/r^2, x/r^2, 0)$, has the property that the integral of the velocity field around any circle C that surrounds the origin always gives

$$\oint_C \mathbf{u} \cdot d\mathbf{x} = 2\pi \ . \tag{1.65}$$

This is because the velocity field drops off as $1/r$, while the perimeter of the circle grows as r. But, by Stokes' theorem, we have

$$\oint_C \mathbf{u} \cdot d\mathbf{x} = \int_S \boldsymbol{\omega} \cdot d\mathbf{S} = 2\pi \tag{1.66}$$

where S is a surface with boundary $\partial S = C$. So it can't quite be true that the vorticity $\boldsymbol{\omega}$ vanishes everywhere! Indeed, the flow is singular at the origin $x = y = 0$ (which, in three dimensions, means that it is singular along the entire z-axis). For the above calculation to be consistent, the vorticity must be non-zero along this axis, with

$$\boldsymbol{\omega} = 2\pi \delta(r)\hat{\mathbf{z}} \ . \tag{1.67}$$

This is sufficient for the flow to have rotation around the origin, even though it doesn't have vorticity at any other point. This slightly subtle example will arise in some later applications. In fact, it's not a bad approximation for what happens when you empty the bath, with the (admittedly finite size) plughole taking the place of $r = 0$.

In the examples above, the fluid swirls in the (x, y)-plane, and the vorticity points in the z-direction. That's the intuitive view of vorticity. But anything involving curls in vector calculus can be counterintuitive! To illustrate this, it's possible to set up a situation such that the vorticity is everywhere equal to the velocity. A simple, but rather contrived, example is known as the ABC flow (after Arnold, Beltrami, and Childress)

$$\mathbf{u} = \begin{pmatrix} A \sin z + C \cos y \\ B \sin x + A \cos z \\ C \sin y + B \cos x \end{pmatrix} . \tag{1.68}$$

It's simple to check that $\boldsymbol{\omega} = \nabla \times \mathbf{u} = \mathbf{u}$. The streamlines in this flow can be shown to exhibit chaotic behaviour. In general, any flow $\mathbf{u}$ that is proportional to its vorticity $\boldsymbol{\omega} = \nabla \times \mathbf{u}$ is referred to as a *Beltrami vector field*.

1.3.1 The Biot–Savart Law

We can invert the equation $\boldsymbol{\omega} = \nabla \times \mathbf{u}$ to get an expression for the velocity in terms of the vorticity. In fact, this is a calculation that we've done elsewhere and it's worth taking the opportunity to remind ourselves of this.

In electromagnetism, the magnetic field obeys $\nabla \cdot \mathbf{B} = 0$, which means that it can be written in terms of a vector potential $\mathbf{B} = \nabla \times \mathbf{A}$. In the case of magnetostatics, the magnetic field is given by Ampère's law

$$\nabla \times \mathbf{B} = \mu_0 \mathbf{J} \quad \Longrightarrow \quad \nabla^2 \mathbf{A} = -\mu_0 \mathbf{J} \tag{1.69}$$

with $\mathbf{J}$ the current density. This is just the Poisson equation for each component of $\mathbf{A}$ and can be solved using the Green's function to find

$$\mathbf{A}(\mathbf{x}) = \frac{\mu_0}{4\pi} \int_V d^3 x' \, \frac{\mathbf{J}(\mathbf{x}')}{|\mathbf{x} - \mathbf{x}'|} . \tag{1.70}$$

If we subsequently take the curl of this equation, then we get an expression for the magnetic field $\mathbf{B}$ in terms of the current density

$$\mathbf{B}(\mathbf{x}) = \frac{\mu_0}{4\pi} \int_V d^3 x' \, \frac{\mathbf{J}(\mathbf{x}') \times (\mathbf{x} - \mathbf{x}')}{|\mathbf{x} - \mathbf{x}'|^3} . \tag{1.71}$$

This is the *Biot–Savart law*. You can read more about the steps above, and the uses of the Biot–Savart law, in Volume 2 on Electromagnetism.

But we can now repeat each of these steps for the fluid velocity. If the fluid is incompressible, so $\nabla \cdot \mathbf{u} = 0$, then we can introduce a vector potential $\mathbf{A}$ such that $\mathbf{u} = \nabla \times \mathbf{A}$. This way of writing the velocity is at the heart of the idea of a stream function, as we saw in Section 1.1.4. The curl of the velocity is the vorticity, so we have

$$\nabla \times \mathbf{u} = \boldsymbol{\omega} \quad \Longrightarrow \quad \nabla^2 \mathbf{A} = -\boldsymbol{\omega} . \tag{1.72}$$

Following the same steps that we took above, the vector potential can then be expressed as

$$\mathbf{A}(\mathbf{x}, t) = \frac{1}{4\pi} \int_V d^3 x' \, \frac{\boldsymbol{\omega}(\mathbf{x}', t)}{|\mathbf{x} - \mathbf{x}'|} . \tag{1.73}$$

Again taking the curl gives the fluid analogue of the Biot–Savart law

$$\mathbf{u}(\mathbf{x}, t) = \frac{1}{4\pi} \int_V d^3 x' \, \frac{\boldsymbol{\omega}(\mathbf{x}', t) \times (\mathbf{x} - \mathbf{x}')}{|\mathbf{x} - \mathbf{x}'|^3} . \tag{1.74}$$

So we see that, just as the magnetic field is sourced by currents, so is the velocity field of a fluid sourced by vorticity. In fact, in both cases we can have some additional contribution to the field and the most general form of the fluid velocity is

$$\mathbf{u}(\mathbf{x}, t) = \nabla \phi(\mathbf{x}, t) + \frac{1}{4\pi} \int_V d^3 x' \, \frac{\boldsymbol{\omega}(\mathbf{x}', t) \times (\mathbf{x} - \mathbf{x}')}{|\mathbf{x} - \mathbf{x}'|^3} \tag{1.75}$$

where the $\mathbf{u} = \nabla \phi$ piece doesn't contribute to the vorticity because $\nabla \times \nabla \phi = 0$. We can only reconstruct the velocity field from the vorticity up to this subtlety. In particular, there are situations where all the physics is sitting in the $\mathbf{u} = \nabla \phi$ term. (The flow in (1.59) is governed by $\phi = \alpha(y^2 - x^2)/2$, and we'll meet many more examples in Sections 1.4 and 1.5.)

While the mathematics leading to the electromagnetic and fluidic versions of the Biot–Savart law is identical, there are some differences. The first is conceptual. In electromagnetism, one thinks of the current $\mathbf{J}$ as something fixed and external, which determines the magnetic field $\mathbf{B}$. In contrast, in fluid mechanics the vorticity $\boldsymbol{\omega}$ is, at least initially, viewed as an object derived from the velocity field $\mathbf{u}$. Nonetheless, as we proceed, there will be times in this book when it's useful to think of vorticity as an object in its own right. The Biot–Savart formula (1.75) is perhaps the first place that this change of status suggests itself.

The second difference is more technical. The electromagnetic Biot–Savart law (1.71) holds only for static currents. There is a generalisation to time-dependent currents, but it requires us to take into account the time that it takes light to travel from the current to the place where the magnetic field is measured. (See Chapter 7 of the book on Electromagnetism.) In contrast, as shown, the fluid version (1.75) holds for time-dependent flows, with the velocity and vorticity fields evaluated at the same time. This can be traced to the fact that we're restricting ourselves to incompressible flows, which allows information to be transmitted infinitely quickly through the fluid. As soon as we appreciate that fluids are, in fact, compressible, we must generalise the story above. We would then find that any disturbance actually propagates at the speed of sound.

1.3.2 The Vorticity Equation

It is interesting to ask how the vorticity $\boldsymbol{\omega}$ evolves. We return to the equation (1.44) that we previously found on the way to deriving Bernoulli's formula, again restricted to a conservative force $\mathbf{f} = -\nabla \Phi$

$$\rho \frac{\partial \mathbf{u}}{\partial t} + \frac{1}{2} \rho \nabla |\mathbf{u}|^2 = \rho \mathbf{u} \times \boldsymbol{\omega} - \nabla P - \nabla \Phi \ . \tag{1.76}$$

If we take the curl of this, and use the fact that $\nabla \times (\nabla \text{ anything}) = 0$, we have

$$\frac{\partial \boldsymbol{\omega}}{\partial t} = \nabla \times (\mathbf{u} \times \boldsymbol{\omega}) \ . \tag{1.77}$$

We now use the vector identity

$$\nabla \times (\mathbf{u} \times \boldsymbol{\omega}) = (\nabla \cdot \boldsymbol{\omega})\mathbf{u} + (\boldsymbol{\omega} \cdot \nabla)\mathbf{u} - (\nabla \cdot \mathbf{u})\boldsymbol{\omega} - (\mathbf{u} \cdot \nabla)\boldsymbol{\omega} \ . \tag{1.78}$$

We have $\nabla \cdot \boldsymbol{\omega} = 0$ because the vorticity $\boldsymbol{\omega}$ is itself a curl, and $\nabla \cdot \mathbf{u} = 0$ because we're dealing with an incompressible flow. Rearranging the remaining terms, we have

$$\frac{D\boldsymbol{\omega}}{Dt} = (\boldsymbol{\omega} \cdot \nabla)\mathbf{u} \ . \tag{1.79}$$

This is the *vorticity equation*. It tells us how the vortex lines stretch and twist as the fluid evolves. Note that this equation is inherently three-dimensional: if the velocity swirls in some plane, then vorticity $\boldsymbol{\omega}$ points out of that plane and, correspondingly, $(\boldsymbol{\omega} \cdot \nabla)\mathbf{u}$ is measuring how much the velocity $\mathbf{u}$ changes in the direction out of the plane.

Using $\nabla \cdot \mathbf{u} = \nabla \cdot \boldsymbol{\omega} = 0$, the vorticity equation can be rewritten as

$$\frac{\partial \omega^i}{\partial t} + \frac{\partial}{\partial x^j}\left(u^j \omega^i - u^i \omega^j\right) = 0 \ . \tag{1.80}$$

This is the standard form of a continuity equation, telling us that vorticity is conserved.

To try to get a feel for what the vorticity equation (1.79) is telling us, first suppose that the right-hand side vanished. Then the vorticity would simply drift with the fluid. We can get a sense for what the right-hand side means by considering two nearby points $\mathbf{x}_1(t)$ and $\mathbf{x}_2(t)$ at some time t, separated by a small distance

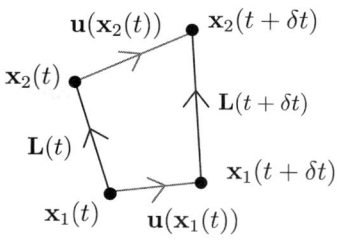

$$\mathbf{L}(t) = \mathbf{x}_2(t) - \mathbf{x}_1(t) \ . \tag{1.81}$$

We'll think about how this material line segment evolves with the flow. At a later time $t + \delta t$, each of these end points has been swept along and now sit at

$$\mathbf{x}_i(t + \delta t) \approx \mathbf{x}_i(t) + \delta\mathbf{x}_i \approx \mathbf{x}_i(t) + \mathbf{u}(\mathbf{x}_i(t))\delta t \ .$$

So the line segment $\mathbf{L}$ has evolved as

$$\begin{aligned} \mathbf{L}(t + \delta t) &\approx \mathbf{x}_2(t + \delta t) - \mathbf{x}_1(t + \delta t) \\ &\approx \mathbf{L}(t) + \big(\mathbf{u}(\mathbf{x}_2(t)) - \mathbf{u}(\mathbf{x}_1(t))\big)\delta t \ . \end{aligned} \tag{1.82}$$

We now Taylor expand $\mathbf{u}(\mathbf{x}_2) = \mathbf{u}(\mathbf{x}_1 + \mathbf{L}) \approx \mathbf{u}(\mathbf{x}_1) + \mathbf{L} \cdot \nabla\mathbf{u}(\mathbf{x}_1)$ to write this as

$$\mathbf{L}(t + \delta t) \approx \mathbf{L}(t) + (\mathbf{L} \cdot \nabla)\mathbf{u}(\mathbf{x}(t))\,\delta t \tag{1.83}$$

where we have evaluated the gradient of the velocity field at $\mathbf{x}$, which could be either $\mathbf{x}_1$ or $\mathbf{x}_2$ or anywhere in between: it doesn't matter as they are close. In the limit $\delta t \to 0$, all the $\approx$ signs become $=$ signs. We see that a small line segment of the fluid evolves as

$$\frac{d\mathbf{L}}{dt} = (\mathbf{L} \cdot \nabla)\mathbf{u} \ . \tag{1.84}$$

But the right-hand-side is the same form as we find in the vorticity equation (1.79). This is telling us that the lines of vorticity are stretched and twisted like the material lines of the fluid itself. We usually say that the vortex lines "move with the fluid".

We can get a more direct expression for the change in the magnitude of the vorticity. First take the dot product of (1.79) with $\boldsymbol{\omega}$. This tells us how the magnitude (squared) of the vorticity $|\boldsymbol{\omega}|^2$ changes

$$\frac{1}{2}\frac{D|\boldsymbol{\omega}|^2}{Dt} = \boldsymbol{\omega} \cdot (\boldsymbol{\omega} \cdot \nabla)\mathbf{u} = \omega_i\omega_j\frac{\partial u_i}{\partial x^j} \tag{1.85}$$

where, in the second term, we've resorted to index notation to clarify what is inner-producted with what. Note, however, that $\omega_i \omega_j$ is symmetric in i and j so this picks out the strain of the flow defined in (1.55). We have

$$\frac{1}{2}\frac{D|\boldsymbol{\omega}|^2}{Dt} = \boldsymbol{\omega} \cdot E\boldsymbol{\omega} \ . \tag{1.86}$$

We learn that vorticity is increased or decreased by the rate of strain in the flow. This, it turns out, is central to turbulence. The integral of $|\boldsymbol{\omega}|^2$ over all of space is known as the *enstrophy*.

This equation is telling us that if, at some time, the vorticity vanishes everywhere, say $\boldsymbol{\omega}(\mathbf{x}, t = 0) = 0$, then it will vanish everywhere at all subsequent times. This holds regardless of any conservative forces that might be at play. This prompts the question: Where does vorticity come from in the first place? The answer is that it comes from non-conservative forces. These include friction forces, as captured through the viscosity of the fluid, and the Coriolis force. Other mechanisms for generating vorticity are at play in stratified fluids. We will devote Chapter 2 to understanding the effects of viscosity and see in a number of explicit examples how it gives rise to vorticity. We will look at the Coriolis force, and a generalisation of vorticity, known as "potential vorticity", in Section 3.3.

An Example: Vortex Stretching

To illustrate how vortex lines stretch and twist, consider the flow

$$\mathbf{u}(\mathbf{x}, t) = \mathbf{u}_{\text{strain}}(\mathbf{x}) + \mathbf{u}_{\text{rot}}(\mathbf{x}, t) \quad \text{with} \quad \begin{cases} \mathbf{u}_{\text{strain}} = \alpha(-x, -y, 2z) \\ \mathbf{u}_{\text{rot}} = f(r, t)(-y, x, 0) \end{cases} \ . \tag{1.87}$$

Both of these flows are similar to the examples given above. The strain flow stretches the fluid in the z-direction, while squeezing it in the (x, y)-plane; the rotational flow clearly rotates in the (x, y)-plane, with an angular velocity determined by the function $f(r, t)$ where $r^2 = x^2 + y^2$.

The vorticity lies in the z-direction, with $\boldsymbol{\omega} = (0, 0, \omega)$ and ω given by (1.64)

$$\omega = \frac{1}{r}\frac{d}{dr}(r^2 f) \ . \tag{1.88}$$

The vorticity equation (1.79) is then a partial differential equation for $\omega(r, t)$

$$\frac{\partial \omega}{\partial t} - \alpha r \frac{\partial \omega}{\partial r} = 2\alpha\omega \ . \tag{1.89}$$

This is solved by

$$\omega(r,t) = e^{2\alpha t} W(re^{\alpha t}) \qquad (1.90)$$

for a function $W(r)$, which is the initial vorticity at time $t = 0$. We see that the strain indeed increases the vorticity, with an exponential growth in time. But the time dependence in the function $W(re^{\alpha t})$ gives a corresponding squeezing of the vorticity in the (x, y)-plane. This effect is known as *vortex stretching*. At heart, this is just the conservation of angular momentum: it is the fluid version of an ice skater who spins faster when they pull in their arms.

In this example, the vorticity is aligned with one of the principal axes of the rate of strain tensor. When this isn't the case, the vortex lines get twisted by the strain.

Bernoulli's Theorem Revisited

It's worth revisiting some of the ideas of Section 1.2.4. First, there is a version of Bernoulli's theorem for the vortex lines, tangent to $\boldsymbol{\omega}$. To see this, we take the inner product of (1.76) with $\boldsymbol{\omega}$ to find that, in a steady flow with $\partial \mathbf{u}/\partial t = 0$, we have

$$\boldsymbol{\omega} \cdot \nabla H = 0 \ . \qquad (1.91)$$

We learn that the Bernoulli function H, defined in (1.49), is constant both along streamlines (as in (1.50)) and along vortex lines.

If the vorticity vanishes everywhere, then the fluid is said to be *irrotational*. In this case, we can say more because, as we already noted in (1.51), a steady, irrotational flow has

$$H = \frac{1}{2}\rho \mathbf{u}^2 + P + \Phi \qquad (1.92)$$

constant everywhere in the fluid, not just along streamlines and vortex lines. We will explore these flows further in Section 1.4.

1.3.3 Kelvin's Circulation Theorem

The *circulation* of a flow around a closed curve C is defined by

$$\Gamma = \oint_C \mathbf{u} \cdot d\mathbf{x} \ . \qquad (1.93)$$

Now consider a *material* curve $C(t)$, meaning that it follows the flow of the underlying fluid elements. We want to understand how the associated circulation $\Gamma(t)$ changes. We have

$$\frac{D\Gamma}{Dt} = \oint_{C(t)} \left(\frac{D\mathbf{u}}{Dt} \cdot d\mathbf{x} + \mathbf{u} \cdot \frac{D(d\mathbf{x})}{Dt} \right) . \qquad (1.94)$$

We can replace $D\mathbf{u}/Dt$ in the first term using the Euler equation (1.29). Assuming a conservative force $\mathbf{f} = -\nabla\Phi$, this gives

$$\oint_{C(t)} \frac{D\mathbf{u}}{Dt} \cdot d\mathbf{x} = \frac{1}{\rho} \oint_{C(t)} (-\nabla P - \nabla\Phi) \cdot d\mathbf{x} = 0 \qquad (1.95)$$

which vanishes because it is the integral of a gradient around a closed path. That leaves us with the second term in (1.94). The notation $D(d\mathbf{x})/Dt$ is a little formal because the material derivative D/Dt was defined to act on fields, while here it's acting on a line element. But the meaning is straightforward: it captures the way that the line element $d\mathbf{x}$ changes under the flow.

To see what this means in practice, we can return to the fundamentals. Consider a small, moving line element $\delta\mathbf{x}(t)$, with end points $\mathbf{x}_1(t)$ and $\mathbf{x}_2(t)$, so $\delta\mathbf{x} \approx \mathbf{x}_2 - \mathbf{x}_1$. We want to know how this line segment evolves. But this is the calculation that we just did when building intuition for the meaning of the vorticity equation: there we called the material line segment $\mathbf{L}(t)$, but it is the same thing as $\delta\mathbf{x}$ in the present context. Importing the result (1.84), we learn how the line element changes and give meaning to the expression $D(d\mathbf{x})/Dt$: it is

$$\frac{D(d\mathbf{x})}{Dt} = (d\mathbf{x} \cdot \nabla)\mathbf{u} . \qquad (1.96)$$

Using this in (1.94), we have

$$\frac{D\Gamma}{Dt} = \oint_{C(t)} \mathbf{u} \cdot (d\mathbf{x} \cdot \nabla)\mathbf{u} = \oint_{C(t)} u_i \frac{\partial u_i}{\partial x^j} dx^j \qquad (1.97)$$

where we've again resorted to index notation to clarify which objects are dotted together. This can be written as

$$\frac{D\Gamma}{Dt} = \frac{1}{2} \oint_{C(t)} \nabla(\mathbf{u} \cdot \mathbf{u}) \cdot d\mathbf{x} = 0 \qquad (1.98)$$

which again vanishes because it is the integral of a gradient around a closed path. The upshot is that the circulation around any closed loop $C(t)$ does not change when we follow this loop with the flow

$$\frac{D\Gamma}{Dt} = 0 . \qquad (1.99)$$

This is *Kelvin's circulation theorem.*

To see the consequences of this result, first note that the circulation is related to the vorticity by Stokes' theorem

$$\Gamma = \int_S \boldsymbol{\omega} \cdot d\mathbf{S} \tag{1.100}$$

where S is any surface with boundary $\partial S = C$. (It's worth remembering at this point that Stokes learned about Stokes' theorem from his friend William Thomson, later known as Lord Kelvin!) So the circulation theorem again tells us that a fluid that starts off as irrotational, with $\boldsymbol{\omega} = 0$, will remain irrotational.

More intuition comes if we focus on flows in which vorticity is localised. To this end, suppose that $\boldsymbol{\omega}$ is non-vanishing only in some region of the fluid. Find a surface S such that the circulation defined in (1.100) is non-vanishing. As we vary the surface S, Γ can't change. This means that the vor-

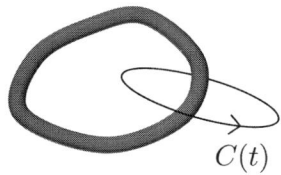

$C(t)$

ticity can't be localised in a lump of space: it must be extended along a tube-like region. This tube might extend to infinity, which is the case in the example of vorticity that we saw earlier in this section. Or it might form a vortex ring, as shown in the figure. In either case, the vorticity can't just end.

We learned previously that the magnitude of the vorticity can change due to the strain in the fluid (1.86). Now we see that, in a certain sense, vorticity must be conserved. There's no contradiction here. As the magnitude of the vorticity increases, the area of the flux tube must decrease so that the vortex flux (1.100) remains unchanged. Indeed, we saw precisely this effect at play in the vortex (1.90).

An Historical Aside

I think it's fair to say that Kelvin got a little carried away with his results on vortices. He was so taken with the stability of vortices, and of smoke rings in particular, that he proposed that they may form the basis of all matter, with different atoms arising as different knots of vortices. Some pictures from one of Kelvin's original papers are shown in Figure 1.5.

With hindsight, Kelvin's idea looks overly optimistic. Nonetheless, modern ideas in physics suggest that they may contain a grain of truth. In quantum

Fig. 1.5 Taken from the 1867 paper "On Vortex Motion" by Sir William Thomson, better known by his later name Lord Kelvin.

field theories, certain particles arise as so-called "solitons" in which the fields wrap themselves in some stable configuration that has some similarities to vortices in fluids. (There are, admittedly, differences. The vortices in fluids are not really solitons, although similar vortices that arise in superfluids and superconductors are.)

From a certain perspective, the proton and neutron can be viewed as solitons of an underlying pion field, known as a *skyrmion*. (Admittedly, the more familiar story of the proton and neutron as made from three quarks is a more fundamental perspective.) Magnetic monopoles, if they exist, would be examples of solitons. Both will be described in later books on Quantum Field Theory and The Standard Model.

1.3.4 Helicity

We've uncovered a number of conserved quantities in the Euler equation already. The equation itself can be viewed as a continuity equation (1.35) capturing the conservation of momentum. In addition, we have also seen continuity equations for energy (1.48) and vorticity (1.80).

There is one further conserved quantity that we highlight. This is the *helicity*, defined to be the integral

$$h = \int d^3x \, \mathbf{u} \cdot \boldsymbol{\omega} \, . \tag{1.101}$$

As the fluid evolves, the helicity remains constant. To see this, we look at the time variation. We use the equation of motion for the velocity,

$$\frac{\partial \mathbf{u}}{\partial t} = -\mathbf{u} \cdot \nabla \mathbf{u} - \frac{1}{\rho} \nabla (P + \Phi) \tag{1.102}$$

and for the vorticity

$$\frac{\partial \boldsymbol{\omega}}{\partial t} = -\mathbf{u} \cdot \nabla \boldsymbol{\omega} + \boldsymbol{\omega} \cdot \nabla \mathbf{u} \tag{1.103}$$

to derive the continuity equation associated to helicity

$$\frac{\partial (\mathbf{u} \cdot \boldsymbol{\omega})}{\partial t} = \nabla \cdot \left(\frac{1}{2} |\mathbf{u}|^2 \boldsymbol{\omega} - (\mathbf{u} \cdot \boldsymbol{\omega}) \mathbf{u} - \frac{1}{\rho} (P + \Phi) \boldsymbol{\omega} \right) . \tag{1.104}$$

The usual arguments then imply that h is conserved, assuming nothing is escaping at infinity.

It's rather nice, and a little surprising, to find yet another conserved quantity in the Euler equation. What is its physical interpretation? Well that, it turns out, is also rather nice: it counts the linking number of intertwined vortex tubes.

Take two vortex rings. We will assume that the vorticity of each ring is localised in some region and the two do not overlap. We denote this vorticity as $\boldsymbol{\omega}_1$ and $\boldsymbol{\omega}_2$ respectively, and the associated circulation carried by each ring as Γ_1 and Γ_2. We will show that, evaluated on this configuration, the helicity counts the linking number $n \in \mathbb{Z}$ of the two vortices,

$$h = 2n\Gamma_1\Gamma_2 . \tag{1.105}$$

For example, the two vortices shown in the figure have linking number $n = 1$.

To prove (1.105), first note that the helicity integral only picks up contributions in the vicinity of the rings. We call these two curves C_1 and C_2. Locally, for each curve, we split the volume element into a line element $d\mathbf{x}$ tangent to the curve and an area element $d\mathbf{S}$ that spans the directions perpendicular to the curve. 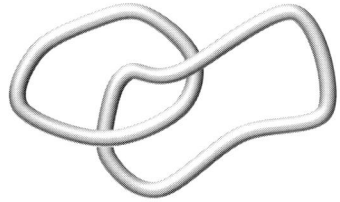 But the vector area element $d\mathbf{S}$ points in the direction perpendicular to the plane, so $d\mathbf{S}$ is actually parallel to $d\mathbf{x}$ which, in turn, is parallel to $\boldsymbol{\omega}$. We then write the helicity as

$$
\begin{aligned}
h &= \int_{V_1} d^3x \; \mathbf{u} \cdot \boldsymbol{\omega}_1 + \int_{V_2} d^3x \; \mathbf{u} \cdot \boldsymbol{\omega}_2 \\
&= \oint_{C_1} \mathbf{u} \cdot d\mathbf{x}_1 \int_{S_1} \boldsymbol{\omega}_1 \cdot d\mathbf{S}_1 + \oint_{C_2} \mathbf{u} \cdot d\mathbf{x}_2 \int_{S_2} \boldsymbol{\omega}_2 \cdot d\mathbf{S}_2 \\
&= \Gamma_1 \oint_{C_1} \mathbf{u} \cdot d\mathbf{x}_1 + \Gamma_2 \oint_{C_2} \mathbf{u} \cdot d\mathbf{x}_2 \; .
\end{aligned}
\tag{1.106}
$$

But, from (1.100) the line integral $\oint \mathbf{u} \cdot d\mathbf{x}$ counts the vorticity passing through a surface spanned by C. For the curve C_1, this vorticity is provided by the other vortex ring and counts the number of times (with sign) that the second curve C_2 links the first. We have

$$
\int_{C_1} \mathbf{u} \cdot d\mathbf{x}_1 = n\Gamma_2 \quad \text{and} \quad \int_{C_2} \mathbf{u} \cdot d\mathbf{x}_2 = n\Gamma_1 \; .
\tag{1.107}
$$

This gives the claimed result (1.105).

The result above mimics an analogous calculation in magnetism, where the Chern–Simons integral $\int d^3x \; \mathbf{A} \cdot \mathbf{B}$ counts the linking number of magnetic flux tubes. This was explained in the book on Electromagnetism (see the section entitled "A Topological Interlude").

1.4 Potential Flows in 3d

In this section we restrict ourselves to flows that are steady, so $\partial \mathbf{u}/\partial t = 0$, incompressible, and irrotational. These latter two properties mean that

$$
\nabla \cdot \mathbf{u} = 0 \quad \text{and} \quad \nabla \times \mathbf{u} = 0 \; .
\tag{1.108}
$$

This suggests two different vector calculus routes to attack the problem. We could use the first condition to write $\mathbf{u} = \nabla \times \mathbf{A}$. This was our previous stream function approach. However, it turns out to be more useful to use the irrotational property. If the domain of the flow is simply connected, then a vector field that obeys $\nabla \times \mathbf{u} = 0$ can be written in terms of a potential ϕ such that

$$
\mathbf{u} = \nabla \phi \; .
\tag{1.109}
$$

The requirement that the flow is incompressible, $\nabla \cdot \mathbf{u} = 0$, then tells us that

$$
\nabla^2 \phi = 0 \; .
\tag{1.110}
$$

This is very familiar: it is just the Laplace equation. A flow that is steady, incompressible, and irrotational is called a *potential flow*. Importantly, the Laplace equation is linear. That means that if we have two solutions then we can simply superpose them to get a third. The non-linearity of the Euler equation disappeared by virtue of the irrotational assumption.

To understand potential flows, all we have to do is solve the Laplace equation. The devil in the details is, as we shall see, all about the boundary conditions that we impose on the flow. Indeed, the further we get with understanding fluids, the more devilish these boundary conditions will become.

(To briefly look forwards, we will later see that that boundary conditions can allow non-linearities to sneak back into the physics, undoing all our good work in finding a simple, linear equation to describe potential flow. In this chapter we will only deal with linear boundary conditions but we will see examples of non-linear, and surprisingly complicated, boundary conditions in Chapter 3 when we discuss surface waves.)

1.4.1 Boundary Conditions

In many courses in theoretical physics, boundary conditions are relatively unimportant beyond the usual requirement that things fall off asymptotically. (There are, of course, exceptions such as the study of electromagnetic waves in materials that we described in the book on Electromagnetism.) For fluids, however, many of the most important results come from imposing the right boundary conditions.

We'll meet various kinds of boundary conditions in this book. For example, later when we come to discuss waves we'll think about dynamical interfaces between two fluids. But, for now, we will restrict to the simplest kind: a solid boundary.

Suppose that the fluid comes into contact with a solid object. Maybe there's a wall at the edge of the container. Or maybe there's some object, like the wing of an aircraft, sitting in the fluid flow. What boundary conditions should we impose?

Our first condition is completely obvious. The fluid can't flow into the solid. To describe this mathematically, we introduce a normal vector $\mathbf{n}(\mathbf{x})$ at each point $\mathbf{x}$ on the boundary. If the boundary is flat, then $\mathbf{n}$ is constant, independent of $\mathbf{x}$. If the boundary curves in some way, then $\mathbf{n}$ changes accordingly. Provided that the boundary itself does not move, we must have

$$\mathbf{n} \cdot \mathbf{u} = 0 \tag{1.111}$$

at each point of the boundary. This is the statement that nothing seeps into the solid. It is also the statement that the boundary of a fluid is a streamline.

We will also be interested in situations in which the boundary does move, with some velocity $\mathbf{U}$. In this case, we place ourselves in the frame of the moving boundary, where the fluid velocity is $\mathbf{u}' = \mathbf{u} - \mathbf{U}$ and the boundary condition is $\mathbf{n} \cdot \mathbf{u}' = 0$. Back in the original frame, we have

$$\mathbf{n} \cdot \mathbf{u} = \mathbf{n} \cdot \mathbf{U} \ . \tag{1.112}$$

This simple statement that the solid is impermeable is an example of a *kinematic boundary condition*. It fixes the component of the fluid velocity perpendicular to the boundary. We'll meet other examples of kinematic boundary conditions in Chapter 3 where we describe waves on the surface of a fluid.

We haven't yet said anything about the component of the velocity that is tangential to the boundary. For example, we might think that a "no-slip" boundary condition should be imposed, which says that the layer of fluid right next to the boundary is stationary. Indeed, this will be important in certain fluid flows (actually, very important!) but these kinds of boundary conditions arise only when we take the viscosity of the fluid into account. For that reason we postpone their discussion to Chapter 2.

1.4.2 Flow Around a Sphere

Perhaps the most familiar solution to the Laplace equation (and certainly the one most useful in the book on Electromagnetism) is the spherically symmetric potential

$$\phi(r) = -\frac{q}{r} \tag{1.113}$$

for some constant q. This corresponds to a radial, three-dimensional flow

$$\mathbf{u} = \frac{q}{r^2}\hat{\mathbf{r}} \ . \tag{1.114}$$

Strictly speaking, this doesn't satisfy the Laplace equation everywhere. Instead, it is the Green's function, obeying

$$\nabla^2\phi = 4\pi q\delta^3(\mathbf{x}) \ . \tag{1.115}$$

The delta function should be thought of as a source (for $q > 0$) or a sink (for $q < 0$) for the fluid.

This radially symmetric solution is simple, but of little immediate utility in the context of fluid dynamics because it's hard to think of a situation in which a fluid spews out radially in 3d from some source. (Explosions do not usually obey the requirements of potential flow.) So instead we turn to slightly more complicated solutions. Our strategy is going to be a little bit cheap: rather than trying to solve a particular problem, we'll instead write down some simple potentials and then try to interpret the results in terms of some fluid flow that might be of interest. We then declare success at having solved something important!

To make progress, we work with spherical polar coordinates

$$x = r \sin\theta \cos\varphi \ , \quad y = r \sin\theta \sin\varphi \ , \quad z = r \cos\theta \ . \quad (1.116)$$

In these coordinates, the Laplacian takes the form

$$\nabla^2 = \frac{1}{r^2}\frac{\partial}{\partial r}\left(r^2\frac{\partial}{\partial r}\right) + \frac{1}{r^2 \sin\theta}\frac{\partial}{\partial \theta}\left(\sin\theta\frac{\partial}{\partial \theta}\right) + \frac{1}{r^2 \sin^2\theta}\frac{\partial^2}{\partial \varphi^2} \ . \quad (1.117)$$

We look for solutions that are independent of the coordinate φ. As explained in Volume 2 on Electromagnetism (and again in Volume 3 on Quantum Mechanics), the most general such solution can be written in terms of Legendre polynomials $P_n(\cos\theta)$,

$$\phi(r,\theta) = \sum_{n=0}^{\infty}\left(A_n r^n + \frac{B_n}{r^{n+1}}\right)P_n(\cos\theta) \ . \quad (1.118)$$

The radial solution that we saw above corresponds to the $n = 0$ term (with $P_0(\cos\theta) = 1$). The next simplest is the $n = 1$ term. Recalling that $P_1(\cos\theta) = \cos\theta$, this solution depends on two constants A and B,

$$\phi(r,\theta) = \left(Ar + \frac{B}{r^2}\right)\cos\theta \ . \quad (1.119)$$

Both of these terms have a natural interpretation in terms of fluid flow. The first term can be rewritten as $\phi = Az$, which tells us that it's simply a straight, constant flow in the z-direction. This is shown on the left of Figure 1.6. The flow runs left to right in the figure, which means that I've made the slightly disorienting choice of the taking the z-axis to lie horizontally. At large distances, this term dominates so we identify

$$A = U \quad (1.120)$$

as the asymptotic velocity of the fluid.

The second term can be viewed, in the language of electromagnetism and

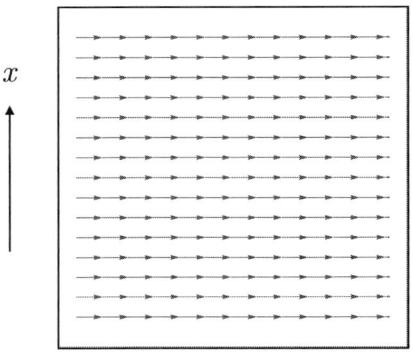
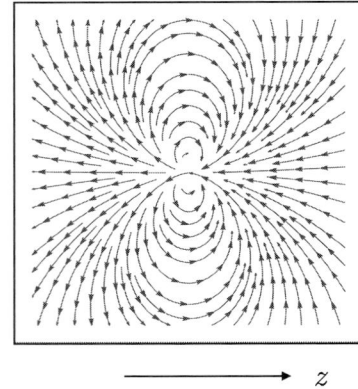

x

z

Fig. 1.6 On the left, the constant flow with $A \neq 0$ and $B = 0$. On the right, the dipole flow with $A = 0$ and $B \neq 0$.

of fluid mechanics, as a dipole. To see this, consider a source and sink of the form (1.113) displaced slightly in some direction $\mathbf{d}$. The potential is

$$\phi = \frac{q}{r} - \frac{q}{|\mathbf{r} + \mathbf{d}|} \ . \tag{1.121}$$

We then look at $\phi(\mathbf{x})$ at distances $r \gg |\mathbf{d}|$. We Taylor expand the second term as

$$\frac{1}{|\mathbf{r} + \mathbf{d}|} \approx \frac{1}{r} + \mathbf{d} \cdot \nabla \frac{1}{r} + \dots = \frac{1}{r} - \frac{\mathbf{d} \cdot \mathbf{r}}{r^3} + \dots \ . \tag{1.122}$$

The potential (1.121) then becomes

$$\phi \approx q \frac{\mathbf{d} \cdot \mathbf{r}}{r^3} + \dots \ . \tag{1.123}$$

If we take the displacement to be aligned with the z-direction, so $\mathbf{d} = d\hat{\mathbf{z}}$ and $\mathbf{d} \cdot \mathbf{r} = dr \cos\theta$, and subsequently take the limit $|\mathbf{d}| \to 0$ keeping the product qd fixed, then we get the second term in (1.119) with $B = qd$. The velocity field can be computed in spherical polar coordinates,

$$\mathbf{u} = \frac{\partial \phi}{\partial r} \hat{\mathbf{r}} + \frac{1}{r} \frac{\partial \phi}{\partial \theta} \hat{\boldsymbol{\theta}} \ . \tag{1.124}$$

The resulting fluid flow is shown on the right in Figure 1.6.

Because the Laplace equation is linear, we can simply add these two flows together for any choice of $A = U$ and B. The result is shown on the left of Figure 1.7. So far it's not immediately obvious that we've constructed something useful. However, if we look at the velocity, we find something

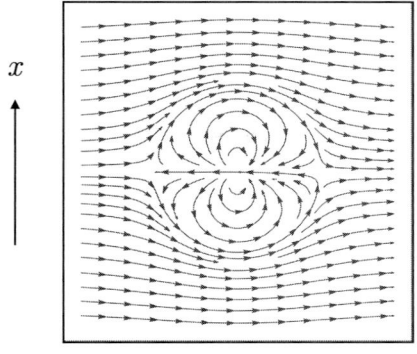 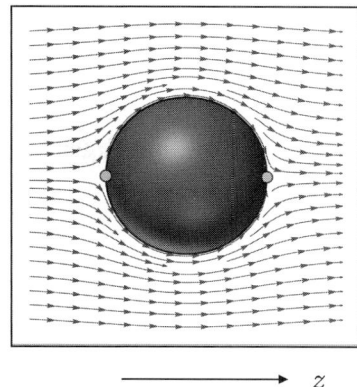

x

z

On the left, the constant flow superposed with the dipole flow. On the right, a well-placed solid sphere, hiding the messy bit.

interesting. The radial and angular velocity are given by

$$u_r = \frac{\partial \phi}{\partial r} = \left(U - \frac{2B}{r^3} \right) \cos \theta$$

$$u_\theta = \frac{1}{r} \frac{\partial \phi}{\partial \theta} = - \left(U + \frac{B}{r^3} \right) \sin \theta \; . \tag{1.125}$$

Crucially, the radial velocity vanishes at a radius R where

$$R^3 = \frac{2B}{U} \; . \tag{1.126}$$

This means that the flow has the appropriate boundary conditions to hold if there is a solid sphere of radius R at the origin. Nothing is flowing into the sphere! We then just ignore the flow inside the sphere with $r < R$ completely. It is only what sits outside that matters. This is shown on the right of Figure 1.7. The point $\theta = 0$ sits on the right of the sphere, and the point $\theta = \pi$ sits on the left, where the fluid comes from.

The upshot is that the potential

$$\phi = U \left(r + \frac{R^3}{2r^2} \right) \cos \theta \tag{1.127}$$

describes a flow of asymptotic velocity U past a solid sphere of radius R. Standard uniqueness theorems then tell us that it is *the* flow with these properties.

We've chosen to describe a flow with asymptotic velocity U and a stationary sphere. Alternatively, we could boost by U. This means that we remove the constant U term in (1.127) to describe a fluid that is asymptotically stationary, but with a sphere moving through it at speed U.

The velocity perpendicular to the sphere vanishes, but the velocity u_θ tangent to the surface of the sphere does not vanish when $r = R$. We may wonder how realistic this is for actual fluids and the answer, in many situations, is not very! We'll revisit this when we come to discuss viscosity.

There are a number of interesting features of the flow (1.127). First, there are two points where the flow stops completely and $\mathbf{u} = 0$. This happens on the surface of the sphere, $r = R$, at $\theta = 0$ (on the right) and $\theta = \pi$ (on the left) as depicted by the small dots in Figure 1.7. This occurs when the fluid comes in with vanishing impact parameter and, on symmetry grounds, can't tell whether to go up or down. So instead it stops. Points where the local fluid velocity vanishes are called *stagnation points*.

Next, we can look at the top and bottom of the sphere with $\theta = \pm\pi/2$. From (1.125), we see that the velocity on the boundary of the sphere is

$$|\mathbf{u}_{\text{top}}| = \frac{3}{2}U \ . \tag{1.128}$$

In other words, the fluid speeds up as it moves past the sphere. In fact, this follows from Bernoulli's principle as we explain below. Relatedly, you can see that the streamlines get squeezed together at the top and bottom of the sphere. This is familiar in other situations: stand at the top of a hill and it's windier than it was at the bottom.

1.4.3 d'Alembert's Paradox

Next we calculate the pressure that the fluid exerts on the sphere. For this we use Bernoulli's principle which says that the function H defined in (1.49) remains constant along streamlines (and, because the flow is irrotational, throughout the fluid). Asymptotically

$$H = \frac{1}{2}\rho U^2 + P_\infty \tag{1.129}$$

where P_∞ is the asymptotic pressure of the flow. Meanwhile, on the surface of the sphere

$$H = \frac{9}{8}\rho U^2 \sin^2\theta + P(\theta) \ . \tag{1.130}$$

So the pressure on the surface of the sphere is

$$P(\theta) = P_\infty + \frac{1}{2}\rho U^2 \left(1 - \frac{9}{4}\sin^2\theta\right) . \tag{1.131}$$

The pressure is highest at the two stagnation points, $\theta = 0$ and $\theta = \pi$. But here's the weird thing: the pressure depends only on $\sin^2\theta$. This means that the pressure exerted on the sphere at the front, where $\pi/2 < \theta \leq \pi$ (this is on the left in the figure) is identical to the pressure exerted behind, where $0 \leq \theta < \pi/2$ (on the right in the figure). And that doesn't sound right at all! We know from experience that an object placed in a stream will suffer a *drag force* which, in this case, should serve to carry the sphere along with the flow. But that's not what we find! Instead the flow finds a way to move seamlessly around the object, exerting no force.

Said differently, we can always boost our solution by speed U so that the fluid is stationary and the sphere is moving through it with speed U. The result above says that the sphere experiences no friction. It just glides through the fluid unimpeded. Tantalising as this sounds, it's simply not right. The fact that the maths differs so wildly from observation is known as the d'Alembert paradox, after the mathematician Jean le Rond d'Alembert who first uncovered this puzzle in 1752.

Another Historical Aside

d'Alembert concluded his paper with:

> It seems to me that the theory, developed in all possible rigour, gives, at least in several cases, a strictly vanishing resistance, a singular paradox which I leave to future Geometers to elucidate.

Future geometers (and physicists) took their time. Even though the Euler equation was replaced by the Navier–Stokes equation, which includes the effects of viscosity, there are arguments that suggest that, at least for fast fluid flows, the effects of viscosity are negligible. Which, if true, would mean that the solution we've described here provides a good approximation for fast-moving fluids and the drag force should be close to zero. Needless to say, that's not what's observed.

As the years wore on, this curious mathematical puzzle turned into something of an embarrassment. And this embarrassment, in turn, grew into a sense of genuine shame as controlled, powered flight became a reality. The Wright brothers made their famous first flight in December 1903, a century and a half after d'Alembert's work, yet the basic resolution of his paradox was still not fully understood, leaving theoretical physicists at something of a disadvantage in explaining the most important technology of the day. The breakthrough came only in 1905 (a good year for physics) and the work of Prandtl on boundary layers. We will describe this in Section 2.5.

Yet Prandtl's discovery was far from a proof and its full importance took some time to seep in. Even as late as 1915, the great Rayleigh finished a review of a book on hydrodynamics with:

> We may hope that before long [artificial flight] may be co-ordinated and brought into closer relation with theoretical hydrodynamics. In the meantime one can scarcely deny that much of the latter science is out of touch with reality.

Part of the goal of this book is to explain that, happily, theoretical hydrodynamics is very much in touch with reality. It's just a little more subtle than the simple approach we've taken here.

1.4.4 A Bubble Rising

A small variation on the calculation allows us to understand how bubbles rise to the surface in a fluid, at least in the inviscid approximation that underlies this section.

Consider a sphere of radius R and mass M moving through a stationary fluid with speed U. At the time when the bubble is centred at the origin, the flow is described by the potential

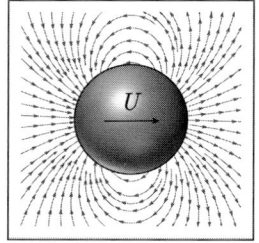

$$\phi = -\frac{UR^3}{2r^2}\cos\theta \ . \qquad (1.132)$$

To see that this is the appropriate flow function, note that $\mathbf{u} \to 0$ as $r \to \infty$, so the fluid is indeed asymptotically stationary. Moreover, on the surface of the sphere $\mathbf{u} = U\cos\theta\,\hat{\mathbf{r}} + \frac{1}{2}U\sin\theta\,\hat{\boldsymbol{\theta}}$ so $\mathbf{u}\cdot\hat{\mathbf{r}} = U\cos\theta$, which is the relevant boundary condition (1.112) for a sphere moving with speed U. The flow is shown in the figure above.

We can calculate the kinetic energy of the fluid

$$
\begin{aligned}
T_{\text{fluid}} &= \frac{1}{2}\rho\int_{r>R}(\nabla\phi)^2\,dV \\
&= \frac{1}{2}\rho\int_{r>R}\left(\nabla\cdot(\phi\nabla\phi) - \phi\nabla^2\phi\right)dV \ . \qquad (1.133)
\end{aligned}
$$

The second term vanishes because the potential obeys $\nabla^2 \phi = 0$. The first term can be evaluated by the divergence theorem and gives

$$T_{\text{fluid}} = \frac{1}{2}\rho \int \phi \, \mathbf{u} \cdot d\mathbf{S} = \frac{1}{4}\rho R^3 U^2 \int_0^{2\pi} d\varphi \int_0^{\pi} d\theta \, \cos^2\theta \sin\theta$$
$$= \frac{\pi}{3}\rho R^3 U^2 \tag{1.134}$$

where $d\mathbf{S}$ is the inward-pointing normal on the sphere, reflecting the fact that the bulk integral is over the exterior region $r > R$. To this we should add the kinetic energy $\frac{1}{2}MU^2$ of the sphere itself, so that the total kinetic energy of the sphere and the fluid is

$$T_{\text{total}} = \frac{1}{2}MU^2 + T_{\text{fluid}} = \frac{1}{2}M_{\text{eff}}U^2 \ . \tag{1.135}$$

Here we've introduced the effective mass M_{eff} as the combined mass of the sphere, together with the surrounding fluid,

$$M_{\text{eff}} = M + \frac{2\pi}{3}\rho R^3 \ . \tag{1.136}$$

Note that the additional contribution $\frac{2\pi}{3}\rho R^3$ is precisely half the mass of the fluid displaced by the sphere.

Now consider the case of a bubble, with $M = 0$. The effective mass is just $M_{\text{eff}} = \frac{2\pi}{3}\rho R^3$. But this bubble is an absence of water. This means that if we raise it by some height z then the potential energy is *lowered*! This is most easily seen because moving a bubble upwards is the same as moving the displaced water downwards. The potential energy is then

$$V = -\left(\frac{4\pi}{3}\rho R^3\right)gz \tag{1.137}$$

where the factor of $\frac{4\pi}{3}\rho R^3 = 2M_{\text{eff}}$ is the mass of the displaced water. The total energy of the bubble is then

$$E = \frac{1}{2}M_{\text{eff}}U^2 - 2M_{\text{eff}}gz \ . \tag{1.138}$$

The minus sign means that the bubble rises, rather than falls, due to gravity. Of course, we knew that anyway. The factor of 2 is perhaps more surprising: it says that the bubble accelerates upwards at twice the usual gravitational acceleration

$$\dot{U} = 2g \ . \tag{1.139}$$

The idea that the absence of something can itself be viewed as a new object – the bubble – is rather intuitive in this context. A more quantum version of the same idea also arises in the theory of solids where the absence of an electron acts very much like a particle with positive electric charge, known as a "hole".

Looking (Far) Forwards: Renormalisation

Before we go on, it's worth pausing to point out that, hidden inside this simple calculation, is an idea that will later blossom into something rather deep. This is the observation that, when immersed in a fluid, an object acts as if it has an effective mass M_{eff}, which is a combination of its original mass M together with the mass of the fluid that it drags along with it.

This same phenomenon occurs at a much more fundamental level for elementary particles such as electrons, quarks, and the Higgs boson. This arises because our universe is filled with different fields, each of which acts in many ways like a fluid. These fields include the familiar electric and magnetic fields as well as many others. When a particle moves through space, these fields become excited and get dragged along with the particle, much like a ball moving through water. The upshot is that the mass of an elementary particle has two contributions: an inherent mass, analogous to M, that is called the "bare mass" in particle physics, and an additional contribution from the other fields. These combine to give the mass M_{eff} which is what we measure in experiment.

The calculations that give rise to this shift in the mass go by the name of *renormalisation*. They can be rather challenging and you will have the pleasure of meeting them later in the book on Quantum Field Theory. While the underlying mathematics can seem daunting, it's worth keeping in mind that what's really going on is little different from the effective mass of a ball moving through water.

1.5 Potential Flows in 2d

In this section, we again look at potential flows with $\mathbf{u} = \nabla \phi$ and $\nabla^2 \phi = 0$, but this time in two dimensions. Equivalently, you can think of these as three-dimensional axisymmetric flows. You might think that things are simpler for 2d flows. However, as we will explain, there is a novelty that doesn't arise in the 3d case. This comes from topology.

We work in 2d polar coordinates (r, θ) so Laplace's equation takes the form

$$\nabla^2 \phi = \frac{1}{r} \frac{\partial}{\partial r} \left(r \frac{\partial \phi}{\partial r} \right) + \frac{1}{r^2} \frac{\partial^2 \phi}{\partial \theta^2} = 0 \ . \tag{1.140}$$

We will explore the different solutions to this equation.

Radial and Angular Flows

Once again, the simplest rotationally invariant solution is not particularly useful: it is the radial, planar flow

$$\phi = q \log \left(\frac{r}{r_0} \right) \quad \Longrightarrow \quad \mathbf{u} = \frac{q}{r} \hat{\mathbf{r}} \ . \tag{1.141}$$

This is again the Green's function, now obeying the 2d Poisson equation

$$\nabla^2 \phi = 2\pi q \, \delta^2(\mathbf{x}) \ . \tag{1.142}$$

However, this time there is a second, rotationally invariant flow. It arises from the potential

$$\phi = \frac{\Gamma}{2\pi} \theta \tag{1.143}$$

with Γ a constant. First note that this is *not* a single-valued potential because $\phi(\theta + 2\pi) \neq \phi(\theta)$ and you may wonder about the validity of such potentials. To see the consequence, we can simply compute the velocity field

$$\mathbf{u} = \nabla \phi = \frac{\Gamma}{2\pi r} \hat{\boldsymbol{\theta}} \ . \tag{1.144}$$

We've met this velocity field before! It was given in (1.63) (where you should substitute $f(r) = \Gamma/2\pi r^2$ in (1.63)). This is the flow that has the property that it swirls around the origin, even though it is irrotational, with $\nabla \times \mathbf{u} = 0$, at least away from $r = 0$. The parameter Γ measures the circulation of the flow

$$\Gamma = \oint_C \mathbf{u} \cdot d\mathbf{x} \tag{1.145}$$

where the integral is taken around any curve that surrounds the origin. Usually the integral of any conservative vector field like $\mathbf{u} = \nabla \phi$ around a closed curve is necessarily vanishing. The reason that it's not the case here is because ϕ is not single-valued. It is the ability to have circulation in 2d flows that adds some extra spice to the proceedings. We'll now see how this manifests itself in a simple example.

1.5.1 Circulation Around a Cylinder

We consider the flow around an infinite cylinder, aligned along the y-direction. This ensures that the flow is effectively two-dimensional; we care only about the velocity in the (x, z)-plane.

The start of our story is the same as the flow around a sphere that we saw in the previous section. The most general solution to the 2d Laplace equation is

$$\phi(r, \theta) = A_0 \log r + B_0 \theta + \sum_{n=1}^{\infty} \left(A_n r^n \cos(n\theta + \alpha_n) + \frac{B_n}{r^n} \cos(n\theta + \beta_n) \right) .$$

Here we focus on the flows with $n = 1$. The integration constants A_0 and B_0 will play no role for now, so we set them to zero and look at

$$\phi = U \left(r + \frac{R^2}{r} \right) \cos \theta . \tag{1.146}$$

This is very similar to the 3d potential (1.119). Again, the first term describes a constant flow with asymptotic velocity U, a fact that we've anticipated in labelling the overall coefficient. The second term is now a two-dimensional dipole. Combined, they give rise to the velocity field

$$\mathbf{u} = \frac{\partial \phi}{\partial r} \hat{\mathbf{r}} + \frac{1}{r} \frac{\partial \phi}{\partial \theta} \hat{\boldsymbol{\theta}}$$
$$= U \left(1 - \frac{R^2}{r^2} \right) \cos \theta \, \hat{\mathbf{r}} - U \left(1 + \frac{R^2}{r^2} \right) \sin \theta \, \hat{\boldsymbol{\theta}} . \tag{1.147}$$

We see that the radial component has the property that

$$u_r = U \left(1 - \frac{R^2}{r^2} \right) \cos \theta = 0 \quad \text{when } r = R . \tag{1.148}$$

This means that this potential describes the flow past a solid cylinder of radius R. The velocity field $\mathbf{u}$ is shown in Figure 1.8, with the two stagnation points shown as the two small circles. The details are slightly different, but the qualitative features are the same as for the sphere.

Adding Circulation

Things get more interesting if we add some circulation. Because the Laplace equation is linear, we can superpose the flow around the cylinder (1.146) with the rotation (1.143),

$$\phi = U \left(r + \frac{R^2}{r} \right) \cos \theta + \frac{\Gamma}{2\pi} \theta . \tag{1.149}$$

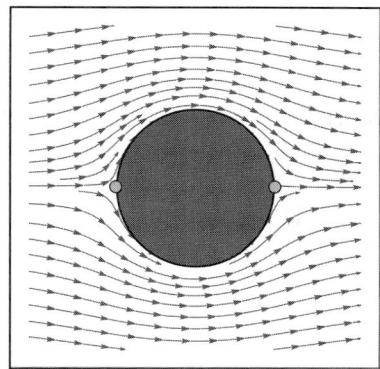

Fig. 1.8   Two-dimensional potential flow around a cylinder.

The extra term affects only the angular part of the velocity, which now takes the form

$$\mathbf{u} = U\left(1 - \frac{R^2}{r^2}\right)\cos\theta\,\hat{\mathbf{r}} + \left[\frac{\Gamma}{2\pi r} - U\left(1 + \frac{R^2}{r^2}\right)\sin\theta\right]\hat{\boldsymbol{\theta}}\ . \quad (1.150)$$

You can check that the associated stream function (1.18) is

$$\psi = Ur\left(1 - \frac{R^2}{r^2}\right)\sin\theta - \frac{\Gamma}{2\pi}\log\left(\frac{r}{r_0}\right)\ . \quad (1.151)$$

To understand the effect on the flow, we can search for the stagnation points at which $\mathbf{u} = 0$. Clearly $u_r = 0$ provided that we sit at radius $r = R$. The angular velocity then vanishes at the angle θ such that

$$\Gamma = 4\pi U R \sin\theta\ . \quad (1.152)$$

But this has a solution only when $|\Gamma| < 4\pi U R$ (where we're taking $U > 0$). This suggests that the flow will be different for small and large circulation Γ.

We start by looking at small $|\Gamma| < 4\pi U R$ so that there are two stagnation points on the surface of the cylinder at $\sin\theta = \Gamma/4\pi U R$. The corresponding flow is shown on the left of Figure 1.9. (I've taken $\Gamma < 0$ in this figure for reasons that will become apparent below.) Note that the stagnation point plays an important role: this is where the fluid separates, with streamlines on either side taking different paths, one above the cylinder and the other below.

Meanwhile, when $|\Gamma| > 4\pi U R$, there is no stagnation point on the surface of the cylinder. Instead it now occurs at $\theta = \pi/2$ (which ensures that $u_r = 0$)

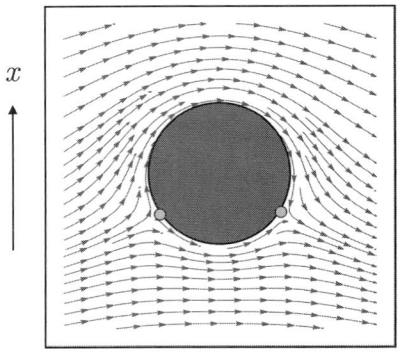
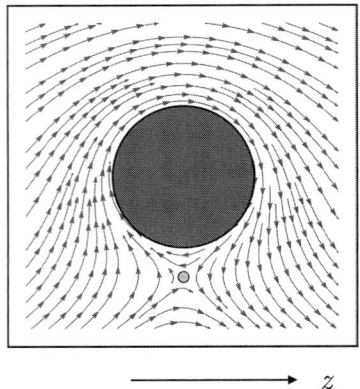

Fig. 1.9 On the left, the flow around a cylinder when the circulation is small; on the right, when the circulation is big.

and a distance r from the centre, given by the solution to the quadratic

$$r^2 - \frac{\Gamma r}{2\pi U} + R^2 = 0 \ . \tag{1.153}$$

This ensures that $u_\theta = 0$. The quadratic is guaranteed to have one positive root sitting outside the sphere provided that $|\Gamma| > 4\pi U R$. The flow is shown on the right of Figure 1.9, again with the stagnation point shown as a dot below the cylinder.

1.5.2 Lift and the Magnus Force

Now we can repeat the calculation that we performed for the sphere to answer the question: What's the pressure that the fluid exerts on the cylinder? We use Bernoulli's principle and the conservation of H throughout the flow. At infinity we have

$$H = \frac{1}{2}\rho U^2 + P_\infty \tag{1.154}$$

while, on the surface of the cylinder, it is

$$H = \frac{1}{2}\rho \left(\frac{\Gamma}{2\pi R} - 2U \sin\theta \right)^2 + P(\theta) \ . \tag{1.155}$$

So the pressure on the surface of the cylinder is

$$P(\theta) = P_\infty + \frac{1}{2}\rho U^2 \left(1 - 4\sin^2\theta\right) + \frac{U\Gamma\rho}{\pi R} \sin\theta - \frac{\Gamma^2\rho}{8\pi^2 R^2} \ . \tag{1.156}$$

The pressure acts radially on the cylinder. We want to decompose this force to compute the component forces F_z in the z-direction (horizontal in the flow diagrams in Figure 1.9) and F_x in the x-direction (vertically in Figure 1.9). From the figure below, we see that

$$F_z = \int_0^{2\pi} P(\theta) R \cos\theta \, d\theta = 0 \ . \ (1.157)$$

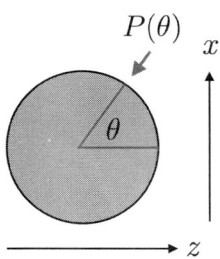

So there is no force in the direction of the flow. Or, said differently, there is no drag force. This is the same result that we saw for the sphere and leads to d'Alembert's paradox. The novelty is that the force perpendicular to the asymptotic flow is non-vanishing: it receives a contribution from the $\sin\theta$ term in (1.156)

$$F_x = -\int_0^{2\pi} P(\theta) R \sin\theta \, d\theta = -\frac{U\Gamma\rho}{\pi}\int_0^{2\pi} \sin^2\theta \, d\theta = -U\Gamma\rho \ . \ (1.158)$$

The minus sign means that, for $\Gamma < 0$ as shown in Figure 1.9, the force is upwards. This makes sense: if you look at Figure 1.9, you see that the streamlines are closer together at the top of the cylinder. This means that the fluid is travelling faster at the top and, correspondingly, there is a lower pressure. Hence the upwards force. This force is called *lift*. We took $\Gamma < 0$ in Figure 1.9 to save ourselves the embarrassment of having a force called "lift" that acts downwards. Although, in that case, it has a different name and is called simply *downforce*. It is what keeps Formula One cars on the track.

In the calculation above, we took the fluid to be circulating and the cylinder to be stationary. However, the same effect occurs if the cylinder rotates while the fluid has no circulation.

There is a related effect that has all the hallmarks of the lift described above but is not, in fact, the same. This is what happens when you kick a football and put spin on it, making it swerve in the air. Think Roberto Carlos and his wonderful banana kick in the 1997 World Cup. This effect is known as the *Magnus force*. However, a little thought will convince you that the Magnus force actually works in the opposite direction to lift. If you kick the ball on the right-hand side, then it will bend to the left. But that means that the force on the ball is in the direction in which the relative air flow is slowest! The physics behind the Magnus force is something else entirely: turbulence. The spin of the ball delays the separation of the boundary layer and leads to a deflected wake, and this is what causes the ball to bend.

These words will be explained in the next chapter, although the full magic of Roberto Carlos' foot will not.

1.6 A Variational Principle

The laws of physics can be expressed using the principle of least action. What about the laws of fluid mechanics?

The action principle is best suited to fundamental laws of physics where there is no friction at play. The full Navier–Stokes equation for fluids (that we will meet in Chapter 2) includes a friction term and so isn't immediately amenable to a formulation using an action. But the Euler equation that we've studied in this section has no such friction term which suggests that it should be possible to write down an action that gives rise to the Euler equation. The question is: How?

This, it turns out, is not quite as straightforward as one might think. But it is possible and, moreover, gives some insight into the mathematical structure of the Euler equation. The purpose of this section is to describe this.

This section is something of a tangent and we won't be returning to the action principle later in the book, not least because we'll be embracing the full Navier–Stokes equation. Also, the terminology in this section can be a little confusing simply because Euler and Lagrange were rather impressive mathematicians. To give you a sense of this, our goal is to work in the Eulerian framework of fluid mechanics, rather than the Lagrangian framework, and then write down a Lagrangian and derive the Euler–Lagrange equations to reproduce the Euler equation. All clear? Good.

1.6.1 The Principle of Least Action

We start by giving a review of the principle of least action, both in the framework of classical mechanics and also in classical field theory. You can read more about this in the books on Classical Mechanics and Electromagnetism.

First, Newtonian mechanics. We consider a single particle with a position given by $\mathbf{x} \in \mathbb{R}^3$. The position changes with time, so the trajectory of a particle traces out a curve $\mathbf{x}(t)$. Of all these possible trajectories, there is typically one that obeys the laws of physics. We want to know which one.

If the particle has mass m then its kinetic energy is $T = \frac{1}{2}m\dot{\mathbf{x}}^2$. We'll assume that the particle experiences a potential energy $V(\mathbf{x})$. We then define the *Lagrangian*

$$L(\mathbf{x}, \dot{\mathbf{x}}) = T - V \tag{1.159}$$

and, from this, the *action*

$$S[\mathbf{x}(t)] = \int dt \; L(\mathbf{x}, \dot{\mathbf{x}}) = \int dt \; \left[\frac{1}{2}m\dot{\mathbf{x}}^2 - V(\mathbf{x}) \right] . \tag{1.160}$$

The action assigns a number S to each trajectory $\mathbf{x}(t)$. Strictly speaking, we should consider the action for all trajectories with certain boundary conditions specified, such as $\mathbf{x}(t_0) = \mathbf{x}_0$ and $\mathbf{x}(t_1) = \mathbf{x}_1$. This is important, but we'll sweep it under the rug in what follows.

The principle of least action states that the true trajectory $\mathbf{x}(t)$ followed by the particle is the one that extremises the action S. Mathematically, this means the following. Suppose that you have a putative trajectory $\mathbf{x}(t)$ with some action S. We look at all neighbouring trajectories $\mathbf{x}(t) + \delta\mathbf{x}(t)$ and compute their action $S + \delta S$. The original trajectory is the one taken by the particle if $\delta S = 0$ for *all* variations $\delta\mathbf{x}(t)$.

For our action (1.160), we have

$$S[\mathbf{x}(t) + \delta\mathbf{x}(t)] = \int dt \; \left[\frac{1}{2}m(\dot{\mathbf{x}} + \delta\dot{\mathbf{x}})^2 - V(\mathbf{x} + \delta\mathbf{x}) \right] \tag{1.161}$$

$$\approx \int dt \; \left[\frac{1}{2}m(\dot{\mathbf{x}}^2 + 2\dot{\mathbf{x}} \cdot \delta\dot{\mathbf{x}}) - V(\mathbf{x}) - \nabla V \cdot \delta\mathbf{x} \right] = S + \delta S$$

where, in going to the second line, we've ignored all terms of order $\delta\mathbf{x}^2$ and higher. This gives us an expression for the variation of the action δS which we can now play with

$$\delta S = \int dt \; [m\dot{\mathbf{x}} \cdot \delta\dot{\mathbf{x}} - \nabla V \cdot \delta\mathbf{x}] = \int dt \; [-m\ddot{\mathbf{x}} - \nabla V] \cdot \delta\mathbf{x} . \tag{1.162}$$

In the second equality we've integrated by parts and thrown away the boundary terms. (We've been careless about why one can throw away boundary terms after integration by parts; that's the bit we're sweeping under the rug.) The principle of least action states that the true trajectory has $\delta S = 0$ for all possible variations $\delta\mathbf{x}$. This can only be true if the expression in square brackets vanishes, meaning

$$m\ddot{\mathbf{x}} = -\nabla V . \tag{1.163}$$

This, of course, is the Newtonian equation of motion. The principle of least action is just a recasting of this familiar result.

The action for a given equation of motion is not necessarily unique. Here, for example, is a different action that yields the same equation of motion (1.163). We will initially think of the position $\mathbf{x}(t)$ and velocity $\mathbf{v}(t)$ of the particle as independent quantities. We'll then enforce the requirement $\mathbf{v} = \dot{\mathbf{x}}$ through the use of a Lagrange multiplier. The upshot is that we can write down the action

$$S[\mathbf{x}(t), \mathbf{v}(t), \boldsymbol{\beta}(t)] = \int dt \ \left[\frac{1}{2} m \mathbf{v}^2 - V(\mathbf{x}) - \boldsymbol{\beta} \cdot (\mathbf{v} - \dot{\mathbf{x}}) \right] \ . \quad (1.164)$$

The equation of motion for $\boldsymbol{\beta}$ reproduces the constraint $\mathbf{v} = \dot{\mathbf{x}}$, while the equation of motion for $\mathbf{v}$ tells us that we should identify the Lagrange multiplier with the velocity: $m\mathbf{v} = \boldsymbol{\beta}$. Finally, the equation of motion for $\mathbf{x}$ is $\dot{\boldsymbol{\beta}} = -\nabla V$. Combining these reproduces (1.163).

For the Newtonian particle, there's clearly no advantage to writing the action (1.164) over (1.160). Indeed, it seems a little perverse to do so. But these kinds of tricks can prove useful in other contexts and one of these turns out to be fluid dynamics.

An Action for Fields

The next conceptual step is to move from particles to fields. Consider a scalar field $\varphi(\mathbf{x}, t)$ which associates a number to each point in space and time. Note, in particular, that the role of the spatial coordinate $\mathbf{x}$ has changed. In the context of Newtonian mechanics, $\mathbf{x}$ was the dynamical degree of freedom, something that evolves over time. But in field theory that's no longer the case. Now $\mathbf{x}$ is just a label, like time t, and the field φ is the dynamical degree of freedom whose value depends on both space and time.

We would like to write down an action for this field. This means that we want to associate a number S to each possible field configuration $\varphi(\mathbf{x}, t)$. We start by defining the *Lagrangian density* $\mathcal{L}$ (although everyone simply refers to it as the "Lagrangian"). A natural choice, which is the analogue of (1.159), is

$$\mathcal{L}(\varphi, \dot{\varphi}, \nabla\varphi) = \frac{1}{2}\dot{\varphi}^2 - \frac{1}{2}c^2(\nabla\varphi)^2 - V(\varphi) \ . \quad (1.165)$$

We have a kinetic-energy-type term, $\dot{\varphi}^2$, but now we have two different kinds of potential energy. The first, proportional to $\nabla\varphi^2$, is an energy arising from spatial gradients of the field. It comes with a constant coefficient c which has dimension $[c] = LT^{-1}$. In many situations, this is the speed of ripples of the field. In addition, we have a second potential energy $V(\varphi)$ which depends only on φ and not on its derivatives. We pick different potentials $V(\varphi)$ to model the situation that we're interested in, just like $V(\mathbf{x})$ in Newtonian mechanics. Typically one picks $V(\varphi)$ so that it penalises large values of φ, e.g. $V(\varphi) \sim \varphi^2$. Here we'll keep $V(\varphi)$ general.

We associate an action S to a given field configuration $\varphi(\mathbf{x}, t)$ by integrating the Lagrangian over both space and time,

$$S = \int dt \, d^3x \, \mathcal{L} = \int dt \, d^3x \, \left[\frac{1}{2}\dot{\varphi}^2 - \frac{1}{2}c^2(\nabla\varphi)^2 - V(\varphi) \right] . \quad (1.166)$$

It's worth stressing, for the second time, the different roles that the spatial coordinate plays in (1.160) and (1.166). It has been demoted from its role as a dynamical degree of freedom in the former to a mere integration variable in the latter.

At this point, we proceed in much the same way as for the Newtonian particle. We take a reference field configuration $\varphi(\mathbf{x}, t)$ and compute its action S. Then we look at all nearby field configurations $\varphi(\mathbf{x}, t) + \delta\varphi(\mathbf{x}, t)$ and compute their action $S + \delta S$. The original field configuration obeys the classical equations of motion if and only if $\delta S = 0$ for all $\delta\varphi$. In equations, we have

$$S[\varphi + \delta\varphi] = \int dt \, d^3x \, \left[\frac{1}{2}(\dot{\varphi} + \delta\dot{\varphi})^2 - \frac{1}{2}c^2(\nabla\varphi + \nabla\delta\varphi)^2 - V(\varphi + \delta\varphi) \right]$$

$$\approx \int dt \, d^3x \, \left[\frac{1}{2}(\dot{\varphi}^2 + 2\dot{\varphi}\,\delta\dot{\varphi}) - \frac{1}{2}c^2(\nabla\varphi^2 + 2\nabla\varphi \cdot \nabla\delta\varphi) \right.$$

$$\left. - V(\varphi) - \frac{\partial V}{\partial \varphi}\delta\varphi \right] \quad (1.167)$$

where, as before, we've truncated our expansion at leading order in $\delta\varphi$ in the second line. From this we can extract the variation of the action

$$\delta S = \int dt \, d^3x \, \left[\dot{\varphi}\,\delta\dot{\varphi} - c^2\nabla\varphi \cdot \nabla\delta\varphi - \frac{\partial V}{\partial \varphi}\delta\varphi \right]$$

$$= \int dt \, d^3x \, \left[-\ddot{\varphi} + c^2\nabla^2\varphi - \frac{\partial V}{\partial \varphi} \right] \delta\varphi . \quad (1.168)$$

Here we've again integrated by parts, now with respect to both temporal and spatial derivatives, so that all terms are proportional to $\delta\varphi$. We've also thrown away the boundary terms. Requiring that $\delta S = 0$ for all possible $\delta\varphi$

tells us that the expression in square brackets must vanish, so

$$\frac{\partial^2 \varphi}{\partial t^2} - c^2 \nabla^2 \varphi = -\frac{\partial V}{\partial \varphi} \ . \tag{1.169}$$

This is the *Klein–Gordon equation*. It is the simplest equation of motion for a classical scalar field.

The equation of motion (1.169) doesn't play a particularly prominent role in classical physics, where our heads are more likely to be turned by more sophisticated theories such as electromagnetism or general relativity. It does, however, arise in various cameo roles and we will meet it briefly in Section 3.3.2 when discussing a certain kind of wave that is driven by the Coriolis force. The Klein–Gordon equation only really comes to the fore when we turn to the world of Quantum Field Theory where it plays more of a starring role.

1.6.2 An Action Principle for Fluids

Now we are in a position to construct an action principle for fluids. Our goal is to write down an action that reproduces the Euler equation for an incompressible fluid

$$\rho \frac{D\mathbf{u}}{Dt} = -\nabla P \quad \text{and} \quad \nabla \cdot \mathbf{u} = 0 \ . \tag{1.170}$$

We could also include further forces, such as gravity, but since these don't add extra conceptual issues we will just ignore them and focus on the simplest equations above.

The first question that we should ask is: What are the dynamical degrees of freedom for a fluid? Until now, we have viewed (1.170) as four equations for four variables, $\mathbf{u}$ and P. But we might suspect that these aren't quite the right variables to construct an action. After all, when writing down an action for the Newtonian particle, it's important that we vary with respect to the position $\mathbf{x}$ rather than the velocity $\dot{\mathbf{x}}$. And the same is true for a fluid. To build an action, we need to start thinking about the "position" of the fluid.

To this end, we will think of the configuration of the fluid as a map from $\mathbb{R}^3 \mapsto \mathbb{R}^3$

$$\mathbf{x} \mapsto \alpha^i(\mathbf{x}, t) \quad i = 1, 2, 3 \ . \tag{1.171}$$

Here $\mathbf{x}$ labels the fixed positions in space, while $\alpha^i(\mathbf{x}, t)$ labels the position of parcels of the fluid. This is the Eulerian (as opposed to Lagrangian) description of a fluid. We will refer to α^i as the embedding coordinate of the fluid.

We will think of $\alpha^i(\mathbf{x}, t)$ as the fields of our Lagrangian although, as we will see, they will need to be augmented by several more. But even before we get going, it's worth pointing out that $\alpha^i(\mathbf{x}, t)$ aren't quite like other fields. This is because the map from $\mathbb{R}^3 \mapsto \mathbb{R}^3$ that describes our fluid must be invertible. For example, there's no configuration of the fluid with, say, $\alpha^i(\mathbf{x}, t) = 0$ for all $\mathbf{x}$. That would describe the entire fluid as sitting at a single point and that's not allowed. In fact, because our fluid is incompressible, we should require that the map from $\mathbb{R}^3 \mapsto \mathbb{R}^3$ is volume preserving. This is assured if

$$\det\left(\frac{\partial \alpha^i}{\partial x^j}\right) = 1 \ . \tag{1.172}$$

We will have to find a way to impose a constraint like this on our map.

(As an aside: these kinds of constraints are not entirely unfamiliar in other contexts. In general relativity, the dynamical degree of freedom is a metric $g_{\mu\nu}(\mathbf{x}, t)$ but, as with a fluid, we're not allowed to set $g_{\mu\nu} = 0$. Instead, we must have $\det(g_{\mu\nu}) \neq 0$.)

We describe the fluid by the maps α^i. How do we define the velocity? You might naively guess that it's just $\dot{\alpha}^i$, but that's not right. Instead, we need to think more physically. Suppose that you focus on one particular parcel of fluid, say the one labelled by $\alpha^i = (3, 7, 4)$. Then we can follow this parcel through the fluid. If $\alpha^i(\mathbf{x}, t)$ changes then the parcel of fluid must have moved to a some neighbouring point, which means that the velocity $\mathbf{u}$ is non-zero. This velocity is defined implicitly as

$$\frac{\partial \alpha^i}{\partial t} + \mathbf{u} \cdot \nabla \alpha^i = 0 \ . \tag{1.173}$$

Because the map from $\mathbb{R}^3 \mapsto \mathbb{R}^3$ is invertible, we can get an explicit expression for $\mathbf{u}$ in terms of α^i. Using the fact that the map preserves volumes (1.172), this is given by

$$u^i(\mathbf{x}, t) = -\frac{1}{2} \epsilon^{ijk} \epsilon_{lmn} \frac{\partial \alpha^l}{\partial t} \frac{\partial \alpha^m}{\partial x^j} \frac{\partial \alpha^n}{\partial x^k} \ . \tag{1.174}$$

To see that this solves (1.173), you just need to use the definition of the determinant in terms of ϵ^{ijk}. It's also straightforward to show that the condition (1.172) ensures that $\nabla \cdot \mathbf{u} = 0$ as expected.

Note that for these incompressible flows, with $\nabla \cdot \mathbf{u} = 0$, the equation (1.173) takes the form of a conservation law $\partial \alpha^i / \partial t + \nabla \cdot (\mathbf{u}\alpha^i) = 0$. There is a simple physical intuition for this: it is just the statement that you can trace the evolution of a given parcel of fluid, a kind of "conservation of particle identity" if you like.

Now we've set up the basic kinematical structure for fluids, our next job is to write down the action. Here a number of choices await us. It is possible to write down an action just for the embedding coordinates $\alpha^i(\mathbf{x}, t)$, with the constraint (1.172) imposed by a Lagrange multiplier. While it's possible, it's also a little messy. It turns out to be more straightforward to write down an action for α^i and u^i, together with a collection of Lagrange multipliers. This is analogous to the slightly daft action (1.164) that we introduced for the Newtonian particle.

We take as our action

$$S[\boldsymbol{\alpha}, \mathbf{u}, \phi, \boldsymbol{\beta}] = \int dt\, d^3x \left[\frac{1}{2}\rho\mathbf{u}^2 + \phi\nabla \cdot \mathbf{u} - \beta_i \left(\frac{\partial \alpha^i}{\partial t} + \mathbf{u} \cdot \nabla\alpha^i \right) \right] . \quad (1.175)$$

The equations of motion arise from varying the action with respect to $\alpha^i(\mathbf{x}, t)$, $u^i(\mathbf{x}, t)$, and the Lagrange multipliers $\phi(\mathbf{x}, t)$ and $\beta^i(\mathbf{x}, t)$.

The Lagrange multipliers are easiest to deal with. Varying with respect to ϕ gives the incompressibility condition $\nabla \cdot \mathbf{u} = 0$, now directly in terms of velocity rather than the more abstract (1.172). Meanwhile, varying with respect to β_i gives us the relation (1.173) between the embedding coordinate and velocity. That leaves us with the equations of motion that come from varying the action with respect to α^i and u^i. If we vary with respect to α^i, we have

$$\frac{\partial \beta_i}{\partial t} + \mathbf{u} \cdot \nabla\beta_i = 0 . \quad (1.176)$$

We see that the Lagrange multipliers β_i obey the same equation (1.173) as the embedding coordinates. Meanwhile, varying with respect to the components of the velocity $\mathbf{u}$ gives the expression

$$\rho\mathbf{u} = \nabla\phi + \beta_i\nabla\alpha^i . \quad (1.177)$$

This is a curious equation, relating the velocity to ϕ, $\boldsymbol{\alpha}$, and $\boldsymbol{\beta}$. Note that the first term is familiar: it is just the kind of potential flow that we met in Section 1.4, with the Lagrange multiplier playing the role of the potential. But the second term is less familiar and it's not immediately obvious how this is related to the Euler equation. In particular, we haven't yet seen how the pressure emerges in this framework.

To make progress, we compute $D\mathbf{u}/Dt$ using the expression (1.177). There's a little bit of algebra involved, but it's not too hard to show that

$$\rho \frac{Du^i}{Dt} = \rho \left(\frac{\partial u^i}{\partial t} + u^j \frac{\partial u^i}{\partial x^j} \right) \tag{1.178}$$

$$= \frac{\partial}{\partial x^i} \left(\frac{\partial \phi}{\partial t} + \beta_j \frac{\partial \alpha^j}{\partial t} + \frac{1}{2} \rho \mathbf{u}^2 \right) + \frac{D\beta_j}{Dt} \frac{\partial \alpha^j}{\partial x^i} - \frac{D\alpha^j}{Dt} \frac{\partial \beta_j}{\partial x^i} .$$

But the last two terms vanish by virtue of (1.173) and (1.176). We're left with,

$$\rho \frac{D\mathbf{u}}{Dt} = -\nabla P \quad \text{where} \quad P = -\frac{\partial \phi}{\partial t} - \beta_j \frac{\partial \alpha^j}{\partial t} - \frac{1}{2} \rho \mathbf{u}^2 + \text{constant} \tag{1.179}$$

with the pressure given, as shown, by a combination of the velocity and Lagrange multipliers. This is the promised Euler equation, now derived from an action principle. Throughout this chapter, we've been mentioning that pressure is rather like a Lagrange multiplier, ensuring that $\nabla \cdot \mathbf{u} = 0$. The above variational principle makes the relationship clearer.

A Slightly Simpler Action

As we mentioned above, there are slightly simpler versions of the fluid action. Here we describe one that succeeds in eliminating the need for embedding coordinates completely. Instead, it uses the fact that a general velocity field $\mathbf{u}(\mathbf{x}, t)$ in $\mathbb{R}^3$ can be written as

$$\mathbf{u} = \nabla \phi + \beta \nabla \alpha \tag{1.180}$$

for some functions ϕ, β and α. (These functions are not unique.) This is sometimes known as the *Clebsch representation*. Note that it's very similar to the form of the velocity (1.177) that arose from our previous variational principle, except now there is just a single α and β function rather than a triplet.

There is a nice way of visualising the form of the velocity field (1.180). The first term is clearly the irrotational, potential flow that we met previously. The second term gives vorticity

$$\boldsymbol{\omega} = \nabla \times \mathbf{u} = \nabla \beta \times \nabla \alpha . \tag{1.181}$$

This is telling us that vortex lines (i.e. integral curves of $\boldsymbol{\omega}$) lie on the intersection of surfaces of constant α and constant β.

Now consider the action

$$S = \int dt \, d^3x \left[-\beta \frac{\partial \alpha}{\partial t} - \frac{1}{2} (\nabla \phi + \beta \nabla \alpha)^2 \right] . \tag{1.182}$$

This is closely related to our previous action (1.175): it's what you get if you substitute the expression (1.177) for $\mathbf{u}$ into the action and drop the $i = 1, 2, 3$ indices on α^i and β_i.

Now when varying the action, we must remember that the velocity $\mathbf{u}$ is defined by (1.180). The equation of motion for ϕ then tells us that $\nabla \cdot \mathbf{u} = 0$. Meanwhile, the equations of motions for α and β are, respectively,

$$\frac{D\beta}{Dt} = 0 \quad \text{and} \quad \frac{D\alpha}{Dt} = 0 \ . \tag{1.183}$$

We can now repeat our previous calculation to once again find the Euler equation

$$\rho \frac{D\mathbf{u}}{Dt} = -\nabla P \quad \text{with} \quad P = \rho \left(-\frac{\partial \phi}{\partial t} - \beta \frac{\partial \alpha}{\partial t} - \frac{1}{2} \left(\nabla \phi + \beta \nabla \alpha \right)^2 \right) \ . \tag{1.184}$$

There is one rather pretty consequence of this: the Lagrangian that appears in (1.182) is recognised as the pressure,

$$S[\phi, \alpha, \beta] = \frac{1}{\rho} \int dt \, d^3 x \ P[\phi, \alpha, \beta] \ . \tag{1.185}$$

If you're thermodynamically inclined, this is perhaps less surprising. In an appropriate ensemble, the pressure is equal to the free energy and there are situations where the action and free energy sit on the same footing.

So far we've built up some intuition for how fluids flow. But we've missed an ingredient and it is, it turns out, a rather important one. That ingredient is the internal friction experienced by a fluid, known as *viscosity*. We'll present a careful discussion of how viscosity arises shortly in Section 2.1, but first we give a quick, slightly hand-wavy derivation of the relevant term in the equation.

The Euler equation describing fluid motion is, as we've seen, just the continuum version of "$F = ma$". It is

$$\rho \frac{D\mathbf{u}}{Dt} = -\nabla P + \mathbf{f} \tag{2.1}$$

where $\mathbf{f}$ describes any other forces experienced by the fluid. We want to write down the force that arises from friction.

We already met friction briefly in our first book on Classical Mechanics. There we explained that friction is not a fundamental force, but one that arises from the underlying interactions of many (say, 10^{23}) particles. We also stated, without proof, that particles moving slowly with velocity v through very viscous fluids experience linear drag, with $F \sim -v$, while particles moving more quickly through less viscous fluids experience quadratic drag, $F \sim -v^2$. (For what it's worth, we'll recover the equations for linear drag in Section 2.4, while quadratic drag often involves turbulent flows and is not so easy to derive from first principles.) The first question that we want to ask is: What kind of friction force does the fluid experience when it rubs against itself?

The answer is that the viscosity is *linear* in the fluid velocity $\mathbf{u}(\mathbf{x}, t)$. At heart, this statement follows simply from a Taylor expansion. The underlying atomic interactions are complicated and they surely result in a friction force that is some arbitrarily complicated function of $\mathbf{u}$. But, for suitably small velocities, the linear term is always larger than the quadratic term. This simple but powerful argument is known as *linear response*. It's the same argument that leads to Ohm's law with the current given by $I = V/R$ rather than some more complicated function of voltage. Those fluids for which the linear approximation is a good one are called *Newtonian*.

So the viscosity should be a force $\mathbf{f}$ that is linear in the fluid velocity $\mathbf{u}$. What can we write down? First, the force can't be proportional to $\mathbf{u}$ itself: that's in contradiction with Galilean relativity, which says that the equations of fluid mechanics must be invariant under a boost of the whole fluid (and any container) by a constant velocity. This tallies with the idea that viscosity is associated to the friction force experienced by the fluid when one part rubs up against another. That means that the different parts of the fluid should be moving at different speeds for viscosity to kick in. Or, in other words, the friction force must depend on spatial changes of $\mathbf{u}$.

The simplest possibility is that the friction force depends on the first derivative of $\mathbf{u}$, but there is only one vector that we can form in this way, the vorticity $\boldsymbol{\omega} = \nabla \times \mathbf{u}$, and this isn't a good candidate for a force. That's because of the symmetry of *parity* or reflections, $\mathbf{x} \to -\mathbf{x}$. Under parity, $\mathbf{u} \to -\mathbf{u}$ and each term in the Euler equation is odd. This should continue to be true of any force $\mathbf{f}$ that acts on the fluid. But vorticity is even: $\boldsymbol{\omega} \to \boldsymbol{\omega}$. It cannot arise as a force.

This means that we must look to two-derivative terms if we are to build a force linear in velocity. There are two possibilities: $\nabla^2 \mathbf{u}$ and $\nabla(\nabla \cdot \mathbf{u})$. But, given that we are dealing with an incompressible fluid, the second of these necessarily vanishes. We're left with just one choice for our friction force

$$\mathbf{f}_{\text{viscous}} = \mu \nabla^2 \mathbf{u} \ . \tag{2.2}$$

The coefficient μ is a constant known as the *dynamic (shear) viscosity*. The resulting equation describing the motion of fluids is

$$\rho \frac{D\mathbf{u}}{Dt} = -\nabla P + \mu \nabla^2 \mathbf{u} + \mathbf{f} \tag{2.3}$$

where, in $\mathbf{f}$, we retain the option to include any other external force such as gravity. This is the famous *Navier–Stokes equation*. Its solutions will occupy us for a large part of this book.

In what follows, we will also frequently come across the ratio

$$\nu = \frac{\mu}{\rho} \ . \tag{2.4}$$

This arises so frequently that it is given its own name: it is called the *kinematic viscosity*.

Simple as the Navier–Stokes equation is, there is a great deal about it that remains to be understood. This includes the most basic questions about the existence and uniqueness of solutions. We will touch upon some of these issues in this book although much our focus will be on the things that are known rather than those that are not.

Viscosity Causes Diffusion of Momentum

As this book progresses, we'll develop some intuition for what viscosity does. But even before we get going, we can gain some insight through analogy. If we focus on two terms in the Navier–Stokes equation, ignoring the others (and also ignoring why we might be allowed to ignore them!) we have

$$\rho \frac{\partial \mathbf{u}}{\partial t} \sim \mu \nabla^2 \mathbf{u} \; . \tag{2.5}$$

It's useful to compare this to the heat equation which describes how temperature changes

$$\frac{\partial T}{\partial t} = \alpha \nabla^2 T \tag{2.6}$$

where α is a constant known as the *thermal diffusivity*. (This equation will be derived later in this book. There is some further discussion in Section 3.4, and a full derivation in Chapter 6.) Solutions to the heat equation are well studied: if we start with a hot spot somewhere, a place where there is a localised increase in temperature, then it will spread out increasing in size as $L \sim \sqrt{t}$. This kind of behaviour is called *diffusion*. It is the characteristic behaviour of any conserved quantity when undergoing random bombardment by some microscopic process. We will devote Chapter 7 to a more comprehensive exploration of the properties of diffusion.

This also tells us how to think of the viscosity term in the Navier–Stokes equation. It is causing momentum $\rho \mathbf{u}$ to diffuse. It doesn't cause the momentum to change direction. But if there was some place in the fluid where the momentum density was higher, then the viscosity term will cause this to spread out, much like temperature in the heat equation. This also makes physical sense: if the momentum density is higher in one region of the fluid then that region will be rubbing against its neighbouring regions. Viscosity is the friction force induced by this rubbing and results in a transfer of momentum from one region into neighbouring regions.

Viscosity Breaks Time Reversal

More intuition comes from the observation that the additional viscosity term breaks the symmetry of *time reversal*. This acts as

$$T : t \to -t \,, \quad T : \mathbf{x} \to \mathbf{x} \,, \quad T : \mathbf{u} \to -\mathbf{u} \,. \tag{2.7}$$

You can check that all the terms in the Euler equation are invariant under this transformation, at least if the external force is time-reversal invariant, so $T : \mathbf{f} \to \mathbf{f}$. But the extra viscosity term is not invariant under time reversal: it transforms as

$$T : \nabla^2 \mathbf{u} \to -\nabla^2 \mathbf{u} \,. \tag{2.8}$$

This is important. If we're given any solution to the Euler equation, then we can always run it backwards in time and we will get another solution. This is not true of solutions to the Navier–Stokes equation which exhibits an arrow of time. This is because, as advertised above, the viscosity is a friction force which, like other friction forces in classical mechanics, causes the system to lose energy. We'll see this explicitly in Section 2.1.2 where we compute the energy lost due to viscosity.

2.1 Stress, Strain, and Viscosity

We now give a slightly more involved derivation of the Navier–Stokes equation which will allow us to unpick the meaning of viscosity. To do this, we go back to basics and start thinking afresh about the kinds of forces that act on a fluid. Recall from (1.26) that, for a volume V of fluid, the equation $F = ma$ becomes

$$\rho \int_V \frac{D\mathbf{u}}{Dt} \, dV = -\int_S P \, d\mathbf{S} + \text{other forces} \tag{2.9}$$

with $S = \partial V$ the surface of the volume. Importantly, the pressure P acts on the surface of the volume and this ensures that it appears in the Navier–Stokes equation as the gradient ∇P when the surface integral is converted to a volume integral by the divergence theorem. Similarly, the friction forces also naturally act on the surface of the volume as a neighbouring piece of fluid brushes past. So our first task is to better understand what the general kind of force acting on a surface looks like.

Consider a small cubic volume V as shown in the following figure. There are six sides, and there may be a force acting on each. The pressure force is special because it acts parallel to the normal on each side. But that's not necessarily the case for all forces. In general, the force might act in any direction. Moreover, the direction of the force will generally depend on the orientation of the surface. For example, this is true of pressure which is parallel to the normal.

The figure shows the normal $\mathbf{n}$ to upper and rightmost faces, together with the force per unit area, which we've denoted $\mathbf{T}$, acting on each of these faces. This force per unit area is called the *stress*. Note that stress is, like pressure, a force per unit area, as opposed to the force density $\mathbf{f}$ that we introduced previously which is the force per unit volume. As the figure illustrates, both the magnitude and direction of $\mathbf{T}$ can differ from one face to another.

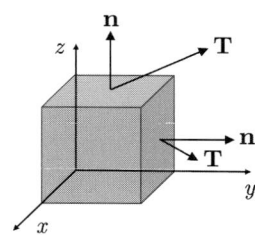

These considerations mean that to specify the force that acts on a surface, we first have to specify the orientation to the surface. This is achieved through the introduction of the *stress tensor*, σ_{ij}. It is defined so that the force $\mathbf{F}$ acting on a small surface with area δA and normal $\mathbf{n}$ is given by

$$\mathbf{F} = \mathbf{T}\,\delta A = (\sigma \mathbf{n})\delta A \ . \tag{2.10}$$

In index notation, we have

$$T_i = \sigma_{ij} n_j \ . \tag{2.11}$$

For pressure, the stress tensor takes the simple diagonal form

$$\sigma_{ij} = -P\,\delta_{ij} \ . \tag{2.12}$$

This is just the statement that the pressure force on a surface always points in the same direction as the normal vector. But, in general, σ_{ij} may take a more complicated form. Furthermore, for a fluid the stress tensor is itself a field $\sigma_{ij}(\mathbf{x}, t)$ that may vary in both space and time. This means that the forces acting on various parts of the fluid depend both on the position in the fluid and on the orientation of the surface that is considered.

The stress tensor has an important property: it is symmetric

$$\sigma_{ij} = \sigma_{ji} \ . \tag{2.13}$$

We will now show this. Consider the (slightly messy) Figure 2.1 depicting a small cube with sides of length L. The leftmost and rightmost faces lie

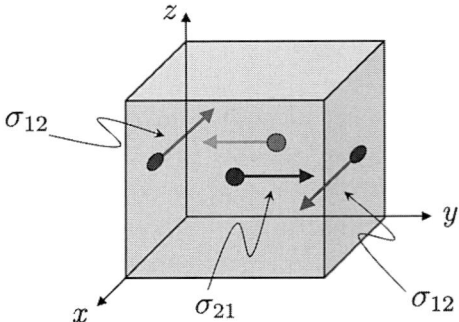

Fig. 2.1   The torque experienced by a small parcel of fluid.

normal to the y-direction. On these faces, we've depicted the component of the stress in the x-direction. From (2.11), this stress is σ_{12}. Note that the stress can depend on position, so $\sigma_{12} = \sigma_{12}(\mathbf{x})$, which means that the force on the leftmost face is not necessarily the same as the force on the rightmost face.

Meanwhile, the front and back faces lie normal to the x-direction. On these faces, we've depicted the component of the stress in the y-direction. This is σ_{21}. (We've labelled σ_{21} only on the front face to keep the figure cleaner.)

These four stresses give rise to a torque. Each σ_{ij} is a force per unit area, so the actual force is $L^2 \sigma_{ij}$. Furthermore, the moment of each force about the centre of the cube is $L/2$. This means that the total torque around a line parallel to the z-axis, through the centre of the cube, is

$$\tau = L^3(\sigma_{12} - \sigma_{21}) + \mathcal{O}\left(\frac{\partial \sigma_{12}}{\partial y}L^4, \frac{\partial \sigma_{21}}{\partial x}L^4\right). \tag{2.14}$$

The leading term comes from the difference between σ_{12} and σ_{21}. The subleading terms come from the difference of, say, σ_{12} on the left- and right-hand faces. (The statement that the cube is small is the assumption that σ_{ij} does not vary much over the length scale L.)

Further torque may come from bulk forces whose strength varies over the inside of the cube. But this torque will always be of order L^4 (times some suitable dimensionful parameter). So, for small cubes, the leading contribution to the torque (2.14) is proportional to the difference $(\sigma_{12} - \sigma_{21})$ and scales as L^3.

But torques that scale as L^3 are bad. To see this, recall some simple facts from the book on Classical Mechanics. A torque τ will increase the angular velocity Ω through the equation $\dot{\Omega} = \tau/I$ where I is the moment of inertia. But the moment of inertia of any object always scales as mass $\times L^2 = \rho L^5$, and so $\dot{\Omega} \sim 1/L^2$. The actual speed of the object is $v \sim \Omega L$ so if the torque scales as L^3, the acceleration will diverge as $\dot{v} \sim 1/L$ for small L. That makes no sense. To avoid this we must have

$$\sigma_{12} = \sigma_{21} \ . \tag{2.15}$$

The same argument works for all other components and so $\sigma_{ij} = \sigma_{ji}$: the stress tensor is necessarily symmetric.

The stress-tensor σ_{ij} is also an object that appears in relativistic field theories where it sits as the spatial components of the energy-momentum tensor. We met one example in the book on Electromagnetism. There we showed that σ_{ij} is necessarily symmetric to ensure conservation of angular momentum, an argument that is closely related to the one given above. A more sophisticated perspective comes from thinking about how the theory behaves when placed on a curved manifold and will be discussed in the book on General Relativity.

2.1.1 Newtonian Fluids

With the technology of the stress tensor, it is straightforward to describe the effect of friction. A *Newtonian fluid* is one where the friction forces are linear in velocity. If we assume that the fluid is isotropic then the form of the stress tensor is pretty much fixed by rotational invariance: it must be a symmetric tensor constructed from ∇ and $\mathbf{u}$ and the only option is $\partial_i u_j + \partial_j u_i$. In fact, a symmetric tensor can be decomposed into its trace and a traceless piece so in general we have, including the pressure term

$$\sigma_{ij} = -P\delta_{ij} + \mu\left(\frac{\partial u_i}{\partial x^j} + \frac{\partial u_j}{\partial x^i} - \frac{2}{3}\nabla \cdot \mathbf{u}\,\delta_{ij}\right) + \zeta\nabla \cdot \mathbf{u}\,\delta_{ij} \tag{2.16}$$

where, as we mentioned previously, the constant μ is called the *dynamic shear viscosity*. This time we've included the extra term proportional to $\nabla \cdot \mathbf{u}$ with a coefficient ζ known as the *bulk viscosity* or sometimes the *volume viscosity*. Importantly, it can be shown that both μ and ζ are necessarily positive. For μ, this follows from energy dissipation and we will give the argument shortly. For ζ it turns out that this follows from considerations of entropy. However, when dealing with incompressible fluids with $\nabla \cdot \mathbf{u} = 0$ we can forget all about the bulk viscosity. We have

$$\sigma_{ij} = -P\delta_{ij} + 2\mu E_{ij} \tag{2.17}$$

where E_{ij} is the rate of strain tensor that we met previously in (1.55)

$$E_{ij} = \frac{1}{2}\left(\frac{\partial u_i}{\partial x^j} + \frac{\partial u_j}{\partial x^i}\right) . \tag{2.18}$$

We can now use this form of the stress tensor in the equation of motion for the fluid. With a general surface force, captured by σ_{ij}, the integrated equation of motion (2.9) for a fluid becomes

$$\rho \int_V \frac{Du_i}{Dt}\, dV = \int_S \sigma_{ij}\, dS^j \tag{2.19}$$

where we have neglected other forces such as gravity. We use the divergence theorem to change the surface integral into a volume integral

$$\rho \int_V \frac{Du_i}{Dt}\, dV = \int_V \frac{\partial \sigma_{ij}}{\partial x^j}\, dV . \tag{2.20}$$

This formula holds for arbitrary volume V, so the equation of motion is

$$\rho \frac{Du_i}{Dt} = \frac{\partial \sigma_{ij}}{\partial x^j} . \tag{2.21}$$

From our equation (2.17), the right-hand side becomes

$$\frac{\partial \sigma_{ij}}{\partial x^j} = -\frac{\partial P}{\partial x^i} + \mu\left(\frac{\partial^2 u_i}{\partial x^j \partial x^j} + \frac{\partial^2 u_j}{\partial x^j \partial x^i}\right) . \tag{2.22}$$

The last term vanishes, again using the incompressibility condition $\nabla \cdot \mathbf{u} = 0$, and we're left with the promised Navier–Stokes equation

$$\rho \frac{D\mathbf{u}}{Dt} = -\nabla P + \mu \nabla^2 \mathbf{u} . \tag{2.23}$$

In what follows, we'll often divide by the density to write the Navier–Stokes equation as

$$\frac{D\mathbf{u}}{Dt} = -\frac{1}{\rho}\nabla P + \nu \nabla^2 \mathbf{u} \tag{2.24}$$

where, as defined earlier, $\nu = \mu/\rho$ is the *kinematic viscosity*. We can also add further forces on the right-hand side to taste.

The derivation of the Navier–Stokes equation that we described above sits entirely within the continuum language that is the essence of fluid mechanics. There is another remarkable, and ultimately better, derivation that really goes back to basics. This is due to Boltzmann. The derivation starts with the underlying $\sim 10^{23}$ atoms and tracks their interactions, albeit in a statistical way. It explains why the variables of the Navier–Stokes equation are the right thing to focus on if you care only about long-time physics and gives a microscopic explanation of the various terms. We present this derivation in Chapter 6.

2.1.2 Momentum and Energy Conservation Revisited

For inviscid fluids, the Euler equation is simply the statement that momentum is conserved, while energy conservation (1.48) led to Bernoulli's principle. What becomes of these conservation laws in the presence of viscosity?

First momentum. Here there is no problem: in the absence of external forces, we can write the Navier–Stokes equation in the form of a continuity equation, telling us that momentum is conserved. The only difference from the Euler equation is that we get an extra term in the momentum current, proportional to the viscosity

$$\frac{\partial(\rho u_i)}{\partial t} + \frac{\partial \Pi_{ij}}{\partial x^j} = 0 \quad \text{with} \quad \Pi_{ij} = \rho u_i u_j + P\delta_{ij} - 2\mu E_{ij} \ . \qquad (2.25)$$

As before, we've used the fact that $\nabla \cdot \mathbf{u} = 0$ for incompressible flows. In particular, we've used this to keep Π_{ij} symmetric by taking the extra term to be proportional to the rate of strain tensor (2.18) rather than just $\partial_j u_i$.

This gives us another perspective on the Navier–Stokes equation. It is, like the Euler equation, simply conservation of momentum, but with an additional term in the momentum current coming from gradients of the velocity. The idea that gradients drive currents is something that also occurs in other, perhaps more familiar, contexts where it goes by the name of *Fick's law*. For example, differences in temperature result in a heat current $\mathbf{J} \sim \nabla T$. When you include viscosity, gradients in the velocity result in a momentum current.

What about energy? We will ignore other bulk forces for now. (We've already seen in Section 1.2.4 that conservative forces don't spoil conservation of energy.) However, it's useful to briefly return to the form of the Navier–Stokes equation (2.21) in which we allow for general stress forces σ_{ij}. Taking the inner product with the velocity $\mathbf{u}$, the matter derivative becomes

$$\mathbf{u} \cdot \frac{D\mathbf{u}}{Dt} = u_i \cdot \left(\frac{\partial u_i}{\partial t} + u_j \frac{\partial u_i}{\partial x^j} \right) = \frac{1}{2}\frac{\partial |\mathbf{u}|^2}{\partial t} + \frac{1}{2}\mathbf{u} \cdot \nabla |\mathbf{u}|^2 \qquad (2.26)$$

and our proto-Navier–Stokes equation (2.21) becomes

$$\frac{\rho}{2} \left(\frac{\partial |\mathbf{u}|^2}{\partial t} + \mathbf{u} \cdot \nabla |\mathbf{u}|^2 \right) = u_i \frac{\partial \sigma_{ij}}{\partial x^j} \ . \qquad (2.27)$$

Remember the game that we're playing: we'd like to massage this into the continuity equation to see the conservation of energy. Using the fact that the fluid is incompressible, so $\nabla \cdot \mathbf{u} = 0$, we have

$$\frac{\rho}{2}\frac{\partial |\mathbf{u}|^2}{\partial t} + \frac{\partial}{\partial x^j}\left(\frac{\rho}{2}|\mathbf{u}|^2 u_j - u_i \sigma_{ij}\right) = -\sigma_{ij}\frac{\partial u_i}{\partial x^j} = -\sigma_{ij}E_{ij} \qquad (2.28)$$

where, in the second equality, we've used the fact that $\sigma_{ij} = \sigma_{ji}$ so the contraction picks out the symmetric part of $\partial u_i/\partial x^j$ which is E_{ij}, the rate of strain tensor defined in (2.18). The two terms on the left-hand side take the form of a continuity equation. But now the right-hand side is not zero. This tells us that, in contrast to the Euler equation, energy is *not* conserved in the Navier–Stokes equation.

We can get an expression for how energy is lost. If we integrate over some fixed volume V then we have

$$\frac{\rho}{2}\frac{\partial}{\partial t}\int_V |\mathbf{u}|^2 \, dV + \int_S \left(\frac{\rho}{2}|\mathbf{u}|^2 u_j - u_i \sigma_{ij}\right) dS^j = -\int_V \sigma_{ij}E_{ij} \, dV \quad (2.29)$$

with $S = \partial V$. The volume term on the left-hand side is clearly the change in the kinetic energy in V. The surface term accounts for (some of) this change: the $|\mathbf{u}|^2 u_j$ term captures the energy that flows out through the surface, while the $u_i \sigma_{ij}$ term is the work done by the surface forces on the fluid contained in V. This includes the work done by both the pressure and the viscous forces. All of this is consistent with the conservation of energy. However, the right-hand side of (2.29) doesn't vanish and this is telling us that energy is, in fact, no longer conserved. Instead, the right-hand side tells us the rate at which energy is dissipated

$$\text{dissipation} = \int_V \sigma_{ij}E_{ij} \, dV = 2\mu \int_V E_{ij}E_{ij} \, dV \qquad (2.30)$$

where, in the second equality, we've used the explicit form of the stress tensor (2.17). We see that the pressure doesn't contribute to energy dissipation (because $\delta_{ij}E_{ij} = \nabla \cdot \mathbf{u} = 0$). This, of course, is something that we found when studying the Euler equation. But we now see that one important consequence of viscosity is that we no longer have energy conservation. Correspondingly, Bernoulli's principle no longer holds when the effects of viscosity are important.

The dissipation is the integral of a total square, so it is clearly positive provided that $\mu > 0$. And the minus sign on the right-hand side of (2.29) is telling us that energy is lost to friction, rather than gained. This is the reason why we should take $\mu > 0$.

There is another way to write the dissipation (2.30). Up to a boundary term, we have

$$\text{dissipation} = 2\mu \int d^3x \; E_{ij}E_{ij} = \mu \int d^3x \; |\boldsymbol{\omega}|^2 \; . \qquad (2.31)$$

The integral of $|\boldsymbol{\omega}|^2$ on the right-hand side is what we previously called *enstrophy*. This tells us that dissipation is due to a combination of viscosity and vorticity. This is one of the key ideas that underlies turbulence and will be described in more detail in Chapter 5.

It is natural to ask: Where did the energy go when it was lost? After all, energy is certainly conserved at a fundamental level. The answer is that it went into heat. The dissipation (2.30) is a transfer of energy from the macroscopic, coherent kinetic energy of the fluid, captured by the coarse-grained velocity field $\mathbf{u}$, to some microscopic, incoherent internal motion of the underlying atoms. This internal motion is still kinetic energy, but not with any overall preferred direction. To properly account for this, we should understand how the temperature and entropy of the fluid change due to these dissipative effects. As with friction forces in classical mechanics, we won't attempt to do this here: we will simply count this as lost energy. We will, however, return to the interplay of heat and energy in Section 3.4 when we discuss sound waves.

2.1.3 Claude-Louis Navier (1785–1836)

Navier was a better mathematician than engineer. For much of his life, he held a professorship at École des Ponts et Chaussées in Paris. His primary job was in the construction of bridges, with a particular focus on the newly invented suspension bridge. His work reached its grand conclusion in the design of a magnificent bridge to cross the Seine in the most prominent and prestigious location: from the Champs Elysées to the Hôtel des Invalides. The bridge was to be a monument to the glory of France and Navier boasted that his plans embraced the very latest mathematical techniques. Construction started in 1824 and went on for more than two years until the first cracks appeared. Just weeks before the planned opening ceremony, the bridge was deemed beyond repair and ordered to be dismantled. Fourier analysis, it turns out, is not always a good substitute for concrete.

This episode did not, ultimately, hurt Navier's career prospects. He was later elected to the Chevalier de la Legion d'Honneur and he is one of 72 scientists whose name adorns the Eiffel Tower.

Navier's joint interest in mathematics and engineering naturally led him

Fig. 2.2   Yeah, you might want to make those foundations a bit bigger. Navier's design for his Parisian bridge.

to the study of fluid mechanics. The foundations of the subject (no concrete needed) were laid down by Euler in 1757, after presenting a preliminary version to the Berlin Academy in 1752. Navier was the first, in 1822, to include the $\nabla^2 \mathbf{u}$ term to capture the effects of friction in a fluid. His viewpoint was rather strange from a modern perspective, with particles sitting at points in a lattice experiencing forces related to their relative velocities. He did not identify the proportionality constant in front of his additional term as viscosity, but rather as some kind of molecular spacing.

The Navier–Stokes equations were subsequently derived by Cauchy and by Poisson, each invoking slightly different arguments. A continuum derivation that we would accept today as correct was first given by Barré de Saint-Venant in 1843. Stokes' derivation came two years later, in 1845.

2.2 Some Simple Viscous Flows

Our first task is to explore some very simple solutions to the Navier–Stokes equation (2.3). This will allow us to build some intuition for the role that viscosity plays.

2.2.1 The No-Slip Boundary Condition

We've already seen the importance of boundary conditions in constructing fluid flows. For an inviscid flow, we introduced the obvious "you shall not pass" condition in Section 1.4

$$\mathbf{n} \cdot \mathbf{u} = 0 \tag{2.32}$$

where $\mathbf{n}$ is the normal to a solid surface. This solid surface might be the walls of the container, or an obstacle sitting in the fluid like the sphere or cylinder we studied previously. If the solid object is itself moving with some velocity $\mathbf{U}$ then this condition becomes

$$\mathbf{n} \cdot \mathbf{u} = \mathbf{n} \cdot \mathbf{U} \; . \tag{2.33}$$

For viscous fluids, we introduce a further boundary condition that restricts the flow *tangent* to a solid. This is the *no-slip* condition that states

$$\mathbf{t} \cdot \mathbf{u} = \mathbf{t} \cdot \mathbf{U} \tag{2.34}$$

where $\mathbf{t}$ is now any vector tangent to the boundary. This means that the velocity of the fluid along the boundary must match the velocity of the boundary itself. It is sometimes written as the requirement that $\mathbf{u} - (\mathbf{u} \cdot \mathbf{n})\mathbf{n}$ is continuous at the boundary.

The no-slip condition (2.34) doesn't follow from the Navier–Stokes equation. Instead, it is something additional that we assert. It is, however, physically sensible and arises from the friction forces between the fluid and the boundary. Importantly, it is also the boundary condition that is observed to be correct for most experiments.

The flows that we met in Chapter 1 describing fluids moving around spheres and cylinders do *not* obey the no-slip condition. Of course, they also failed miserably in explaining drag forces. This is our first hint that we should do a better job of describing the flows close to the boundary of an object. You might wonder why we just don't search for other solutions to the Euler equations that include the no-slip condition. The reason is that there simply aren't any such solutions. This is because the Euler equation is first order in spatial derivatives and we only get to impose one boundary condition, namely the impenetrability condition (2.32). In contrast, the Navier–Stokes equation is second order. This means that we must impose an additional boundary condition when solving the equation. The no-slip condition is the boundary condition of choice.

2.2.2 Couette Flow

Take two infinite parallel plates lying in the (x,y)-plane and separated by some distance h in the z-direction. The bottom plate is stationary while the top plate moves with a constant speed U in the x-direction. What happens to fluid trapped between them?

We will look for a steady flow with $\partial \mathbf{u}/\partial t = 0$ and with the velocity lying solely in the x-direction. The speed of the fluid depends only on the

z-direction, meaning

$$\mathbf{u}(\mathbf{x}, t) = (u(z), 0, 0) \ . \tag{2.35}$$

With this ansatz $(\mathbf{u} \cdot \nabla)\mathbf{u} = 0$ so the material time derivative vanishes: $D\mathbf{u}/Dt = 0$. There are no pressure gradients in the fluid, so the only surviving term in the Navier–Stokes equation comes from the viscosity and tells us

$$\mu \frac{d^2 u}{dz^2} = 0 \ . \tag{2.36}$$

The boundary conditions are $u(0) = 0$ and $u(h) = U$. This is an easy equation to solve and the velocity profile must increase linearly to match the speeds of the two plates

$$u(z) = \frac{Uz}{h} \ . \tag{2.37}$$

The result is known as *Couette flow* and is shown in the figure. Flows of this kind, in which adjacent layers of fluids move at different speeds, are collectively referred to as *shear flows*.

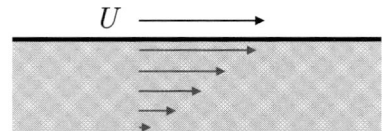

Couette flow is not a potential flow. The simplest way to see this is to note that, even though the flow doesn't look like it's rotating, it has vorticity

$$\boldsymbol{\omega} = \nabla \times \mathbf{u} = (0, U/h, 0) \ . \tag{2.38}$$

This vorticity arises because we've implemented the no-slip boundary condition, ensuring that the upper plate drags the fluid along with it. This suggests that the no-slip boundary condition may be a way to generate vorticity. We will see later that this is an important observation.

It is a simple matter to compute the stress exerted on the fluid using (2.17)

$$\sigma = \begin{pmatrix} -P & 0 & \mu U/h \\ 0 & -P & 0 \\ \mu U/h & 0 & -P \end{pmatrix} \ . \tag{2.39}$$

This tells us that the force per unit area exerted by the top plate with $\mathbf{n} = \hat{\mathbf{z}}$ is

$$\mathbf{T} = (\mu U/h, 0, -P) \tag{2.40}$$

while the bottom plate, with $\mathbf{n} = -\hat{\mathbf{z}}$, exerts an equal and opposite force. We usually think of the bottom plate, and the distance h between the plates, as fixed externally. We then ask what force we have to exert on the upper plate to keep it moving if it has (large) area A. The answer is

$$\frac{F}{A} = \frac{\mu U}{h} \ . \tag{2.41}$$

This is the operational definition of viscosity μ that we met in the book on Classical Mechanics. The work done by this pushing (again, per unit area) is just $\mu U^2/h$. You can check that this agrees with the more formal definition of dissipation given in (2.30).

Circular Couette Flow

The same kind of flow occurs in different geometries. Consider, for example, two concentric, infinite cylinders, aligned along the z-direction. The inner cylinder has radius R_1 and rotates with angular velocity Ω_1. The outer cylinder has radius R_2 and rotates with angular velocity Ω_2.

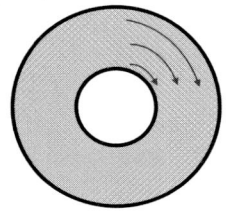

From the geometry, we see that the flow should be rotationally invariant, meaning that it takes the form

$$\mathbf{u} = \Omega(r)\,(y, -x, 0) \tag{2.42}$$

where $r^2 = x^2 + y^2$ and $\Omega(r)$ is the angular velocity of the fluid. The no-slip condition implements the boundary conditions $\Omega(R_1) = \Omega_1$ and $\Omega(R_2) = \Omega_2$.

This time the story is a little different because we can no longer ignore the non-linear term in the Navier–Stokes equation,

$$(\mathbf{u} \cdot \nabla)\mathbf{u} = -r\Omega^2 \hat{\mathbf{r}} \ . \tag{2.43}$$

But this is something familiar, it is just the outward-pointing centrifugal force that comes from the rotation of the fluid. It gives rise to a pressure gradient in the fluid, with the radial pressure $P(r)$ obeying

$$\frac{\partial P}{\partial r} = r\Omega^2 \quad \Longrightarrow \quad \frac{D\mathbf{u}}{Dt} = -\nabla P \ . \tag{2.44}$$

Such a flow obeys the Euler equation for any choice of $\Omega(r)$. But to satisfy the Navier–Stokes equation we must have, in addition,

$$\mu \nabla^2 \mathbf{u} = 0 \ . \tag{2.45}$$

A quick calculation shows that $\nabla^2 \mathbf{u} = (3\Omega'/r + \Omega'')(y, -x, 0)$ so the angular velocity of the flow must take the form

$$\Omega'' = -\frac{3\Omega'}{r} \quad \Longrightarrow \quad \Omega = A + \frac{B}{r^2} \; . \tag{2.46}$$

The first term is just a constant rotation, while the second term corresponds to the irrotational line vortex that we met in Section 1.3. The no-slip boundary conditions fix these coefficients to be

$$A = \frac{\Omega_2 R_2^2 - \Omega_1 R_1^2}{R_2^2 - R_1^2} \quad \text{and} \quad B = (\Omega_1 - \Omega_2)\frac{R_1^2 R_2^2}{R_2^2 - R_1^2} \; . \tag{2.47}$$

This circular Couette flow is also known as *Taylor–Couette flow*. (Taylor gets his name attached because he discovered certain instabilities in the flow that we will describe in Section 4.2.)

2.2.3 Poiseuille Flow

Here's another simple example. Again, take a fluid sitting between two infinite parallel plates, lying in the (x, y)-plane. This time it will be slightly more convenient if we separate the plates by distance $2h$ in the z-direction. We take them to sit at $z = \pm h$.

In contrast to Couette flow, both plates are now stationary. However, this time we induce a constant pressure gradient through the fluid, parallel to the plates

$$\frac{dP}{dx} = \text{constant} \; . \tag{2.48}$$

We again look for a steady, shear flow of the form $\mathbf{u} = (u(z), 0, 0)$. The Navier–Stokes equation is now

$$\mu\frac{d^2 u}{dz^2} = \frac{dP}{dx} = \text{constant} \; . \tag{2.49}$$

With the no-slip boundary conditions $u(z = \pm h) = 0$, the solution is

$$u(z) = -\frac{1}{2\mu}\frac{dP}{dx}(h^2 - z^2) \; . \tag{2.50}$$

This is known as *Poiseuille flow* or, more properly, *planar Poiseuille flow*. The minus sign is sensible: it tells us that if the pressure is greatest to the left, so $dP/dx < 0$, then the fluid moves to the right. Clearly the speed increases as we

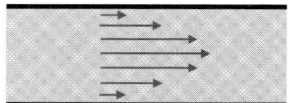

move away from the edges and is maximum in the middle where $z = 0$. Again, the flow has vorticity induced by the no-slip boundary condition.

The stress (2.17) is

$$
\sigma = \begin{pmatrix} -P(x) & 0 & z\,dP/dx \\ 0 & -P(x) & 0 \\ z\,dP/dx & 0 & -P(x) \end{pmatrix} . \tag{2.51}
$$

It is independent of viscosity. The top and bottom plates have normal $\mathbf{n} = \pm\hat{\mathbf{z}}$ and sit at $z = \pm h$, giving a force per unit area

$$
\mathbf{T} = \left(h\frac{dP}{dx}, 0, \mp P(x) \right) . \tag{2.52}
$$

The total force exerted by both plates is now in the negative x-direction, as it should be.

A simple generalisation of this story describes flow down a circular pipe of radius R with a constant pressure gradient. We work in cylindrical polar coordinates, (r, θ, x), which has the slightly unconventional choice of taking the polar coordinates to parametrise the (y, z)-plane. The pressure gradient along the pipe is again constant, while the velocity takes the form

$$
\mathbf{u} = u(r)\hat{\mathbf{x}} . \tag{2.53}
$$

The Navier–Stokes equation is

$$
\mu\nabla^2 u = \frac{dP}{dx} \quad \Longrightarrow \quad \frac{\mu}{r}\frac{d}{dr}\left(r\frac{du}{dr} \right) = \frac{dP}{dx} . \tag{2.54}
$$

The solution with the appropriate boundary conditions is

$$
u(r) = -\frac{1}{4\mu}\frac{dP}{dx}(R^2 - r^2) . \tag{2.55}
$$

This is known as *pipe flow* or *Hagen–Poiseuille flow*.

2.2.4 Vorticity Revisited and the Burgers Vortex

As our final example of a flow, we will look at something that swirls. This gives us the opportunity to revisit vorticity in the presence of viscosity.

Previously we derived the vorticity equation (1.79) from the Euler equation. We can follow the same steps, now taking the curl of the Navier–Stokes equation to find

$$
\frac{D\boldsymbol{\omega}}{Dt} = (\boldsymbol{\omega}\cdot\nabla)\mathbf{u} + \nu\nabla^2\boldsymbol{\omega} . \tag{2.56}
$$

This is the *vorticity equation* for a viscous fluid. The term due to viscosity, naturally written in terms of $\nu = \mu/\rho$, should be viewed as analogous to the diffusion term in the heat equation. Just as viscosity gives rise to diffusion of momentum, so it gives rise to diffusion of vorticity too. It is telling us that if there is some vorticity localised in some region of space, the viscosity will tend to make it diffuse into neighbouring regions. For example, if you blow a smoke ring then the size of the ring will grow over time as the vorticity diffuses into neighbouring regions, provided the ring remains stable.

For inviscid fluids, Kelvin's circulation theorem told us that $\Gamma = \oint_{C(t)} \mathbf{u} \cdot d\mathbf{x}$ doesn't change for curves $C(t)$ that move with the fluid. You can check that the addition of the viscosity term means that the circulation is no longer conserved in the full Navier–Stokes equations.

Burgers Vortex

To highlight how viscosity changes the physics, we can return to the vortex solution that we saw back in Section 1.3. There we looked at a combination of a strain and rotation

$$\mathbf{u}(\mathbf{x}, t) = \mathbf{u}_{\text{strain}}(\mathbf{x}) + \mathbf{u}_{\text{rot}}(\mathbf{x}, t) \quad \text{with} \quad \begin{cases} \mathbf{u}_{\text{strain}} = \alpha(-x, -y, 2z) \\ \mathbf{u}_{\text{rot}} = f(r, t)(-y, x, 0) \end{cases} \quad (2.57)$$

with $r^2 = x^2 + y^2$. The strain part of the flow stretches the fluid in the z-direction, while squeezing it in the (x, y)-plane; the rotational flow clearly rotates in the (x, y)-plane, giving rise to a vorticity $\boldsymbol{\omega} = (0, 0, \omega)$ with ω given by (1.64)

$$\omega = \frac{1}{r}\frac{d}{dr}(r^2 f) \,. \tag{2.58}$$

The vorticity equation (2.56) is a partial differential equation for ω

$$\frac{\partial \omega}{\partial t} - \alpha r \frac{\partial \omega}{\partial r} - 2\alpha\omega = \nu \frac{1}{r}\frac{\partial}{\partial r}\left(r\frac{\partial \omega}{\partial r}\right) \,. \tag{2.59}$$

Previously we solved this equation when $\nu = 0$ to find an example of vortex stretching (1.90). The solution we found was time-dependent, with $\omega(r, t) = e^{2\alpha t} W(re^{\alpha t})$, and shows the magnitude of vorticity increasing exponentially, while being squeezed in the (x, y)-plane so that the overall flux is conserved in a way that is consistent with the circulation theorem.

Now we want to solve the vorticity equation with $\nu \neq 0$ to include the effect of viscosity. We already noted that the contribution $\nu\nabla^2\boldsymbol{\omega}$ to the vorticity equation looks like a diffusion term. This suggests that we might be able to find a time-independent solution in which the squeezing of vorticity is balanced by an outward diffusion caused by the viscosity. For steady solutions, the equation (2.59) becomes

$$\frac{1}{r}\frac{\partial}{\partial r}\left(\alpha r^2\omega + \nu r\frac{\partial\omega}{\partial r}\right) = 0 \ . \tag{2.60}$$

We can integrate once to get

$$\frac{\partial\omega}{\partial r} = -\frac{\alpha r}{\nu}\omega \tag{2.61}$$

where we've set the integration constant to zero by requiring that ω and ω' decay suitably quickly asymptotically. This equation gives an exponentially localised vorticity

$$\omega(r) = \frac{\Gamma\alpha}{2\pi\nu}e^{-\alpha r^2/2\nu} \ . \tag{2.62}$$

Here Γ is a constant that determines the overall magnitude of vorticity. The slightly strange combination of constants that accompany it ensure that Γ can also be identified with the asymptotic circulation of the flow

$$\Gamma = \int_S \boldsymbol{\omega}\cdot d\mathbf{S} = 2\pi\int_0^\infty dr\ r\omega(r) \ . \tag{2.63}$$

We can then solve (2.58) to get the associated profile function for the angular velocity,

$$f(r) = \frac{\Gamma}{2\pi r^2}\left(1 - e^{-\alpha r^2/2\nu}\right) \ . \tag{2.64}$$

This is the *Burgers vortex solution* and is shown in the figure to the right. It is the simplest model for a hurricane.

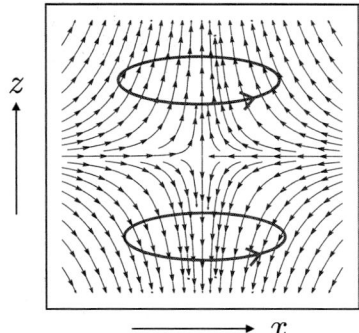

We can compute the dissipation due to the vortex. The integral (2.31) is simple and gives us the dissipation per unit length

$$\text{dissipation per unit length} = 2\pi\mu\int_0^\infty dr\ r\omega^2 = \frac{\Gamma^2\alpha\rho}{4\pi} \tag{2.65}$$

where the density ρ has made an appearance through $\mu = \nu\rho$. Remarkably, for fixed circulation Γ, the dissipation is independent of the viscosity ν. Indeed, in the limit $\nu \to 0$ the vortex becomes a singular delta function localised at $r = 0$, but the dissipation remains finite. This resonates with the so-called zeroth law of turbulence that we will meet in Chapter 5.

Table 2.1 The viscosities of some substances at room temperature.		
Fluid	Kinematic Viscosity $(\mathrm{m^2 s^{-1}})$	Dynamic Viscosity $(\mathrm{kg\, m^{-1} s^{-1}})$
Air	1.5×10^{-5}	1.8×10^{-5}
Water	10^{-6}	10^{-3}
Honey	$\sim 10^{-3}$	~ 10
Pitch	$\sim 10^5$	$\sim 10^8$

2.3 Dimensional Analysis

The Navier–Stokes equation is

$$\frac{\partial \mathbf{u}}{\partial t} + \mathbf{u} \cdot \nabla \mathbf{u} = -\frac{1}{\rho} \nabla P + \nu \nabla^2 \mathbf{u} \ . \tag{2.66}$$

Each term has dimension LT^{-2}. This means that the dimension of the kinematic viscosity ν is

$$[\nu] = L^2 T^{-1} \ . \tag{2.67}$$

Meanwhile, the dimension of dynamic viscosity $\mu = \rho \nu$ is

$$[\mu] = ML^{-1}T^{-1} \ . \tag{2.68}$$

The unit for dynamic viscosity is *poise* (P), with $1\,\mathrm{P} = 0.1\ \mathrm{kg\,m^{-1}s^{-1}}$. Values of the two viscosities for various fluids are shown in Table 2.1. To get a sense of the scales involved, we can do some further dimensional analysis. The kinematic viscosity has dimension of velocity times distance. For a fluid, the relevant internal velocity (as opposed to the velocity of some flow) is the speed of sound, c_s. On dimensional grounds, this is given by

$$c_s \sim \sqrt{\frac{k_B T}{m}} \tag{2.69}$$

where T is the temperature, k_B is Boltzmann's constant and m is the mass of the constituent atom or molecule. (We'll derive this formula, together with the overall coefficient, in Section 3.4.) Meanwhile, the relevant distance scale is the average separation a of atoms in the fluid. This suggests that the viscosity should be of order

$$\nu \sim c_s a \ . \tag{2.70}$$

For water, $c_s \sim 1000\,\mathrm{m\,s^{-1}}$, with a characteristic separation between molecules of $a \sim 10^{-9}$ m. This gives $\nu \sim 10^{-6}$ m^{2}s^{-1}, which is indeed in the right ball-park. Of course, we should add that the Navier–Stokes equation is valid only for incompressible flows, which really means for fluid speeds $|\mathbf{u}| \ll c_s$.

For some fluids, the internal molecular forces are strong, resulting in a much higher viscosity. Honey is a particularly familiar example. One of the most viscous fluids is pitch, also known as tar, which has a viscosity many orders of magnitude higher than that of water. A pitch drop experiment was set up in the University of Queensland, Australia in 1927. The flow of the pitch is ten times slower than continental drift. To date, nine drops have fallen. None have been witnessed. A webcam was set up in the 1990s, but was offline when the eighth drop fell in 2000. The ninth drop was accidently broken off before it fell in 2014. You can watch online and wait for the tenth drop, although there's not likely to be any action until 2030 or so.

By the time you get to geophysical phenomena, even the viscosity of pitch starts to look puny. Mantle rock has a viscosity ten orders of magnitude greater!

At the other end of the spectrum, superfluids such as helium-4 at low temperatures have strictly zero viscosity. This is very much a quantum mechanical effect and a proper description requires us to leave the comfortable classical realm of this book. We will study superfluids later in this series of books.

2.3.1 The Reynolds Number

Solving the Navier–Stokes equation (2.66) in full generality is, to put it mildly, a challenging problem. We make progress only by making some approximation. This involves deciding which terms, if any, can be ignored in any given situation. The obvious thing to do is to ask whether the viscosity is small or large. But this question in itself doesn't make any sense. Viscosity is dimensionful. There's no meaning to it being absolutely small or absolutely large. It can only be small or large relative to something else.

That something else depends on the flow. Suppose that the flow has a characteristic speed U and length L. Here U could be the speed of the fluid relative to some boundary, or the rotational speed of the fluid. Similarly L could be some geometrical distance over which the flow changes. From this we can construct a dimensionless ratio called the *Reynolds number*

$$Re = \frac{UL}{\nu} \ . \tag{2.71}$$

Roughly speaking, this captures the relative importance of the inertial term $\mathbf{u} \cdot \nabla \mathbf{u}$ and the viscosity term $\nu \nabla^2 \mathbf{u}$,

$$\frac{\text{inertial term}}{\text{viscosity term}} = \frac{|\mathbf{u} \cdot \nabla \mathbf{u}|}{|\nu \nabla^2 \mathbf{u}|} \sim \frac{U^2/L}{\nu U/L^2} \sim Re \ . \tag{2.72}$$

We can make this clearer if we rescale all the variables in the Navier–Stokes equation to become dimensionless. We write $\tilde{\mathbf{u}} = \mathbf{u}/U$ and $\tilde{\mathbf{x}} = \mathbf{x}/L$, but also $\tilde{t} = tU/L$ and $\tilde{P} = P/\rho U^2$. The dimensionless Navier–Stokes equation then becomes

$$\frac{\partial \tilde{\mathbf{u}}}{\partial \tilde{t}} + \tilde{\mathbf{u}} \cdot \tilde{\nabla} \tilde{\mathbf{u}} = -\tilde{\nabla} \tilde{P} + \frac{1}{Re} \tilde{\nabla}^2 \tilde{\mathbf{u}} \ . \tag{2.73}$$

Here we see explicitly that Re is the only parameter in the Navier–Stokes equation. So once we fix the geometry of the flow (meaning boundary conditions and so on) you can change the speed U, the length L, and viscosity ν of the fluid, but the only combination that matters is the Reynolds number. This is known as the *similarity principle*.

With very broad brush, fluid flows can be characterised in one of two different types:

- High Reynolds Number, $Re \gg 1$: In this case, the flow is inertia dominated. In many cases, we can drop the viscosity term and return to the Euler equation that we studied in Chapter 1. Flows at high Reynolds number have an associated time scale that comes from equating the kinetic term $\partial \mathbf{u}/\partial t$ with the inertial term. This time scale is simply the time it takes the fluid to move some distance: $T \sim L/U$.

  For example, for the flow past an aircraft wing, $U \sim 100 \ \mathrm{m\,s}^{-1}$, while $L \sim 1$ m is the width of the wing. Using the value $\nu \sim 10^{-5}$, we have $Re \sim 10^7 \gg 1$, which suggests that the viscosity term is unimportant for such flows.

  However, this example also suggests that we should be nervous about such simple arguments. If we can really neglect viscosity at high Reynolds number then we run smack into the d'Alembert paradox that we met previously because, as we saw in Section 1.4, the Euler equation doesn't correctly capture the drag force that a fluid exerts on an object. Indeed, the argument that we can ignore the viscosity term is precisely what led to physicists being unable to understand how planes fly! We'll resolve these issues in Section 2.5 where we will see that, even at high Reynolds number, there can be situations where the viscosity term is important after all because it qualitatively changes certain aspects of the flow, in particular through the introduction of a so-called "boundary layer".

- Low Reynolds Number, $Re \ll 1$: In this case, the flow is dominated by viscosity. If we ignore both the inertial term and the pressure term, then the Navier–Stokes equation becomes

$$\frac{\partial \mathbf{u}}{\partial t} = \nu \nabla^2 \mathbf{u} . \tag{2.74}$$

As we've seen previously, this is the same as the heat equation that describes diffusion. It's telling us that flows at low Reynolds number exhibit diffusive transport of momentum, with the kinematic viscosity understood as a measure of momentum diffusivity. In this regime, the time scale associated to the flow is $T \sim L^2/\nu$.

For example, consider a bug of size $L \sim 10^{-5}$ m moving in water. It could be bombing along at a whopping $U \sim 10^{-5}$ m s^{-1} – that's one body length every second – but the associated Reynolds number is still $Re \sim 10^{-4}$. In the bug's world, viscosity rules. We'll explore the low Reynolds world further in Section 2.4.

Other Dimensionless Ratios

In different circumstances, there are further dimensionless ratios that we can form to characterise a flow and help us formulate good approximations to the equations. For example, if there is some characteristic time scale T to the flow – perhaps because the flow is being forced in some way – then we can form the *Strouhal number*

$$St = \frac{L}{UT} . \tag{2.75}$$

This is also written as $St = L\omega/U$, with ω the frequency of oscillation. The Strouhal number tells us the relative importance of the acceleration term $\partial \mathbf{u}/\partial t \sim U/T$ and the inertial term $\mathbf{u} \cdot \nabla \mathbf{u} \sim U^2/L$, with the acceleration term dominant when $St \gg 1$.

We get further dimensionless numbers when we add further forces. For example:

- The *Euler number* captures the relative importance of pressure gradients to the inertial term

$$Eu = \frac{\Delta P}{\rho U^2} . \tag{2.76}$$

- The *Froude number* captures the relative importance of the inertial term $\sim U^2/L$ to the gravitational force $\sim g$

$$Fr = \frac{U}{\sqrt{gL}} \, . \qquad (2.77)$$

As we will see in Chapter 3, the combination $\sqrt{gL}$ is identified with the speed of surface waves, so the Froude number can alternatively be viewed as the ratio of two speeds.

We'll meet other dimensionless quantities as this book progresses. In Section 3.6, we'll come across the *Mach number*, which measures how fast the flow is compared to the speed of sound, and in Section 4.4 the *Rayleigh number* and *Prandtl number*, both of which play a role when temperature differences are important.

2.3.2 Scaling

We upgraded ourselves from the Euler equation to the Navier–Stokes equation by adding a higher derivative term $\nabla^2\mathbf{u}$. But if we're happy to add a term with two derivatives, why not further terms with four derivatives? Or sixteen derivatives? Why should we stop here?

In fact there is a reason why higher derivative terms are irrelevant, at least if we look on suitably large distance scales. (The term "irrelevant" has a technical meaning in the language of physics, but happily it coincides with the usual meaning in this context!) To see this note that, in the absence of any external force, the Navier–Stokes equation

$$\frac{\partial \mathbf{u}}{\partial t} + \mathbf{u} \cdot \nabla \mathbf{u} = -\frac{1}{\rho}\nabla P + \nu \nabla^2 \mathbf{u} \qquad (2.78)$$

has a novel scaling symmetry

$$t \to \lambda^2 t \, , \quad \mathbf{x} \to \lambda \mathbf{x} \, , \quad \mathbf{u} \to \lambda^{-1}\mathbf{u} \, , \quad P \to \lambda^{-2}P \, . \qquad (2.79)$$

The whole Navier–Stokes equation scales with an overall factor of λ^{-3} under this scaling. But, crucially, all terms scale in the same way. This means that if we find one solution to the Navier–Stokes equations, then we can always rescale by some factor λ and get another solution. Because the spatial coordinate scales as $\mathbf{x} \to \lambda \mathbf{x}$, as we increase λ any features in the flow – for example, vortices – will clearly get bigger. Note that the Reynolds number (2.71) is invariant under this scaling: $Re \to Re$. This, in large part, is why it's important: the Reynolds number is a scale-invariant way of characterising a flow.

Now suppose that, in a fit of excitement, you decide that you'd like to add further terms to the Navier–Stokes equation. You should retain rotational symmetries and Galilean boosts (i.e. constant shifts of $\mathbf{u}$) but otherwise you can write down anything you like. The terms that you add will contain some number of time derivatives, spatial derivatives, and factors of the fields $\mathbf{u}$ and P. Schematically, we might have

$$\frac{\partial \mathbf{u}}{\partial t} + \mathbf{u} \cdot \nabla \mathbf{u} = -\frac{1}{\rho}\nabla P + \nu \nabla^2 \mathbf{u} + \mathcal{O}\left(\partial_t^{n_1}, \nabla^{n_2}, \mathbf{u}^{n_3}, P^{n_4}\right) \qquad (2.80)$$

where the integers n_i, with $i = 1, 2, 3, 4$, tell us the number of the various objects that appear. We can ask how this new term fares under the scaling symmetry (2.79). We have

$$\mathcal{O}\left(\partial_t^{n_1}, \nabla^{n_2}, \mathbf{u}^{n_3}, P^{n_4}\right) \to \lambda^{-(2n_1+n_2+n_3+2n_4)}\,\mathcal{O}\left(\partial_t^{n_1}, \nabla^{n_2}, \mathbf{u}^{n_3}, P^{n_4}\right)\ .$$

The key point is that the Navier–Stokes equation already contains the leading terms, each of which scales as λ^{-3}. Any new term that you try to construct scales away more quickly than the λ^{-3}. This means that if you try to scale a flow to larger length scales, then these additional terms play an increasingly diminished role in determining the form of the solution. In particular, on suitably large length scales they will always be less important than those that appear in the Navier–Stokes equation. This is what we mean when we say that they are irrelevant.

This isn't to say that higher derivative terms are never important under any circumstances. If the gradients of fields are large enough, then higher derivative terms will surely compete with the others. But how large do they need to be? The answer to that is governed by the coefficients of these higher derivative terms which characterise the fluid. On dimensional grounds, these coefficients must have certain length or time dimensions, with the relevant scale set by some microscopic interactions. But these new scales are then likely to be set by microscopic physics – say the mean free path of the underlying molecules – and we certainly don't expect fluid mechanics to be relevant if the fields vary significantly on such scales.

The upshot of this discussion is that the Navier–Stokes equation is special because each of the terms scales as λ^{-3} and any other term is always more irrelevant. For this reason, we never add any higher derivative terms. Instead we go the other way! Much of our work in the remainder of this book will be in figuring out what terms in the Navier–Stokes equation we get to drop in certain circumstances, in the hope that what's left may actually be easy enough to solve.

2.4 Stokes Flow

At low Reynolds number, $Re \ll 1$, the flow is dominated by viscosity. There are situations at low Reynolds number in which we can ignore the material derivative $D\mathbf{u}/Dt$ completely. What remains are the *Stokes equations*

$$\nabla P = \mu \nabla^2 \mathbf{u} \quad \text{and} \quad \nabla \cdot \mathbf{u} = 0 \qquad (2.81)$$

We view these as four equations in four unknowns, $\mathbf{u}$ and P. They should certainly be augmented with the no-slip boundary condition, which is appropriate for our uber-viscous Stokesian world. In some circumstances, we may wish to add further external forces $\mathbf{f}$ to the first of the equations.

Solutions to the Stokes equations are known as *Stokes flows*, or sometimes *creeping flows*. They describe, among many other things, micro-organisms swimming in water.

The lack of time derivatives is unusual when solving dynamical equations. It means that the fluid reacts instantaneously to any imposed force. In some sense, the fluid has no life of its own as there are no propagating waves. Instead, it just does what it's told.

More surprisingly, the lack of any time derivatives means that any flow is reversible. Act with an external force $\mathbf{F}$ for some time and the fluid will evolve. Then act with the opposite force $-\mathbf{F}$ for an equal amount of time and it will evolve back again, returning to its original state. There are dramatic demonstrations of this in which some ink is dropped in a fluid at low Reynolds number. The ink doesn't disperse, but just sits there. The fluid is then stirred, and the ink swirls and mixes with the fluid as expected. But when the stirring is reversed, so too is the mixing until the ink returns to its original starting point. It's the kind of behaviour that the second law of thermodynamics usually prohibits. But life is different at low Reynolds number.

There's something a little disconcerting about this reversible behaviour, not least because, as we've seen above, dissipation in fluids only arises because of viscosity. This means that the increase of entropy in a fluid is also due to viscosity. Yet, when viscosity completely dominates, the dynamics becomes reversible and there is no increase in entropy! Or, said better, there is no dynamics since we have neglected the time derivative term.

There is a very simple variational principle that gives the Stokes equations

(2.81). They follow from the functional

$$\mathcal{E}(\mathbf{u}, P) = \int d^3 x \left(\frac{\mu}{2} \nabla \mathbf{u} \cdot \nabla \mathbf{u} - P \nabla \cdot \mathbf{u} \right) . \tag{2.82}$$

We don't need any of the complicated gymnastics of Section 1.6; we just extremise with respect to $\mathbf{u}$ and P and out pop the Stokes equations. And, finally, we see a context in which P isn't something akin to a Lagrange multiplier for incompressibility. It really is the Lagrange multiplier.

Solving the Stokes Equations

In the remainder of this section, we'll explore Stokes flow in a number of different settings. There are some simple manipulations that we can make to highlight the mathematical structure of the Stokes equations. Taking the divergence of both sides of $\nabla P = \mu \nabla^2 \mathbf{u}$, and using the fact that $\nabla \cdot \mathbf{u} = 0$, tells us that the pressure is necessarily a harmonic function

$$\nabla^2 P = 0 . \tag{2.83}$$

Recall that, in the Euler equation, the pressure always satisfies the Poisson equation (1.31). Now, in this low Reynolds number limit of the Navier–Stokes equation, the pressure satisfies the Laplace equation. Meanwhile, taking the curl of both sides tells us that the vorticity $\boldsymbol{\omega} = \nabla \times \mathbf{u}$ is also harmonic

$$\nabla^2 \boldsymbol{\omega} = 0 . \tag{2.84}$$

Finally, acting with ∇^2 on both sides, and using the fact that $\nabla^2 P = 0$, tells us that the velocity itself is "biharmonic", meaning that

$$\nabla^4 \mathbf{u} := \nabla^2 \nabla^2 \mathbf{u} = 0 . \tag{2.85}$$

In some situations, this is a useful starting point for solving the equations. But, for our first application, we'll take a different route.

2.4.1 Flow Around a Sphere

We want to repeat the calculation that we did for an inviscid fluid flowing around a sphere in Section 1.4. In that case, we found the solution by super-imposing a constant flow with a dipole flow, and then hiding the singularity behind the sphere. Because the Stokes equations are linear, it's perfectly possible that a similar strategy will again work, now for very viscous fluids. We'll see that this is indeed the case, albeit with some of the details changed.

To kick things off, we'll look for the Green's function to the equations

(2.81). This is a velocity field $\mathbf{u}$ and a pressure P obeying

$$\mu \nabla^2 \mathbf{u} - \nabla P = -\mathbf{a}\, \delta^3(\mathbf{x}) \tag{2.86}$$

together with the requirement that $\nabla \cdot \mathbf{u} = 0$. The right-hand side of (2.86) includes an arbitrary constant vector $\mathbf{a}$.

Claim: The Green's function for the Stokes equations is

$$\mathbf{u} = G\mathbf{a} \quad \text{and} \quad P = \frac{\mathbf{x} \cdot \mathbf{a}}{4\pi r^3} \tag{2.87}$$

where G is the matrix

$$G_{ij} = \frac{1}{8\pi\mu}\left(\frac{\delta_{ij}}{r} + \frac{x_i x_j}{r^3}\right) . \tag{2.88}$$

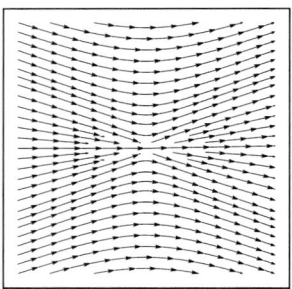

The tensor G is known as the *Stokeslet*. The two terms in G conspire to ensure that $\nabla \cdot \mathbf{u} = 0$. The Stokeslet flow is shown in the figure, with $\mathbf{a} = \hat{\mathbf{z}}$ pointing to the right.

Proof: First, we'll check that the solution (2.87) obeys $\mu \nabla^2 \mathbf{u} = \nabla P$ everywhere except for $r = 0$. Then we'll check that the coefficient of the delta function works out. First look at the velocity term. We have

$$\mu \nabla^2 u_i = \frac{1}{8\pi} \nabla^2 \left(\frac{a_i}{r} + \frac{x_i x_j a_j}{r^3}\right) . \tag{2.89}$$

We recognise the first $1/r$ term as the Green's function for ∇^2 (we already met this interpretation when we discussed potential flows in Section 1.4), with $\nabla^2(1/r) = -4\pi\delta(\mathbf{x})$. Clearly this contributes to the delta function on the right-hand side of (2.86), but only with a coefficient of $\frac{1}{2}$. We'll see that another $\frac{1}{2}$ comes from the other terms. Staying away from $r = 0$ for now, a little bit of algebra is needed to differentiate the second term twice. We have

$$\mu \nabla^2 u_i = \frac{1}{8\pi} \partial_k \partial_k \left(\frac{x_i x_j a_j}{r^3}\right) = \frac{1}{4\pi}\left(\frac{a_i}{r^3} - 3\frac{x_i x_j a_j}{r^5}\right) \quad \text{for } r \neq 0 . \tag{2.90}$$

But now it's simple to check that this is cancelled by the pressure

$$(\nabla P)_i = \frac{1}{4\pi}\partial_i\left(\frac{x_j a_j}{r^3}\right) = \frac{1}{4\pi}\left(\frac{a_i}{r^3} - 3\frac{x_i x_j a_j}{r^5}\right) \quad \text{for } r \neq 0 . \tag{2.91}$$

So we do indeed have a solution to (2.86) away from the origin. Now we just need to check that the $1/8\pi$ normalisation of G gives the correct strength for the delta function. For this we integrate over a ball B_R of radius R centred at the origin, and use the divergence theorem to convert this into an integral over the sphere S_R^2 of radius R

$$\int_{B_R} d^3x \left(\mu \nabla^2 u_k - \partial_k P \right) = \int_{S_R^2} dS_i \left(\mu \partial_i G_{kj} a_j - \frac{1}{4\pi} \delta_{ik} \frac{x_j a_j}{r^3} \right)$$

$$= \frac{a_j}{8\pi} \int_{S_R^2} dS_i \left(\partial_i \left(\frac{\delta_{kj}}{r} + \frac{x_k x_j}{r^3} \right) - 2\frac{\delta_{ik} x_j}{r^3} \right) .$$

Doing the differentiation leaves us with four terms in the integrand

$$\frac{a_j}{8\pi} \int_{S_R^2} dS_i \left(-\frac{\delta_{jk} x_i}{r^3} - \frac{\delta_{ik} x_j}{r^3} + \frac{\delta_{ij} x_k}{r^3} - 3\frac{x_i x_j x_k}{r^5} \right) . \tag{2.92}$$

At this stage, it's all about the placement of indices. The first term is straightforward: it is the usual integral of a radial field over a sphere and gives

$$\frac{a_j}{8\pi} \int_{S_R^2} dS_i \left(-\frac{\delta_{jk} x_i}{r^3} \right) = -\frac{1}{2} a_k . \tag{2.93}$$

This is the same factor of $\frac{1}{2}$ contribution that we noted above. The remaining three terms in the integral must, ultimately, be proportional to δ_{kj} because that's the only invariant tensor available. A standard trick is to take the trace over k and j indices and evaluate the integral: this then gives $3\times$ the coefficient in front of δ_{kj}. If we do this, we find that the second and third terms cancel, while the final term is

$$\frac{a_j}{8\pi} \int_{S_R^2} dS_i \left(-3\frac{x_i x_j x_k}{r^5} \right) = -\frac{1}{2} a_k . \tag{2.94}$$

That's the extra factor of $\frac{1}{2}$ that we were looking for. We learn that our flow and pressure do satisfy (2.86) as promised. □

Given a basic solution like (2.86), we can always generate further solutions by differentiating. These solutions will be more singular at the origin, but drop off quicker asymptotically. This is how the dipole solution is generated for potential flow (and, indeed, for electromagnetism). And it turns out to be what we need for the present problem. The relevant flow is again referred to as a dipole and is given by

$$\mathbf{u}_{\text{dipole}} = (\nabla^2 G)\mathbf{a} \quad \text{with} \quad (\nabla^2 G)_{ij} = \frac{1}{4\pi\mu} \left(\frac{\delta_{ij}}{r^3} - 3\frac{x_i x_j}{r^5} \right) . \tag{2.95}$$

Here the expression for $\nabla^2 G$ comes from (2.90). The associated pressure field is simply $P_{\text{dipole}} = 0$ because, as we saw in (2.83), the original pressure (2.86) is necessarily a harmonic function.

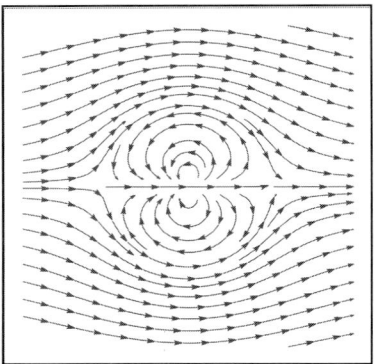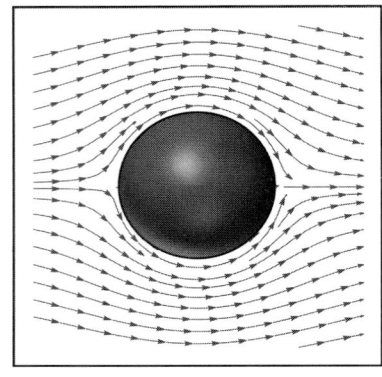

Fig. 2.3   Stokes flow around a sphere. The left-hand figure shows that the flow in the middle moves in the opposite direction to the flow outside. This means that we must have $\mathbf{u} = 0$ on some surface. This is the surface of the solid sphere, as shown on the right.

We now have all the ingredients to solve our problem of interest: a Stokes flow around a sphere of radius R. Importantly, this flow must satisfy the no-slip condition which means that $\mathbf{u} = 0$ for all $|\mathbf{x}| = R$.

We start with a superposition of the different flows that we've found. We take a constant flow $\mathbf{u} = \mathbf{U}$, together with some combination of the Stokeslet and dipole flows. Both the latter flows involve some constant vector $\mathbf{a}$ and, on symmetry grounds, this must be proportional to the asymptotic velocity $\mathbf{U}$. We're left with

$$\mathbf{u} = \mathbf{U} + 4\pi\mu\alpha \left(G + \beta \nabla^2 G \right) \mathbf{U}$$
$$= \mathbf{U} \left(1 + \frac{\alpha}{2r} + \frac{\alpha\beta}{r^3} \right) + (\mathbf{U} \cdot \mathbf{x})\mathbf{x} \left(\frac{\alpha}{2r^3} - \frac{3\alpha\beta}{r^5} \right) \qquad (2.96)$$

where α and β are constants that are fixed by the boundary condition on the sphere. As we've seen, this requires that $\mathbf{u} = 0$ when $|\mathbf{x}| = R$, which is achieved only if both terms are individually vanishing. So we must have

$$\beta = \frac{R^2}{6} \quad \text{and} \quad \alpha = -\frac{3R}{2} \ . \qquad (2.97)$$

So our final flow for a very viscous fluid around a sphere is then

$$\mathbf{u} = \mathbf{U} \left(1 - \frac{3R}{4r} - \frac{R^3}{4r^3} \right) + (\mathbf{U} \cdot \mathbf{x})\mathbf{x} \left(-\frac{3R}{4r^3} + \frac{3R^3}{4r^5} \right) \ . \qquad (2.98)$$

This is shown in Figure 2.3. By eye, the flow outside the sphere doesn't look wildly different from the potential flow that we saw in Section 1.4. But there is a key difference that is clear if you look closely at the left-hand figure, before we place the sphere over it. In Figure 2.3, the fluid inside is moving in the opposite direction to the flow outside. This is what ensures the existence of a surface $r = R$ for which the flow is strictly vanishing, as befits the no-slip boundary condition. In contrast, for the potential flow shown in Figure 1.7, the fluid inside moves in the same direction as the fluid outside.

The fact that $\mathbf{u} = 0$ at $r = R$ makes a big difference. This first manifests itself in the pressure field, which is

$$P = P_\infty - \frac{3}{2} R\mu \frac{\mathbf{U} \cdot \mathbf{x}}{r^3} \ . \tag{2.99}$$

If we take $\mathbf{U} = U\hat{\mathbf{z}}$ and work in spherical polar coordinates, the pressure on the surface of the sphere is

$$P = P_\infty - \frac{3U\mu\cos\theta}{2R} \ . \tag{2.100}$$

This means that the pressure is bigger than P_∞ on the front of the sphere (which is the left side of the sphere Figure 2.3) where $\pi/2 < \theta \le \pi$ and $\cos\theta < 0$. The pressure is less than P_∞ at the back of the sphere where $0 \le \theta < \pi/2$. This, of course, sounds very reasonable: it's simply because the flow is exerting pressure on the sphere. But it was this simple physics that was noticeably absent in the potential flow of Section 1.4. (See equation (1.131) for the analogous equation in that case.) This is the first hint that we may be on the way to finally understand the drag force.

Such a Drag

The pressure is not the only force that the sphere experiences. The technology to compute the drag force comes from the stress tensor (2.17)

$$\sigma_{ij} = -P\delta_{ij} + 2\mu E_{ij} \tag{2.101}$$

where E_{ij} is the rate of strain tensor. We can compute this for the flow (2.98). It simplifies somewhat when evaluated on the surface of the sphere:

$$E_{ij}(|\mathbf{x}| = R) = \frac{3}{4R^2}(U_i x_j + U_j x_i) - \frac{3}{2R^4}(\mathbf{U} \cdot \mathbf{x})x_i x_j \ . \tag{2.102}$$

To compute the force experienced by any point on the sphere, we consider $\sigma_{ij}n_j = \sigma_{ij}(x_j/R)$ where $\mathbf{n} = \mathbf{x}/R$ is the unit normal to the surface of the sphere. Using our expressions above, we have (ignoring the asymptotic pressure P_∞ which has no net effect on the sphere)

$$\sigma_{ij}n_j = \frac{3\mu(\mathbf{U} \cdot \mathbf{x})x_i}{2R^3} + 2\mu\left(\frac{3U_i}{4R} - \frac{3(\mathbf{U} \cdot \mathbf{x})x_i}{4R^3}\right) \ . \tag{2.103}$$

We see that, rather nicely, the first term from the pressure cancels the final term from the strain. This means that the force acting on any point of the sphere is constant, and in the direction $\mathbf{U}$ of the asymptotic flow

$$\sigma \mathbf{n} = \frac{3\mu}{2R}\mathbf{U} \ . \tag{2.104}$$

It's now very easy to compute the drag force: we just integrate this over the whole sphere, getting an additional factor of the surface area $4\pi R^2$. The total drag force acting on the sphere is then

$$\text{drag force} = 6\pi\mu R\mathbf{U} \ . \tag{2.105}$$

This is a famous formula, known as *Stokes' law*. It is the drag experienced by a sphere moving at very low Reynolds number. The force is linear in the velocity U. This was the result that we quoted in the book on Classical Mechanics for a slow-moving object. Here we now know what "slow" really means: it is $Re \ll 1$.

We see that, in contrast to the Euler equations, the Stokes equations do give rise to a calculable drag force. That's both nice and useful, but it's not really a resolution to the d'Alembert paradox of Section 1.4. That's because the paradox still holds at high Reynolds number, where we can seemingly neglect viscosity, and the Navier–Stokes equation would appear to reduce to the Euler equation.

In fact, Stokes flow comes with its own peculiarities. In particular, the $1/r$ term in the velocity profile (2.98) should worry you. If we boost to a frame where the sphere is moving, and the fluid is stationary, then this is the leading term in the velocity at large distance. But it drops off so slowly that the kinetic energy of the fluid, $\int d^3x \, \mathbf{u}^2$, diverges logarithmically. In the language of quantum field theory, we would call this an "infrared divergence". So apparently it takes infinite energy to move a sphere through a fluid at very low Reynolds number, which doesn't really sound right! It turns out that this is sort of a low Reynolds version of the d'Alembert paradox in the sense that, for certain questions, neither the Euler equation nor the Stokes equations are good approximations to the full Navier–Stokes equation. We'll address the high Reynolds number story in Section 2.5.

2.4.2 Uniqueness and the Minimum Dissipation Theorem

We found a solution for the flow around the sphere. But it turns out that it is *the* solution: there is no other with the same boundary conditions. This follows from a uniqueness theorem that is proven in the same way as the uniqueness of solutions to the Laplace equation.

Suppose that we have two solutions, $\mathbf{u}_1$ and $\mathbf{u}_2$, both obeying no-slip boundary conditions on the surface. Then the difference $\mathbf{v} = \mathbf{u}_1 - \mathbf{u}_2$ necessarily vanishes on the boundary. With $\tilde{P} = P_1 - P_2$ the difference in the pressure fields, we have

$$
\begin{aligned}
0 &= \int_V \mathbf{v} \cdot (\mu \nabla^2 \mathbf{v} - \nabla \tilde{P})\, dV \\
&= \int_V \partial_i \left(\mu v_j \partial_i v_j - v_i \tilde{P} \right) dV - \int_V (\partial_i v_j)^2 \, dV \ .
\end{aligned}
\tag{2.106}
$$

The first term on the right-hand side vanishes because it's a total derivative and $\mathbf{v} = 0$ on the boundary ∂V. Moreover, the second term is the integral of a total square so this can be zero only if the integrand vanishes: $\partial_i v_j = 0$. Hence $v_j = 0$ everywhere and our original solutions $\mathbf{u}_1$ and $\mathbf{u}_2$ were the same.

Stokes Flow Dissipates Less Than Any Other Flow

Here's a cute mathematical result. Among all the incompressible flows with the same boundary conditions, the Stokes flow dissipates the least energy.

To prove this, suppose that we have a solution $\mathbf{u}$ and P to the Stokes equations with no external force (2.81), and a second flow $\tilde{\mathbf{u}}$ that satisfies the same boundary conditions but is otherwise arbitrary. Recall from (2.30) that the energy dissipated by an arbitrary flow $\tilde{\mathbf{u}}$ is (2.30)

$$
\begin{aligned}
\text{dissipation} &= 2\mu \int_V \tilde{E}_{ij}\tilde{E}_{ij} \, dV \\
&= 2\mu \int_V \left[E_{ij}E_{ij} + (E_{ij} - \tilde{E}_{ij})^2 + 2E_{ij}(\tilde{E}_{ij} - E_{ij}) \right] dV \\
&\geq 2\mu \int_V \left[E_{ij}E_{ij} + 2E_{ij}(\tilde{E}_{ij} - E_{ij}) \right] dV \\
&= \text{Stokes dissipation} + 4\mu \int_V E_{ij}(\tilde{E}_{ij} - E_{ij}) \, dV \ .
\end{aligned}
\tag{2.107}
$$

We'll now show that this second integral actually vanishes. To see this, recall that the stress tensor for the Stokes flow is

$$
\sigma_{ij} = -P\delta_{ij} + 2\mu E_{ij} \ .
\tag{2.108}
$$

Importantly, the stress tensor is divergence-free for the Stokes flow,

$$
\partial_i \sigma_{ij} = -\partial_j P + \mu \nabla^2 u_j = 0
\tag{2.109}
$$

where the other term in E_{ij} vanishes because it involves $\partial_i u_i = 0$ and the whole thing is equal to zero by virtue of the Stokes equations. This is the special property of the Stokes flow that we need. If we now contract the Stokes stress with the strain tensor $\tilde{E}_{ij}$ for any other flow, we have

$$\sigma_{ij}\tilde{E}_{ij} = 2\mu E_{ij}\tilde{E}_{ij} \qquad (2.110)$$

where the other term $-P\tilde{E}_{ii}$ vanishes because the flow is incompressible and $\tilde{E}_{ii} = \nabla \cdot \tilde{\mathbf{u}}$. We now have

$$4\mu \int_V E_{ij}(\tilde{E}_{ij} - E_{ij})\, dV = 2\int_V \sigma_{ij}(\tilde{E}_{ij} - E_{ij})\, dV$$

$$= 2\int_V \sigma_{ij}\left(\partial_i \tilde{u}_j - \partial_i u_j\right)\, dV$$

$$= 2\int_V \partial_i\left[\sigma_{ij}(\tilde{u}_j - u_j)\right]\, dV = 0 \qquad (2.111)$$

where, in the second line, we've used the fact that σ_{ij} is symmetric and, in the final line, we've used the special property of the Stokes flow (2.109), together with the divergence theorem, which means that the integral only cares about the boundary where, by assumption, $\tilde{\mathbf{u}} = \mathbf{u}$. The upshot is that for any flow $\tilde{\mathbf{u}}$ that is *not* a Stokes flow, we necessarily have

$$\int_V \tilde{E}_{ij}\tilde{E}_{ij}\, dV > \int_V E_{ij}E_{ij}\, dV \ . \qquad (2.112)$$

The dissipation from other flows is always greater than the corresponding Stokes flow. This is the *Helmholtz minimum dissipation theorem*.

There is, it turns out, a deep relationship between drag and dissipation, known as the *fluctuation-dissipation theorem*. The fact that the Stokes flow has the smallest dissipation translates into the statement that it also results in the smallest drag. This means that, as we increase the Reynolds number, the drag on the sphere will only increase beyond that given by Stokes' law (2.105). Indeed, one can set up a perturbation expansion to understand the effects of the terms in the Navier–Stokes equation that we neglected. This is an expansion in the Reynolds number $Re \ll 1$ and the leading order term turns out to be

$$\text{drag force} = 6\pi\mu R\mathbf{U}\left(1 + \frac{3}{8}Re + \dots\right) \ . \qquad (2.113)$$

We will learn more about the fluctuation-dissipation theorem in Chapter 7.

2.4.3 Eddies in the Corner

As you might imagine, there are many different Stokes flows that exhibit interesting properties. Here is another one. We simply look at fluid passing around a corner. This corner has an opening angle that we denote as 2α. We want to know what happens.

This problem is effectively two-dimensional and can be solved quite straight-forwardly by working in cylindrical polar coordinates and introducing a stream function $\psi(r,\theta)$. Recall from Section 1.1.4 that the stream function allows us to construct a vector field $\mathbf{A} = \psi\hat{\mathbf{z}}$ and, from that, an incompressible flow $\mathbf{u} = \nabla \times \mathbf{A}$. In cylindrical polar coordinates, the resulting flow is

$$\mathbf{u} = \frac{1}{r}\frac{\partial\psi}{\partial\theta}\hat{\mathbf{r}} - \frac{\partial\psi}{\partial r}\hat{\boldsymbol{\theta}} \ . \tag{2.114}$$

The associated vorticity is

$$\boldsymbol{\omega} = \nabla \times \mathbf{u} = -(\nabla^2\psi)\hat{\mathbf{z}} \tag{2.115}$$

with

$$\nabla^2\psi = \frac{1}{r}\frac{\partial}{\partial r}\left(r\frac{\partial\psi}{\partial r}\right) + \frac{1}{r^2}\frac{\partial^2\psi}{\partial\theta^2} \ . \tag{2.116}$$

But we've seen in (2.84) that the vorticity $\boldsymbol{\omega}$ is harmonic for Stokes flows, which means that the stream function must be biharmonic

$$\nabla^4\psi = \left(\frac{1}{r}\frac{\partial}{\partial r}\left(r\frac{\partial}{\partial r}\right) + \frac{1}{r^2}\frac{\partial^2}{\partial\theta^2}\right)^2\psi = 0 \ . \tag{2.117}$$

The form of the equation suggests that it might be fruitful to look for scale-invariant, separable solutions of the form

$$\psi(r,\theta) = r^\lambda f(\theta) \tag{2.118}$$

for some exponent λ and some function $f(\theta)$. The biharmonic condition then becomes a differential equation for f,

$$\nabla^4\psi = r^{\lambda-4}\left(\frac{\partial^2}{\partial\theta^2} + \lambda^2\right)\left(\frac{\partial^2}{\partial\theta^2} + (\lambda-2)^2\right)f(\theta) = 0 \ . \tag{2.119}$$

The solution is simply

$$f(\theta) = A\sin(\lambda\theta) + B\cos(\lambda\theta) + C\sin((\lambda-2)\theta) + D\cos((\lambda-2)\theta) \tag{2.120}$$

with four integration constants as well
as the exponent λ still to be determined.
At this point we bring out some boundary conditions. We'll arrange the geome-
try so that the boundaries lie at $\theta = \pm\alpha$.
An example of the kind of flow we're
looking for is shown in the figure, which
shows the streamlines of the flow. Note,
in particular, that the fluid comes in
parallel to one boundary, and out par-

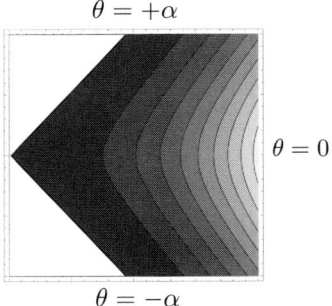

allel to the other. This means that the radial component $\mathbf{u} \cdot \hat{\mathbf{r}}$ of the flow
should be an odd function of θ. The expression (2.114) then tells us that the
stream function should be an even function of θ, so $A = C = 0$.

We now have two further boundary conditions since both components of
$\mathbf{u}$ must vanish along the boundary. The requirement that no fluid moves
into the boundary is

$$\left.\frac{\partial\psi}{\partial r}\right|_{\theta=\pm\alpha} = 0 \quad\Longrightarrow\quad B\cos(\lambda\alpha) + D\cos((\lambda-2)\alpha) = 0 \ . \quad (2.121)$$

Meanwhile, the no-slip condition tells us that

$$\left.\frac{\partial\psi}{\partial\theta}\right|_{\theta=\pm\alpha} = 0 \quad\Longrightarrow\quad B\lambda\sin(\lambda\alpha) + D(\lambda-2)\sin((\lambda-2)\alpha) = 0 \ . \quad (2.122)$$

Or, combined,

$$\lambda\sin(\lambda\alpha)\cos((\lambda-2)\alpha) = (\lambda-2)\cos(\lambda\alpha)\sin((\lambda-2)\alpha) \ . \quad (2.123)$$

This equation always has the solution $\lambda = 1$, but the conditions above tell
us that if $\lambda = 1$ then $B = -D$ and, correspondingly, $\psi = 0$. This is not what
we want. So we'll look for solutions with $\lambda \neq 1$. Expand each sin and cos
above in terms of $e^{i(\text{whatever})}$ and rearrange to get

$$\frac{\sin(2(\lambda-1)\alpha)}{\lambda-1} = -\sin 2\alpha \ . \quad (2.124)$$

This equation determines the exponent λ in terms of the opening angle of
the corner 2α, admittedly in a slightly opaque form. To understand what
it's telling us, write $x = 2(\lambda-1)\alpha$, so the equation becomes

$$\frac{\sin x}{x} = -\frac{\sin 2\alpha}{2\alpha} \ . \quad (2.125)$$

Suppose that the opening angle α is small. Then, as you can see from Figure
2.4, the value of $\sin 2\alpha/2\alpha$ is large. But there is no value of x for which

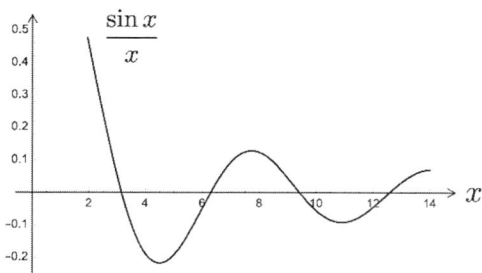

Fig. 2.4 The graph of $\sin x / x$ with the value at the first minimum around -0.217.

$\sin x / x$ has the equal negative value. So for small opening angles, we can't solve (2.125), at least not for real x.

As the opening angle gets bigger, we do get solutions. The smallest value of $\sin x / x$ occurs at the first minimum as shown in Figure 2.4, which sits at

$$ x \approx 1.43\pi \quad \Longrightarrow \quad \frac{\sin x}{x} \approx -0.217 \ . \tag{2.126} $$

This corresponds to a value of 2α given by

$$ 2\alpha \approx 0.813\pi \approx 146° \quad \Longrightarrow \quad \frac{\sin 2\alpha}{2\alpha} \approx +0.217 \ . \tag{2.127} $$

We learn that there is a critical value of the opening angle, given by

$$ 2\alpha_{\mathrm{crit}} \approx 146° \ . \tag{2.128} $$

For opening angles larger than this, we can find solutions to (2.125). A contour plot of the stream function for $2\alpha = 160°$ is shown in the figure to the right. The lines of constant value are the streamlines and they simply flow around the corner undisturbed.

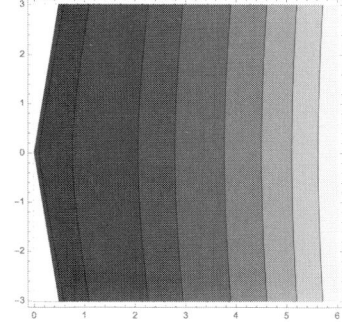

What happens when the opening angle is smaller than 146°? Now, no solutions to (2.125) exist. Or, said more precisely, no *real* solutions exist! There are, however, always complex solutions. For example, suppose that we have a right-angle corner, with $2\alpha = \pi/2 < 2\alpha_{\mathrm{crit}}$. Then there is an infinite sequence of complex solutions to (2.125), starting with

$$2\alpha = \frac{\pi}{2} \quad \Longrightarrow \quad \lambda \approx 3.74 + 1.12i$$

$$\lambda \approx 7.84 + 1.66i \ \dots \ . \quad (2.129)$$

What is the interpretation of these solutions? If we have a solution with

$$\lambda = \lambda_1 + i\lambda_2 \quad (2.130)$$

then, because the velocity (2.114) is a linear function of ψ, we can take the real part of the stream function to get

$$\psi(r,\theta) = \mathrm{Re}\left[r^\lambda f(\theta)\right]$$
$$= r^{\lambda_1}\left[\cos(\lambda_2 \log r)\,\mathrm{Re}f(\theta) - \sin(\lambda_2 \log r)\,\mathrm{Im}f(\theta)\right] . \quad (2.131)$$

That $\cos\log r$ behaviour is striking! For a fixed angle θ, it gives rise to increasingly wild oscillations as $r \to 0$, albeit with decreasing amplitude because of the overall r^{λ_1} scaling. You can check that this means that the angular velocity $\mathbf{u}\cdot\hat{\boldsymbol{\theta}}$ is also oscillating in sign as $r \to 0$. This is telling us that the flow no longer takes the simple form, as shown in the figure for large opening angle, but instead develops eddies. In fact, there are an infinite number of these eddies, becoming increasingly small as $r \to 0$. These are known as *Moffatt eddies*. Some special Moffat eddies are shown in Figure 2.5.

The stream function for a right-angle corner is shown in the figure to the right, clearly exhibiting one such eddy. The logarithm means that both the size of the eddies, and the amplitude of the stream function, decrease exponentially as we head into the corner. The centres of consecutive eddies lie at

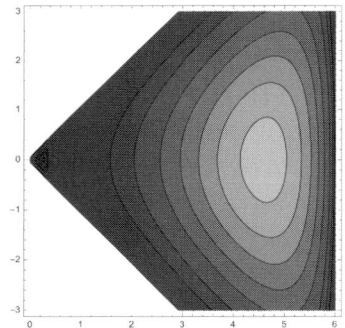

$$\lambda_2 \log r_{n+1} = \lambda_2 \log r_n - \pi$$
$$\Longrightarrow \quad \frac{r_{n+1}}{r_n} = e^{-\pi/\lambda_2} \quad (2.132)$$

and this also characterises the size of the eddies. (If you squint, you can just see a second eddie in the figure centred around $x \approx 0.2$.) Meanwhile, the size of the stream function, which determines the magnitude of the velocity, scales as

$$\frac{|\psi(r_{n+1})|}{|\psi(r_n)|} \sim \left(\frac{r_{n+1}}{r_n}\right)^{\lambda_1} = e^{-\lambda_1 \pi/\lambda_2} . \quad (2.133)$$

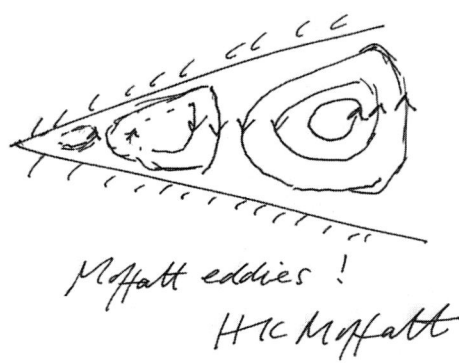

Fig. 2.5 Not just any old Moffatt eddies, but Moffatt's Moffatt eddies. My thanks to Keith Moffatt for graciously humouring my fanboy request to sketch these.

The magnitude of velocities involves a derivative of the stream function, $u \sim \partial \Psi / \partial r$, and so scale as $(r_{n+1}/r_n)^{\lambda_1 - 1} = e^{-(\lambda_1 - 1)\pi/\lambda_2}$. For the right-angle corner shown in the figure, this ratio is around 2000. This exponential scaling doesn't just make it difficult to plot the eddies; it also makes it difficult to experimentally observe more than two or three.

Although the eddies get smaller as you approach the vertex, the flow also becomes slower so it takes significantly longer for a particle to orbit the smaller eddies than the larger ones.

Something rather pretty has happened here: the solution exhibits a discrete scale invariance, with an infinite number of ever-decreasing eddies, each reduced by a constant factor. Such discrete scale invariance arises in other areas of physics too, albeit as something of a rarity. The classic example in quantum mechanics is the *Efimov state*, in which three bosons form an infinite series of bound states, with the ratio of energies similar to the ratio of distances (2.132).

2.4.4 Hele-Shaw Flow

In this short section, we look at a particular way of restricting Stokes flow to two dimensions. However, rather than simply solving the 2d version of the Stokes equations, we instead do something more physical. We trap the fluid between two parallel, stationary plates, separated by a distance h. This scale will be much smaller than any other scale, such as the size of any object that the fluid moves around.

We separate the plates in the z-direction and consider situations in which the fluid flows only in the (x, y)-plane

$$\mathbf{u} = (u(x, y, z), v(x, y, z), 0) \tag{2.134}$$

and we now solve the Stokes equation

$$\nabla P = \mu \nabla^2 \mathbf{u} . \tag{2.135}$$

The first thing to realise is that gradients in the z-direction are of order $\partial/\partial z \sim 1/h$ and so are much bigger than anything else. These gradients can't vanish because the no-slip condition means that $\mathbf{u}$ vanishes at $z = 0$ and $z = h$ but we want it to be non-vanishing in the middle. We work in the approximation that these z-gradients are entirely accounted for by the pressure

$$\mu \frac{\partial^2 \mathbf{u}}{\partial z^2} = \nabla P \quad \Longrightarrow \quad u = \frac{1}{2\mu} \frac{\partial P}{\partial x} z(z - h)$$

$$\text{and} \quad v = \frac{1}{2\mu} \frac{\partial P}{\partial y} z(z - h) \tag{2.136}$$

where the boundary conditions have been chosen so that the no-slip condition is satisfied. This is the same kind of velocity profile that we saw for Poiseuille flow (2.50), but now in 2d rather than 1d. In the present context, it is known as *Hele-Shaw* flow. (One person, not two! He chose, I think rather unusually, to adopt both his father's and his mother's names.)

But Hele-Shaw flow is something very familiar: we have a situation where the 2d velocity field $\mathbf{u}_{2d} = (u, v)$ is given by

$$\mathbf{u}_{2d} = \nabla_{2d}\phi \quad \text{with} \quad \phi(x, y; z) = \frac{1}{2\mu} P(x, y)z(z - h) \tag{2.137}$$

and with $\nabla_{2d} = (\partial_x, \partial_y)$. In other words, we're back in the realm of 2d potential flow that we solved in Section 1.5. This means, for example, that if you place a cylinder between the plates, with its axis pointing in the z-direction, then the velocity flow around it coincides with the velocity (1.147) that we previously calculated.

There is an irony here. We originally introduced potential flow as a description of completely inviscid fluids. Yet the same solutions also describe extremely viscous fluids when sandwiched between plates! In fact the irony runs deeper. If you attempt to go to a regime where viscosity can be neglected – which means high Reynolds number – then another effect, known as the boundary layer, kicks in and the flows don't look at all like potential flows near objects. (We will describe this in Section 2.5.) So, in fact, the only

Fig. 2.6   Hele-Shaw flow. Credit: S.T. Thoroddsen.

way to genuinely manufacture the inviscid potential flows of Section 1.5 is
to work with very viscous fluids. The Hele-Shaw flow around a cylinder is
shown in Figure 2.6, a realisation of the 2d potential flow that we previously
plotted in Figure 1.8.

There is, however, a difference between Hele-Shaw flows and the general 2d
potential flow. Hele-Shaw flows can have no circulation in the (x, y)-plane,

$$\Gamma = \oint \mathbf{u} \cdot d\mathbf{x} = 0 \ . \tag{2.138}$$

This is because, as we showed in Section 1.5, circulation arises only from
potentials that are not single-valued. In contrast, the potential for Hele-Shaw
flows is effectively the pressure $P(x, y)$ and this is certainly single-valued.
The upshot is that Hele-Shaw flows don't include those shown in Figure 1.9
which induce a lift force on the obstacle.

2.4.5 Swimming at Low Reynolds Number

Given the constraints of their biology, scallops are remarkably elegant swim-
mers. They open their shells, then quickly close them, forcing water out
through the hinges to propel themselves forward.

This strategy works in the ocean. But it would be hopeless at low Reynolds
number. This is because, as we mentioned at the beginning of this section,
the lack of time derivatives in the Stokes equations means that motion at
low Reynolds number is reversible. When friction dominates, the speed at
which a scallop opens or closes its shell is irrelevant. It moves in one direction
when the shell opens, and comes back the same amount when it closes. A
scallop dropped in honey can no longer swim. Although it surely tastes nice.

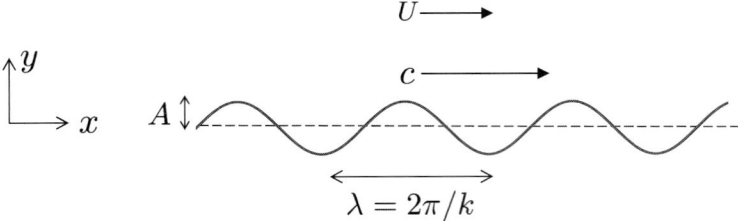

Fig. 2.7 A waving plate in a viscous fluid. Ripples travel down the plate at speed c, resulting in it "swimming" through the fluid at the lower speed $U = \frac{1}{2}A^2k^2c$.

To swim at low Reynolds number, you need a different strategy. You can't just flap your arms (or your legs or your fins) back and forth, because what is done by the forward flap is undone by the backward one. Instead, you need to change your shape in some way that is not, itself, time reversible. In other words, you need something like breaststroke.

Such non-reversible strategies have been developed by micro-organisms living in water, for whom life is lived at low Reynolds number. For example, the bacterium *E. coli* has a helical flagellum which rotates to make it swim. The underlying mathematics is closely related to the rotation of deformable (as opposed to rigid) bodies described in the book on Classical Mechanics.

In this section, we describe a very simple model that captures the essence of swimming at low Reynolds number. It also captures the tension between finding a mathematical model that is easy to solve, and finding one that looks vaguely like a living creature.

An Infinite, Wavy Plate

As our proxy micro-organism, we take an infinite thin plate, lying in the (x, z)-plane, as shown in Figure 2.7. (Admittedly, this object is unlikely to make a good pet.) The plate "swims" by wriggling so that a wave passes down in the x-direction, meaning that the position of the plate in the y-direction, which is perpendicular to the flat plate, is

$$y(x) = A \sin\left(kx - \omega t\right) . \tag{2.139}$$

Here A is the amplitude of the wave, which has wavenumber k and frequency ω. Said another way, the wave has wavelength $\lambda = 2\pi/k$ and travels with speed

$$c = \frac{\omega}{k} . \tag{2.140}$$

We want to understand the flow that results from this wriggling. Our goal is to show that the wriggling induces a constant asymptotic velocity in the fluid. This isn't quite swimming of course: it's staying still and making all the universe move around you. But, by Galilean relativity, this is equivalent to the fluid staying still and the plate moving. And that's what we mean by swimming.

To proceed, we introduce a stream function $\psi(x, y)$ so that $\mathbf{u} = (u, v, 0)$ with

$$u = \frac{\partial \psi}{\partial y} \quad \text{and} \quad v = -\frac{\partial \psi}{\partial x} . \tag{2.141}$$

Repeating the argument that we saw for corner flows, we know that $\boldsymbol{\omega} = \nabla \times \mathbf{u} = -\nabla^2 \psi\, \hat{\mathbf{z}}$ and $\nabla^2 \boldsymbol{\omega} = 0$, which, combined, tell us that

$$\nabla^4 \psi = 0 . \tag{2.142}$$

We must solve this, subject to the no-slip requirement that the velocity of the flow matches that of the plate

$$u = 0 \quad \text{and} \quad v = -A\omega \cos{(kx - \omega t)} \quad \text{on} \quad y = A \sin{(kx - \omega t)} . \tag{2.143}$$

We'll also impose suitable boundary conditions asympotically. We'll flag these up as we go along.

Simple as the equations above are, it's not straightforward to solve them because the boundary condition (2.143) is evaluated on the waving plate. To proceed, we need an approximation. Roughly speaking, we want the amplitude of the wave A to be small in the hope that the boundary condition is easier to implement. But A is dimensionful, so it has to be small relative to something else and the only other length scale we have is the wavelength. So the relevant dimensionless expansion parameter is

$$\epsilon = Ak \ll 1 . \tag{2.144}$$

To understand how to write our equations in terms of a Taylor expansion, it's useful to introduce dimensionless distances and a dimensionless stream function

$$\tilde{x} = kx , \quad \tilde{y} = ky , \quad \tilde{\psi} = \frac{k\psi}{A\omega} . \tag{2.145}$$

The boundary conditions (2.143) then become

$$\frac{\partial \tilde{\psi}}{\partial \tilde{y}} = 0 \quad \text{and} \quad \frac{\partial \tilde{\psi}}{\partial \tilde{x}} = \cos{(\tilde{x} - \omega t)} \quad \text{on} \quad \tilde{y} = \epsilon \sin{(\tilde{x} - \omega t)} . \tag{2.146}$$

Now we can see that, for $\epsilon \ll 1$, we do indeed impose the boundary condition on a value of $\tilde{y}$ that is small. It means that we can Taylor expand the function $\tilde{\psi}$ around $\tilde{y} = 0$, so these boundary conditions read

$$\left.\frac{\partial\tilde{\psi}}{\partial\tilde{y}}\right|_{\tilde{y}=\epsilon\sin(\tilde{x}-\omega t)} = \left.\frac{\partial\tilde{\psi}}{\partial\tilde{y}}\right|_{\tilde{y}=0} + \epsilon\sin(\tilde{x}-\omega t)\left.\frac{\partial^2\tilde{\psi}}{\partial\tilde{y}^2}\right|_{\tilde{y}=0} + \ldots = 0 \quad (2.147)$$

and

$$\left.\frac{\partial\tilde{\psi}}{\partial\tilde{x}}\right|_{\tilde{y}=\epsilon\sin(\tilde{x}-\omega t)} = \left.\frac{\partial\tilde{\psi}}{\partial\tilde{x}}\right|_{\tilde{y}=0} + \epsilon\sin(\tilde{x}-\omega t)\left.\frac{\partial^2\tilde{\psi}}{\partial\tilde{x}\partial\tilde{y}}\right|_{\tilde{y}=0} + \ldots$$
$$= \cos(\tilde{x}-\omega t) . \quad (2.148)$$

We now expand the stream function itself in powers of ϵ

$$\tilde{\psi} = \tilde{\psi}_0 + \epsilon\tilde{\psi}_1 + \ldots . \quad (2.149)$$

Each $\tilde{\psi}_n$ is biharmonic, meaning that it satisfies $\nabla^4\tilde{\psi}_n = 0$ for $n = 0, 1, 2, \ldots$. But each $\tilde{\psi}_n$ obeys different boundary conditions at $\tilde{y} = 0$. We start with $\tilde{\psi}_0$ which has boundary conditions

$$\frac{\partial\tilde{\psi}_0}{\partial\tilde{y}} = 0 \quad \text{and} \quad \frac{\partial\tilde{\psi}_0}{\partial\tilde{x}} = \cos(\tilde{x}-\omega t) \quad \text{on} \quad \tilde{y} = 0 . \quad (2.150)$$

The biharmonic function obeying these boundary conditions in the region above the plate, $\tilde{y} > 0$, is

$$\tilde{\psi}_0 = (1+\tilde{y})e^{-\tilde{y}}\sin(\tilde{x}-\omega t) \quad (2.151)$$

which obeys $\tilde{\nabla}^2\tilde{\psi}_0 = -2e^{-\tilde{y}}\sin(\tilde{x}-\omega t)$ and so $\tilde{\nabla}^4\tilde{\psi}_0 = 0$. Note that we've thrown away a similar solution that scales as $e^{+\tilde{y}}$ on the grounds that it gives an unbounded velocity field as $\tilde{y} \to +\infty$. A solution of this kind is relevant below the plate for $\tilde{y} < 0$.

The first correction to this solution is $\tilde{\psi}_1$ which, from (2.147) and (2.148), obeys

$$\frac{\partial\tilde{\psi}_1}{\partial\tilde{y}} + \sin(\tilde{x}-\omega t)\frac{\partial^2\tilde{\psi}_0}{\partial\tilde{y}^2} = 0$$
$$\text{and} \quad \frac{\partial\tilde{\psi}_1}{\partial\tilde{x}} + \sin(\tilde{x}-\omega t)\frac{\partial^2\tilde{\psi}_0}{\partial\tilde{x}\partial\tilde{y}} = 0 . \quad (2.152)$$

Both boundary conditions (2.147) and (2.148) should again be imposed at $\tilde{y} = 0$. Using our solution (2.151) for $\tilde{\psi}_0$, these become

$$\frac{\partial\tilde{\psi}_1}{\partial\tilde{y}} = \sin^2(\tilde{x}-\omega t) \quad \text{and} \quad \frac{\partial\tilde{\psi}_1}{\partial\tilde{x}} = 0 \quad \text{on} \quad \tilde{y} = 0 . \quad (2.153)$$

The $\sin^2$ term is where our interest lies. We decompose this into Fourier modes by using the double angle formula $\sin^2 \tilde{x} = \frac{1}{2}(1 - \cos 2\tilde{x})$. The $\cos 2\tilde{x}$ term is just telling us that the second harmonic is excited. That's little surprise. The constant term is more interesting as it tells us that there must be a constant component to the fluid motion. Indeed, you can check that the biharmonic function obeying these boundary conditions is

$$\tilde{\psi}_1 = \frac{1}{2}\tilde{y} - \frac{1}{2}\tilde{y}e^{-2\tilde{y}} \cos(2(\tilde{x} - \omega t)) \ . \tag{2.154}$$

Again, this solution holds above the plate for $\tilde{y} > 0$ and there is an analogous solution with an $e^{+\tilde{y}}$ below the plate but with the same constant $\frac{1}{2}\tilde{y}$ term. That linear term is what we're after. Putting the various constants back in, it gives a contribution to the stream function that looks like $\psi = \frac{1}{2}A^2k^2cy + \ldots$ where the $\ldots$ are the oscillatory terms that drop off as $e^{-k|y|}$ or $e^{-2k|y|}$ as we move away from the plate. The constant term is telling us that, far away from the plate, there is necessarily a constant fluid velocity

$$\mathbf{u} \to \frac{1}{2}A^2k^2c\,\hat{\mathbf{x}} \quad \text{as } y \to +\infty \ . \tag{2.155}$$

Alternatively, if we boost to another frame so that the fluid is asymptotically stationary, then the plate must be moving to the left with speed $U = \frac{1}{2}A^2k^2c$. In other words, the plate is swimming. The speed is proportional to the speed c with which waves propagate down the plate but, at least in this approximation, suppressed by $\epsilon^2 = A^2k^2 \ll 1$.

2.5 The Boundary Layer

In the previous section, we focussed on very viscous flow at low Reynolds number. Now we turn to the opposite regime of high Reynolds number. We're going to revisit the question of flows around some fixed object, like a sphere or the wing of an aircraft.

When the Reynolds number is large, the inertia term in the Navier–Stokes equation should dominate over the viscosity term

$$Re = \frac{\text{inertial term}}{\text{viscosity term}} = \frac{|\mathbf{u} \cdot \nabla \mathbf{u}|}{|\nu \nabla^2 \mathbf{u}|} \gg 1 \ . \tag{2.156}$$

For example, for a plane flying we have $Re \sim 10^7$. Given this, it's tempting to think that we can drop the viscosity term completely. But this brings us back to the Euler equation and, as we have seen in Chapter 1, inviscid flows do not give rise to any drag on an object. Something is amiss! In fact it turns out that, no matter how small the viscosity, it still plays an important role.

Mathematically this is because the character of the Navier–Stokes equation changes if we set $\nu = 0$. With $\nu \neq 0$, we have an equation that is second order in spatial derivatives. When $\nu = 0$, it changes to an equation that is first order. As we have commented previously, this means that we must impose two boundary conditions when solving the $\nu \neq 0$ Navier–Stokes equation, but only a single boundary condition when solving the Euler equation. The boundary condition that is expendable is the no-slip condition and, in its absence, solutions exhibit no drag force. However, as soon as we have ν, no matter how small, we're back in business and we can impose the no-slip condition to our heart's content.

Physically, a continuous flow with a no-slip boundary condition must have a layer of almost stationary fluid sitting next to the object. This is the *boundary layer*. The purpose of this section is to understand some of its properties.

We can make progress with some simple dimensional analysis, coupled with a little intuition built on what we've learned so far. For example, one of the most basic questions that we can ask is: What is the width of the boundary layer? It seems plausible that when the fluid first hits the leading edge of the object, only those molecules immediately in contact know about its existence. But, as we look further down the flow, more and more of the fluid should be affected. How much?

Suppose that our object has length L, and travels with speed U relative to the fluid. The Reynolds number is then

$$Re = \frac{UL}{\nu} \ . \tag{2.157}$$

By assumption, $Re \gg 1$.

Any fluid element takes a time $T = L/U$ to move past the object. Close to the object, the fluid will be affected by the no-slip condition and it is reasonable to think that the near-boundary behaviour mimics that of Couette or Poiseuille flow. One of the simple, yet important facts about these flows is that they have vorticity, as the fluid near the boundary travels at different speeds. And we know from the vorticity equation (2.56) that viscosity causes vorticity to diffuse, with diffusion constant ν. Importantly, things that diffuse spread as $\sqrt{\text{time}}$ rather than linearly in time. This means that in the

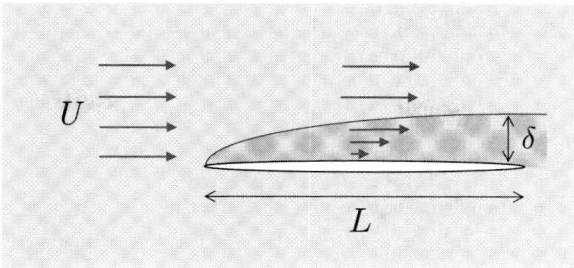

Fig. 2.8 An exaggerated picture of the boundary layer, of width δ, forming over a thin plate of length L. A second boundary layer will, of course, also form below.

time scale T, the vorticity will diffuse a distance

$$\delta \sim \sqrt{\nu T} \sim \sqrt{\frac{\nu L}{U}} \sim \frac{L}{\sqrt{Re}} \ . \tag{2.158}$$

This is the result for the width of the boundary layer that we wanted. It suggests that, at high Reynolds number, there are actually two length scales in the game. The first is the size L of the object. The second, $\delta \ll L$, is the width of a boundary layer that surrounds the object where the effects of both viscosity and vorticity are important. The existence of this thin boundary layer is the 1905 insight of Prandtl.

Outside of the boundary layer, we may neglect viscosity and the fluid is well described by the Eulerian flows of Section 1.4. But much of the physics is dictated by what happens inside the boundary layer where there are large velocity gradients. We want to better understand the properties of this boundary layer.

2.5.1 Prandtl's Boundary Layer Equation

As usual, we don't want to attack the full Navier–Stokes equations. Instead, we will extract the relevant equations that will suffice to model the boundary layer.

We'll set things up as follows. We consider a two-dimensional flow in the (x, y)-plane. As shown in Figure 2.8, we take a thin plate that extends in the x-direction sitting at $y = 0$. The flow is two-dimensional and we write

$$\mathbf{u} = (u, v) \ . \tag{2.159}$$

Asymptotically, $\mathbf{u} \to (U, 0)$. We impose the boundary conditions $u = v = 0$

on the plate at $y = 0$. Incompressibility tells us that

$$\frac{\partial u}{\partial x} + \frac{\partial v}{\partial y} = 0 . \tag{2.160}$$

We'll also look only at steady flows, so there are no time derivatives. The full Navier–Stokes equations then read

$$u\frac{\partial u}{\partial x} + v\frac{\partial u}{\partial y} = -\frac{1}{\rho}\frac{\partial P}{\partial x} + \nu\left(\frac{\partial^2 u}{\partial x^2} + \frac{\partial^2 u}{\partial y^2}\right) , \tag{2.161}$$

$$u\frac{\partial v}{\partial x} + v\frac{\partial v}{\partial y} = -\frac{1}{\rho}\frac{\partial P}{\partial y} + \nu\left(\frac{\partial^2 v}{\partial x^2} + \frac{\partial^2 v}{\partial y^2}\right) . \tag{2.162}$$

We want to ask: Which of these terms can we safely ignore? And which should we keep in the boundary layer?

We look at how the flow changes over a horizontal scale L. We start with the assumption that velocities vary in the x-direction only over the scale L, but may vary in the y-direction on the much smaller scale $\delta \ll L$. Our goal is to construct a consistent truncation of (2.161) and (2.162) such that the terms we're omitting are systematically smaller by a factor of the dimensionless parameter δ/L.

Our first piece of information comes from the incompressibility condition (2.160), with the terms scaling as

$$\frac{\partial u}{\partial x} \sim \frac{U}{L} \quad \text{and} \quad \frac{\partial v}{\partial y} \sim \frac{v}{\delta} \quad \Longrightarrow \quad v \sim \frac{\delta}{L}U . \tag{2.163}$$

So the vertical velocity v is much smaller than the horizontal velocity U. This equation is telling us that the fluid flow is deflected only through a small angle $\sim \delta/L$.

Now let's look to the Navier–Stokes equations (2.161) and (2.162). Both terms on the left-hand side of (2.161) scale as U^2/L, while both terms on the left-hand side of (2.162) scale as $U^2\delta/L^2$. This means that the equation (2.162) is significantly less important than (2.161). In particular, if we assume that the pressure terms have the same order of magnitude then this tells us that

$$\left|\frac{\partial P}{\partial y}\right| \sim \frac{\delta}{L}\left|\frac{\partial P}{\partial x}\right| . \tag{2.164}$$

So, to leading order, pressure becomes a function only of the horizontal distance: $P = P(x)$.

Now we turn to the second order terms on the right-hand side of (2.161). We have

$$\frac{\partial^2 u}{\partial x^2} \sim \frac{U}{L^2} \quad \text{and} \quad \frac{\partial^2 u}{\partial y^2} \sim \frac{U}{\delta^2} \, . \tag{2.165}$$

The second of these is clearly the more important, and we may ignore the $\partial^2 u/\partial x^2$ term. Moreover, assuming that the $\partial^2 u/\partial y^2$ term has the same order of magnitude, U^2/L, as those on the left-hand side of (2.161) tells us that

$$\frac{U^2}{L} \sim \frac{\nu U}{\delta^2} \quad \Longrightarrow \quad \delta \sim \sqrt{\frac{\nu L}{U}} \sim \frac{L}{\sqrt{Re}} \tag{2.166}$$

which confirms our earlier estimate (2.158) and reassures us that the whole approximation scheme is valid at large Reynolds number.

The upshot is that, when solving for the fluid in the boundary layer, we may ignore the y-component of the Navier–Stokes equation (2.162) and the x-component (2.161) simplifies to

$$u\frac{\partial u}{\partial x} + v\frac{\partial u}{\partial y} = -\frac{1}{\rho}\frac{dP}{dx} + \nu\frac{\partial^2 u}{\partial y^2} \, . \tag{2.167}$$

This is the *Prandtl boundary layer equation*. It should be solved in conjunction with the incompressibility condition (2.160).

There is one final finesse. We know that the pressure is approximately a function only of x. This means that we are at liberty to evaluate the pressure $P(x)$ far from the boundary layer, $y \gg \delta$. But here the viscosity terms may be neglected completely, and the flow is governed by the Euler equation. The velocity field takes some profile

$$\mathbf{u} \to (U(x), 0) \quad \text{as} \quad y/\delta \to \infty \tag{2.168}$$

where $U(x) \to U$ as $x \to -\infty$. The Euler equation then tells us that, for a steady flow,

$$-\frac{1}{\rho}\frac{dP}{dx} = U\frac{\partial U}{\partial x} \tag{2.169}$$

which can be substituted into (2.167).

Our next task is to solve (2.167). Far from the plate, the term proportional to ν is unimportant. There is a mathematical framework to solve equations of this kind, whose characteristic form differs in some limit such as $\nu \to 0$. This is the theory of "matched asymptotic expansion". We won't need this in what follows. Instead, we'll look just at some simple examples.

2.5.2 An Infinite Flat Plate

Our simple example is a semi-infinite flat plate. The plate starts at $x = 0$, which we refer to as the leading edge, and continues indefinitely for $x > 0$.

Asymptotically, the flow is constant, $\mathbf{u} \to (U, 0)$ as $y/\delta \to \infty$, so, from (2.169), we have $dP/dx = 0$ and the Prandtl equation becomes

$$u\frac{\partial u}{\partial x} + v\frac{\partial u}{\partial y} = \nu\frac{\partial^2 u}{\partial y^2} \ . \tag{2.170}$$

The flow is two-dimensional so we can again use a stream function $\psi(x, y)$, such that $\mathbf{u} = (u, v)$ with

$$u = \frac{\partial \psi}{\partial y} \quad \text{and} \quad v = -\frac{\partial \psi}{\partial x} \ . \tag{2.171}$$

If we take the stream function to scale as $\psi \sim U\delta$ then, with the scalings described above, we expect to get $u \sim U$ and $v \sim (\delta/L)U$ which is what we want. In looking for a solution, we'll be guided by Figure 2.8. We know that as we move further in the x-direction, the width δ of the boundary layer grows. We will search for "self-similar" solutions in which the velocity profile within the boundary layer remains the same, but gets stretched in the y-direction as the layer grows. Mathematically, this means that we'll search for solutions of the form

$$\psi(x, y) = U \, \delta(x) \, f(\eta) \tag{2.172}$$

where η is the rescaled y coordinate,

$$\eta = \frac{y}{\delta(x)} \tag{2.173}$$

with $\delta(x)$ the characteristic size of the boundary layer (2.158)

$$\delta(x) = \sqrt{\frac{\nu x}{U}} \ . \tag{2.174}$$

(Note: for once $\delta(x)$ has nothing to do with the Dirac delta function!) For our whole approximation to be valid, we required $\delta \ll L$ which in the present context means

$$\delta(x) \ll x \quad \Longrightarrow \quad x \gg \frac{\nu}{U} \ . \tag{2.175}$$

In other words, we can only trust what follows a distance ν/U from the leading edge of the plate. It only gives a good description beyond that point.

The velocity in the x-direction is

$$u = U f' \ . \tag{2.176}$$

Meanwhile, in the y-direction, we have

$$v = -U\delta'f - U\delta f'\frac{\partial \eta}{\partial x} = -U(f - \eta f')\delta' \ . \tag{2.177}$$

Now we can start building the various terms in the Prandtl equation (2.170). We have

$$\frac{\partial u}{\partial x} = -Uf''\frac{\eta}{\delta}\delta' \quad \text{and} \quad \frac{\partial u}{\partial y} = \frac{U}{\delta}f'' \quad \text{and} \quad \frac{\partial^2 u}{\partial y^2} = \frac{U}{\delta^2}f''' \ . \tag{2.178}$$

So putting it all together, the Prandtl equation (2.170) becomes

$$-U^2\eta\frac{\delta'}{\delta}f''f' - U^2\frac{\delta'}{\delta}(f - \eta f')f'' = \nu\frac{U}{\delta^2}f''' \ . \tag{2.179}$$

Two of the terms happily cancel, and we're left with

$$U\delta'\delta f f'' + \nu f''' = 0 \ . \tag{2.180}$$

But, from (2.174), we have $\delta'\delta = \nu/2U$ so our problem reduces to an ordinary, third order differential equation for $f(\eta)$,

$$f''' + \frac{1}{2}ff'' = 0 \ . \tag{2.181}$$

We need to solve this subject to the no-slip boundary condition

$$f = f' = 0 \quad \text{at} \quad \eta = 0 \tag{2.182}$$

and the asymptotic requirement

$$f' \to 1 \quad \text{as} \quad \eta \to \infty \tag{2.183}$$

which ensures that, far from the plate, $\mathbf{u} \to (U, 0)$.

There's no analytic solution to this equation. But it's straightforward to solve numerically. The resulting velocity profile is shown in the figure on the right and is known as the *Blasius boundary layer*. The distance from the plate $y \sim \eta$ is plotted vertically and the velocity $u \sim f'(\eta)$ plotted horizontally. You can see that the velocity interpolates from its zero value on the plate, to the asymptotic value. The graph also gives a more accurate estimate of the 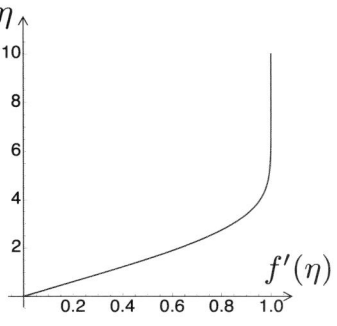 thickness of the boundary layer as something like $\sim 4 - 5$ times δ, by which point the velocity is pretty much at its asymptotic value.

The numerical solution tells us something else. Asymptotically, as $\eta \to \infty$, we find that

$$f(\eta) \approx \eta - 1.72 + \mathcal{O}(1/\eta) \ . \qquad (2.184)$$

This means that, far from the plate, there is vertical component to the velocity (2.177),

$$v \approx 1.72 \sqrt{\frac{\nu U}{4x}} \quad \text{as } y/\delta \to \infty \ . \qquad (2.185)$$

This is capturing what we saw previously in (2.163): the fluid is deflected by an angle $\sim \delta/L$. This angle gets smaller as we get further from the leading edge. This is because the boundary layer increases, and so the velocity gradient – which is always such that the velocity changes from zero to U – decreases as x gets larger, and this fact is reflected by the velocity component in the y-direction infinitely far from the plate. The would-be divergence at $x = 0$ is mitigated by the fact that, as we have seen, our solution only makes sense for distances $x \gg \nu/U$ from the leading edge.

The Drag Force on a Finite Plate

Strictly, the calculation above holds for an infinite plate. We've also seen that it fails within a distance ν/U of the leading edge, and one may expect that it similarly fails near the trailing edge. But we may hope that, for large L, it gives a suitable approximation of the boundary layer over much of a finite plate. With this assumption, we can compute the drag force.

The force on the plate comes from the appropriate component of the stress tensor (2.17). For a single boundary layer, we have

$$\sigma_{xy} = \rho\nu \left(\frac{\partial u}{\partial y} + \frac{\partial v}{\partial x} \right)_{y=0} = \rho\nu \frac{U}{\delta} f''(0) \qquad (2.186)$$

where only the $\partial u/\partial y$ term contributes because $\partial v/\partial x$ vanishes at $y = 0$. We use the numerical solution to evaluate $f''(0) \approx 0.33$. We also need to remember that there are two boundary layers, one on each side. So the total drag force is

$$F_{\text{drag}} = 2 \times 0.33 \times \rho\nu^{1/2}U^{3/2} \int_0^L dx \ \frac{1}{\sqrt{x}} = 1.33\,\rho\nu^{1/2}U^{3/2}L^{1/2} \ . \quad (2.187)$$

Note that the drag force increases as $\sqrt{L}$ rather than proportional to L as one might naively expect. This is because, as the boundary layer thickens, the velocity gradients decrease and, hence, so too does the stress on the plate.

This is our first honest resolution of d'Alembert's paradox: the drag force for an object at high Reynolds number, where one might think that the Euler equation is sufficient, is non-zero. We see explicitly that the drag does vanish if we set $\nu = 0$. If we embed the viscosity in the dimensionless Reynolds number $Re = UL/\nu$, we have

$$F_{\text{drag}} = 1.33\rho \frac{U^2 L}{\sqrt{Re}} \ . \tag{2.188}$$

Taken at face value, this says that the drag force is, in fact, vanishing in the limit $Re \to \infty$. But, sadly, there's another catch awaiting us. The calculation above breaks down at large Reynolds numbers due to the effects of turbulence. Experimentally, this is found to happen at $Re \sim 10^5$ or 10^6.

2.5.3 Boundary Layers with Pressure Gradients

There is a generalisation of the ideas above that exhibits some novel behaviour within the boundary layer. This will be important in the next section when we look at the fate of the boundary layer when it leaves an object.

The generalisation involves looking at boundary layers in flows that are accelerating or decelerating asymptotically. We will again take a semi-infinite flat plate. Far from the boundary layer, the fluid flow takes the form $\mathbf{u} \to (U(x), 0)$, now with

$$U(x) = U \left(\frac{x}{l}\right)^m \tag{2.189}$$

with l some length scale and m a parameter that determines the acceleration. Note that when $m < 0$ our velocity profile (2.189) diverges at $x = 0$. We deal with this by ignoring it: our interest is only in the behaviour of the boundary layer downstream at $x > 0$.

From (2.169), we must have a pressure gradient driving this flow

$$\frac{1}{\rho}\frac{dP}{dx} = -U\frac{dU}{dx} = -\frac{mU^2}{l}\left(\frac{x}{l}\right)^{2m-1} \ . \tag{2.190}$$

There are two distinct cases that will interest us:

- $m > 0$: Accelerating flow with $dP/dx < 0$.

- $m < 0$: Decelerating flow with $dP/dx > 0$.

This is the asymptotic pressure gradient. But, by the arguments of Section 2.5.1, there is no change in the pressure in the y-direction, perpendicular to the plate. This means that the boundary layer also experiences the pressure gradient dP/dx. Our goal is to understand how the boundary layer reacts to this gradient.

The Prandtl equation is

$$u\frac{\partial u}{\partial x} + v\frac{\partial u}{\partial y} = U\frac{dU}{dx} + \nu\frac{\partial^2 u}{\partial y^2} \ . \tag{2.191}$$

We again seek a self-similar solution, now of the form

$$\psi(x,y) = U(x)\,\delta(x)\,f(\eta) \ . \tag{2.192}$$

Here $U(x)$ is given by (2.189), while $\delta(x)$ is a generalisation of our previous expression for the boundary layer thickness

$$\delta(x) = \sqrt{\frac{\nu x}{U(x)}} \tag{2.193}$$

which takes into account the x-dependence of the asymptotic velocity. Note that, for accelerating flows, the boundary layer becomes thinner, relative to the $m = 0$ case, as the flow proceeds. It becomes thicker for decelerating flows. Finally, $\eta = y/\delta(x)$ is the rescaled y-coordinate, as before.

The velocities in the x-direction and y-direction are now

$$u = Uf' \quad \text{and} \quad v = -(U\delta)'f + U\eta f'\delta' \ . \tag{2.194}$$

After a small amount of algebra, the Prandtl equation (2.191) becomes

$$UU'f'^2 - \frac{U}{\delta}(U\delta)'ff'' = UU' + \frac{\nu U}{\delta^2}f''' \ . \tag{2.195}$$

Now we use the explicit expression for the asymptotic velocity (2.189), which tells us that $U \sim x^m$ and $U\delta \sim x^{(m+1)/2}$. Substituting these into the equation above, we see that all terms scale as U^2/x and we may divide by this. Happily, the partial differential equation reduces once again to an ordinary differential equation,

$$mf'^2 - \frac{1}{2}(m+1)ff'' = m + f''' \ . \tag{2.196}$$

This reduces to our previous equation (2.181) when $m = 0$.

Again, we solve this subject to the boundary conditions

$$f = f' = 0 \quad \text{at } \eta = 0 \ \text{and} \ f' \to 1 \ \text{as} \ \eta \to \infty \ . \tag{2.197}$$

The solutions are known as the *Falkner–Skan family of boundary layers*. The velocity profiles $u \sim f'(\eta)$ for a number of different flows are shown in the figure. The curves correspond, from top to bottom, to $m = -0.09$, $m = -0.07$, $m = 0$ (the middle line), $m = 0.2$, and $m = 0.7$.

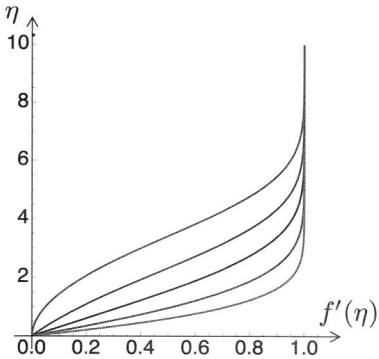

For accelerating flows, with $m > 0$, there isn't a great deal of difference from our previous results. One can show that the solution to the equations is unique and, as you can see from the graph, the velocity profiles all live underneath the $m = 0$ curve, coming in at ever more acute angles at the origin. This can be understood because there is a greater transfer of momentum from the accelerating fluid above. It also has consequence: the angle at which the graph intersects the origin is related to (the inverse of) $f''(0)$. As the acceleration increases, so too does $f''(0)$. But, from (2.186), this means that the force imparted on the plate due to the boundary layer also increases.

At first glance, things don't look too different for decelerating flows with $m < 0$ either. The top two curves in the figure have $m = -0.09$ and $m = -0.07$. Now the graphs come in more steeply at the origin, corresponding to a smaller value of $f''(0)$ and, correspondingly, a smaller stress on the plate. But when we look more closely, there is a surprise waiting us: numerically, we find that for some critical value m_{crit}, the solution actually comes into the origin vertically,

$$m = m_{\text{crit}} \approx -0.0904 \quad \Longrightarrow \quad f''(0) = 0 \ . \tag{2.198}$$

In other words, for a critical deceleration, there is no friction force between the plate and fluid!

What's going on here? Consider an element of fluid near the boundary. It has a force to the right due to the fluid moving above it. But there are also forces to the left, both from the pressure gradient $dP/dx > 0$ and from the viscous force of the boundary. At $m = m_{\text{crit}}$, these precisely cancel. The result is that not only is $u = 0$ on the boundary, but also $du/dy = 0$.

What happens if we decrease m below the value m_{crit}? Naively, one might have thought that one would find solutions with $du/dy < 0$, which would mean the fluid closest to the boundary actually flows in the opposite direction. It turns out that this *doesn't* happen. There are no solutions for $m < m_{\mathrm{crit}}$.

However, there are further solutions that *do* exhibit reverse flows. It turns out that these solutions exist for any $m_{\mathrm{crit}} < m < 0$ where there are two branches of solutions. The first, given above, has $u > 0$ everywhere. The second has a region with $u < 0$ close to the plate. An example is shown in the figure for $m = -0.05$. It has $f''(0) \approx -0.1$. In this case, a fluid element in the region closest to the boundary has a velocity in the opposite direction to the rest of the flow. This reverse flow can be understood as the pressure gradient pushing to the left, while the force from both the fluid above it, and also from the plate, pushes to the right.

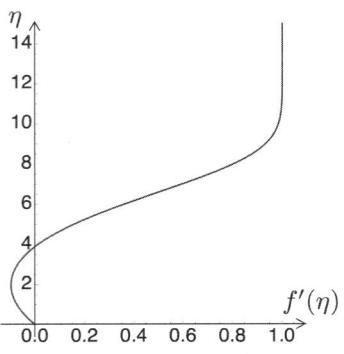

It seems that these boundary solutions with reverse flow cannot be set up in an experiment because they are thought to be unstable in this particular context. Nonetheless, the existence of such reversed boundary layers is crucial to understand the next topic that we turn to. This is the fate of the boundary layer when the boundary ends.

2.5.4 Separation

So far we've understood how the boundary layer develops, but only by restricting to a flat, semi-infinite plate. Needless to say, that's not particularly realistic. Most objects are neither flat, nor semi-infinite. Clearly, we need to understand the physics of the boundary layer for objects that are curved and finite.

This, it turns out, is not so easy. Until now, we've made progress by finding clever ways to reduce the Navier–Stokes equations to an ordinary differential equation which can then easily be solved. But the problem that we're now interested in offers no such simplification. That means that to get a complete handle on the problem we must resort to solving partial differential equations numerically. Which is possible, but challenging, and beyond the scope of this book. Instead we will make do with some rather

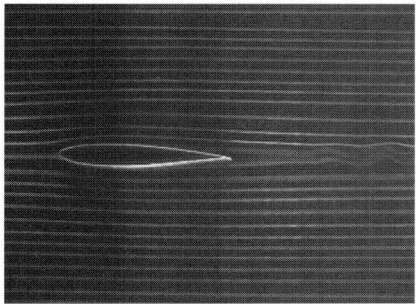 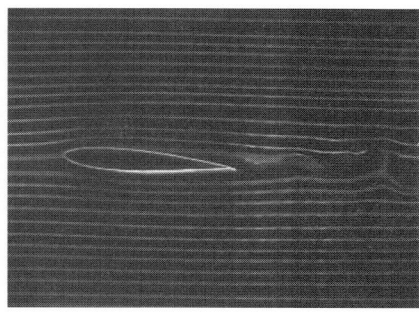

Fig. 2.9 The flow, from left to right, around a streamlined object at high Reynolds number. On the left, the object is aligned with the streamlines and the boundary layer merges smoothly into the flow at the trailing edge. On the right, the object is inclined by 5°. The boundary layer separates from the object on the upper edge. Credit: Stanford University.

qualitative arguments, piecing together various bits of physics that we've learned so far.

First, we can gain some intuition for what's going on by turning to experiment. Figure 2.9 shows the streamlines for a high Reynolds number flow ($Re \approx 7000$) around an elegant pointy object. In the figure on the left, the object is aligned with the streamlines, which glide around much like the flows that we've discussed so far in this book. Such flows, where there is little mixing between adjacent layers of the fluid, are called *laminar*. A boundary layer forms around the object but, at least as far as the photograph shows, appears to merge seamlessly back into the bulk fluid at the tail end.

On the right of Figure 2.9 is the same object, again at $Re \sim 7000$, but now tilted at an angle of 5°. The flow is again laminar at the front and below the object. But you can see that something screwy is happening on the upper trailing edge. There is clearly a streamline that moves away from the object, leaving a swirling indeterminate flow beneath it.

The same phenomenon occurs for less aerodynamic objects. Figures 2.10 and 2.11 show flows moving past a circular cylinder. The first flow, at $Re \approx 10$, clearly shows an anti-symmetry between the front and back of the cylinder as the streamlines separate from the body. This is unsurprising, but sits in stark contrast to the potential flows and Stokes flows that we've seen previously, where it's difficult to see by eye the difference between the front and back of the flow. (See, for comparison, Figure 1.7 or Figure 2.3.) In the second picture in Figure 2.10, the Reynolds number has increased

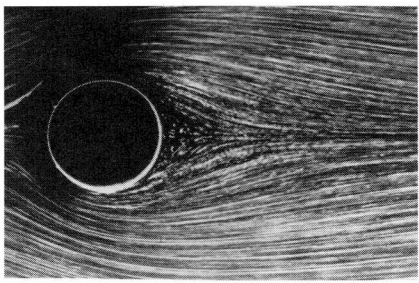

Fig. 2.10   The flow around a circular cylinder with $Re \approx 10$ on the left, and $Re \approx 26$ on the right. In both cases, the flow separates from the cylinder at some point, leaving two trailing eddies in the wake (more visible in the second picture). Credit: S. Taneda.

to $Re \approx 26$ and we again see the flow separating from the body, this time clearly leaving two counter-circulating eddies in its wake.

The Reynolds numbers in Figure 2.10 are fairly low and it's not at all obvious that we can use the theory of boundary layers, which relies on the approximation $Re \gg 1$. But this is surely valid for the picture in Figure 2.11, now at $Re \approx 2000$. Now we clearly see that the laminar flow at the front of the cylinder separates somewhere near the top and bottom, leaving a turbulent flow in its wake.

There are a bunch of things to unpack here. First, how do we extend the theory of a boundary layer to a curved object like those shown in the figures? Second, why does the flow separate from the object at some point? And, finally, how can we understand the physics of the wake left behind? We'll deal with each of these in turn.

Here is a cartoon of the physics. First, extending the theory of the boundary layer to a curved object turns out to be fairly straightforward. We use the same equations as before, but with x and y now curvilinear coordinates: x is the coordinate along the boundary and y the coordinate perpendicular. The boundary layer is so thin that, locally, it barely notices the curvature. All we must do is ensure that the pressure in the boundary layer is given by (2.169)

$$-\frac{1}{\rho}\frac{dP}{dx} = U\frac{\partial U}{\partial x} \ . \tag{2.199}$$

Here, as a first approximation to $U(x)$, we should take the *near*-boundary limit of the flow that surrounds the boundary layer. Provided that this flow isn't turbulent, we can use the near-boundary limit of the inviscid potential

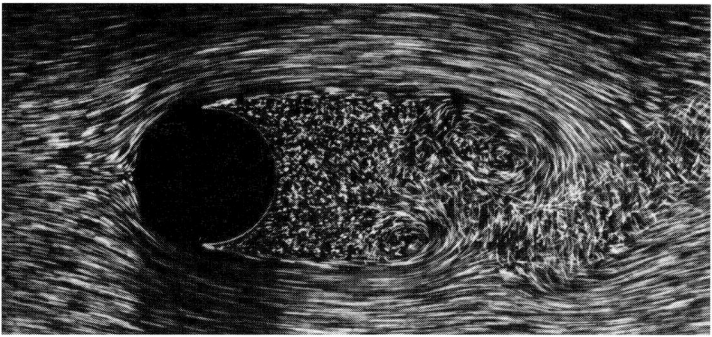

Fig. 2.11 A flow around a circular cylinder, now with $Re \approx 2000$. We're now at values where the boundary layer theory should work. The picture clearly shows laminar flow at the front of the cylinder, where the boundary layer remains attached. It separates somewhere near the top and bottom of the cylinder, leaving a turbulent wake. Credit: Weré and Gannon, ONERA.

flows that we described in Section 1.4. But we know how the pressure changes over the sphere or cylinder due to a potential flow. (The answer for the sphere was given in (1.131) and the result for the cylinder is similar.) There we saw that the pressure directly at the front and back is the same as the asymptotic pressure, but the pressure reduces as you move up or down over the sphere and takes its minimum value at the top and bottom. Crucially, the pressure for an inviscid potential flow is symmetric on the front and back: this, of course, was what led to d'Alembert's paradox.

Now we can see what this means for the boundary layer. On the front edge of the cylinder, the pressure is decreasing, $P' < 0$. This corresponds to an accelerating flow. But on the back edge, the pressure is increasing, $P' > 0$, and the flow is decelerating. This suggests that we might get the kind of behaviour that we observed for decelerating flows in the Falkner–Skan family of boundary layers. In particular, at some point the velocity u tangential to the boundary will obey

$$\left.\frac{\partial u}{\partial y}\right|_{y=0} = 0 \qquad (2.200)$$

where y is the direction perpendicular to the boundary. This is the *separation point*, with the streamline bifurcating and leaving the boundary. Beyond this point, one expects reverse flow close to the boundary. Beyond the separation point, the boundary layer moves off into the bulk of the fluid, leaving behind the wake. A sketch of the scenario is shown in Figure 2.12.

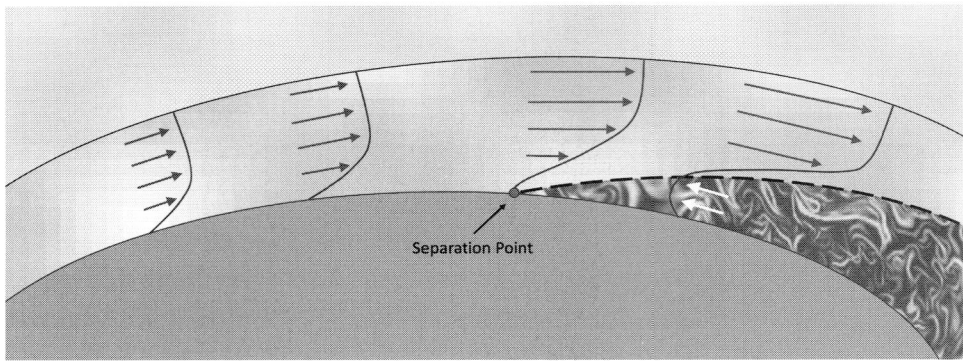

Fig. 2.12 A cartoon of the evolution of the boundary layer and its ultimate separation from the boundary as the flow becomes reversed.

The boundary layer itself cannot just dissolve once it has separated from the boundary. One might reasonably wonder what distinguishes it from the bulk of the fluid. After all, they're made from the same stuff. The answer is that the boundary layer has vorticity, generated by the no-slip condition

$$\boldsymbol{\omega} = \left(\frac{\partial v}{\partial x} - \frac{\partial u}{\partial y} \right) \hat{\mathbf{z}} \approx - \frac{\partial u}{\partial y} \hat{\mathbf{z}} \tag{2.201}$$

where the first term dominates in the boundary layer approximation. For the boundary layers described above, we have $|\boldsymbol{\omega}| = U f''(0)/\delta$. Meanwhile, as we saw previously, the outer laminar flow is irrotational. The vorticity persists in the wake that trails the objects.

For low Reynolds number, the stream flow is low and this vorticity has time to diffuse due to the effects of viscosity. The result is the two large eddies trailing the object seen in Figure 2.10. The flow is steady. These are steady eddies.

But as the Reynolds number is increased to around $Re \sim 100$, something more interesting happens. One of the eddies grows until it peels off from the boundary in a process known as *vortex shedding*. The flow then curls back around the boundary and a new eddy forms. Meanwhile, the eddy on the other side then undergoes the same process. The result is a gorgeous flow pattern of alternating eddies known as the *von Kármán vortex street*. An example is shown in Figure 2.13. At these Reynolds numbers, there is no steady flow of the kind that we've searched for so far. Instead, the flow is time-dependent, but periodic.

There is much that we have swept under the carpet in the discussion

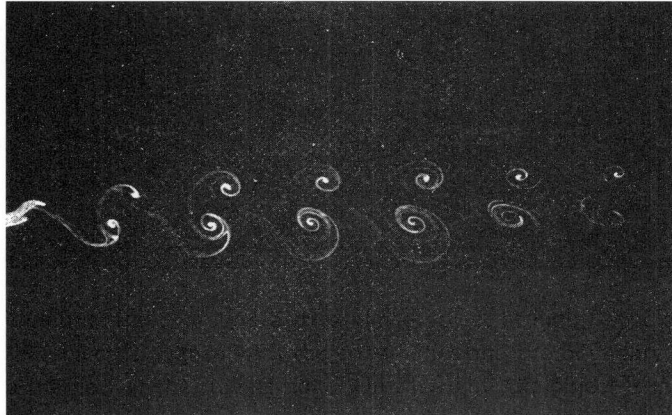

Fig. 2.13 The von Kármán vortex street from a cylinder coated in condensed milk, moving through water at $Re \approx 55$. Credit: M. Rutgers.

above. The elephant in the room is turbulence. As the pictures clearly show, for large Reynolds number the flow is far from laminar. Indeed, the flow is no longer even two-dimensional, but twists and turns in a noisy fashion in three dimensions. This occurs for $Re \gtrsim 10^4$ when the wake becomes turbulent as shown in Figure 2.11. A process known as *turbulent mixing* causes the pressure to be uniform across the turbulent wake, and equal to its value at the point of separation. This means that there is a much lower pressure behind the object and, correspondingly, a much larger drag force.

As the Reynolds number is increased yet further to $Re \gtrsim 10^5$ something novel happens: now the boundary layer itself becomes turbulent. The same turbulent mixing means that vorticity can be transferred vertically much more efficiently, and the result is that the boundary layer gets thicker. This has two competing effects. The first is that the drag due to the turbulent boundary layer increases compared to the laminar boundary layer. The second is that the separation of the boundary layer is delayed, with the reversed flow happening further downstream. This results in a narrower wake which reduces the drag. It turns out that this reduced drag from the narrower wake is more than sufficient to compensate for the increased drag due to the turbulent boundary layer, and the result is that, surprisingly, the drag force actually drops suddenly at this Reynolds number. This goes by the name of the *drag crisis* and is what's ultimately responsible for the Magnus force and Roberto Carlos' magic, which we mentioned in Section 1.5.

Waves

Our story so far has involved the bulk motion of fluids, flowing from one place to another, sometimes trying to negotiate obstacles in their way. But fluids are more subtle and interesting than this. They contain mechanisms to transfer energy through space, but *without* the bulk of the fluid travelling very far. This is achieved through oscillatory behaviour known as *waves*.

Waves are familiar, both from our everyday experience as well as from other areas of physics. Our purpose in this chapter is to explore some of large variety of waves that can occur in fluids. This includes, in Section 3.4, sound waves, which gives us an opportunity to look at some of the novelties that arise with compressible fluids.

3.1 Surface Waves

> Now, the next waves of interest, that are easily seen by everyone and which are usually used as an example of waves in elementary courses, are water waves. As we shall soon see, they are the worst possible example, because they are in no respects like sound and light; they have all the complications that waves can have.
>
> *Richard Feynman*

We start with waves travelling on the surface of a fluid. These include waves on the ocean. As Feynman points out, there are a surprisingly large number of subtleties that arise in understanding these waves.

Viscosity will not play a leading role in our story, so we return to the Euler equation of Chapter 1,

$$\rho \left(\frac{\partial \mathbf{u}}{\partial t} + \mathbf{u} \cdot \nabla \mathbf{u} \right) = -\nabla P + \rho \mathbf{g} \ . \tag{3.1}$$

We've included the effects of gravity on the right-hand side. As we will see, this provides the restoring force needed to create waves.

We will shortly solve the Euler equation using the same techniques that

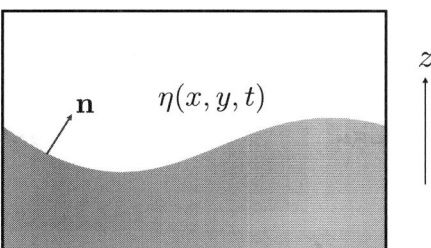

Fig. 3.1 An interface between two fluids.

we met in Chapter 1. All of the novelties come, like so many things in fluid dynamics, from the boundary conditions. So before we get going, we need to think about the kind of boundary condition we should impose on the surface of a fluid.

3.1.1 Free Boundary Conditions

The surface of a fluid is best viewed as the interface between two different fluids. In the case of the ocean, this is the water and the air above. But we could also have a situation where we have two immiscible liquids, like oil and water. The surface is free to move, and so is sometimes referred to as a *free boundary*.

Suppose that the boundary lies close to some $z \approx$ constant surface, as shown in Figure 3.1. Clearly this is appropriate for the surface of the ocean. The surface can fluctuate and, in general, is described by some function

$$F(\mathbf{x}, t) = z - \eta(x, y, t) = 0 . \tag{3.2}$$

The normal to such a surface is parallel to ∇F

$$\mathbf{n} \propto \nabla F = \left(-\frac{\partial \eta}{\partial x}, -\frac{\partial \eta}{\partial y}, 1 \right) . \tag{3.3}$$

Meanwhile, the velocity of the interface is, by construction, in the z-direction and given by

$$\mathbf{U} = \left(0, 0, \frac{\partial \eta}{\partial t} \right) . \tag{3.4}$$

The appropriate boundary condition on the fluid velocity $\mathbf{u}$ is the same as we saw in (1.112) for a solid surface moving with some velocity $\mathbf{U}$: one fluid cannot permeate the other. This means that if we write the fluid velocity as

$\mathbf{u} = (u_x, u_y, u_z)$, then we have

$$\mathbf{n} \cdot \mathbf{u} = \mathbf{n} \cdot \mathbf{U} \quad \Longrightarrow \quad -u_x \frac{\partial \eta}{\partial x} - u_y \frac{\partial \eta}{\partial y} + u_z = \frac{\partial \eta}{\partial t} \ . \tag{3.5}$$

We can equivalently write this as

$$u_z - \mathbf{u} \cdot \nabla \eta = \frac{\partial \eta}{\partial t} \quad \Longrightarrow \quad \frac{D\eta}{Dt} = u_z \ . \tag{3.6}$$

Alternatively, if we return to our original definition of the interface as the surface $F(\mathbf{x}, t) = 0$ given in (3.2), the fact that one fluid cannot invade the other can be written in the elegant form

$$\frac{DF}{Dt} = 0 \ . \tag{3.7}$$

In addition, there is a further dynamical boundary condition that comes from the requirement that forces on the surface are balanced. For an inviscid fluid, there is no tangential stress. In the simplest situations, the only component of the stress tensor perpendicular to the interface is the pressure and the boundary condition is simply that the pressure must be continuous

$$P(x, y, \eta(x, y)) = P_0 \tag{3.8}$$

where, for the waves on the ocean, P_0 is atmospheric pressure. For example, if the water is stationary, so $\mathbf{u} = 0$, with a flat, free boundary at $z = 0$, then the Euler equation, together with this boundary condition, brings us back to the equation of hydrostatic pressure (1.38)

$$P(z) = P_0 - \rho g z \ . \tag{3.9}$$

We would like to understand how to generalise this to the case where waves propagate on the boundary.

There are further complications that we could add to this story. In particular, there is an additional force that acts on the boundary known as *surface tension*. We postpone a discussion of this to Section 3.1.5.

3.1.2 The Equations for Surface Waves

We will look for irrotational flows which, as in Chapter 1, allows us to introduce a velocity potential

$$\nabla \times \mathbf{u} = 0 \quad \Longrightarrow \quad \mathbf{u} = \nabla \phi \ . \tag{3.10}$$

The incompressibility of the fluid then tells us that the potential ϕ must obey the Laplace equation

$$\nabla \cdot \mathbf{u} = 0 \quad \Longrightarrow \quad \nabla^2 \phi = 0 \ . \tag{3.11}$$

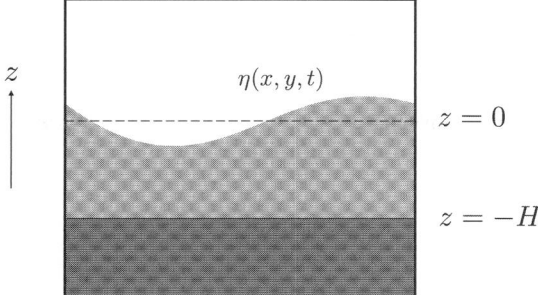

Fig. 3.2 Surface waves, with a boundary on the bottom at $z = -H$ and another, dynamical boundary, on the top at $z = \eta(x, y, t)$.

All the subtleties lie in the boundary conditions.

We'll take the waves to be propagating in an ocean of height H, as shown in Figure 3.2. The bottom of the ocean lies at $z = -H$ while the surface of the ocean is at $z = \eta(x, y, t)$, some height above or below the equilibrium value of $z = 0$. The boundary condition at the bottom of the ocean is straightforward: the water can't flow into the ocean floor so

$$u_z(z = -H) = \left.\frac{\partial \phi}{\partial z}\right|_{z=-H} = 0 \ . \tag{3.12}$$

Meanwhile, on the surface $z = \eta(x, y, t)$, we impose the free boundary condition (3.6)

$$u_z = \frac{D\eta}{Dt} \implies \left.\frac{\partial \phi}{\partial z}\right|_{z=\eta} = \frac{\partial \eta}{\partial t} + \left.\frac{\partial \phi}{\partial x}\right|_{z=\eta} \frac{\partial \eta}{\partial x} + \left.\frac{\partial \phi}{\partial y}\right|_{z=\eta} \frac{\partial \eta}{\partial y} \ . \tag{3.13}$$

This boundary condition is all well and good if we're given the equation of the surface $z = \eta(x, y, t)$. But here, of course, the surface is something that arises dynamically and our goal is to find it.

We need one further equation that relates the velocity potential ϕ and the surface η. This comes from the continuity of pressure (3.8). To implement this, we tread the same steps that we took in Section 1.2.4 when deriving Bernoulli's principle. So, following (1.44), the Euler equation (3.1) can be written as

$$\rho \left(\frac{\partial \mathbf{u}}{\partial t} + \frac{1}{2} \nabla |\mathbf{u}|^2 - \mathbf{u} \times \boldsymbol{\omega} \right) = -\nabla (P + \rho g z) \ . \tag{3.14}$$

But we're dealing with irrotational flows, so $\boldsymbol{\omega} = 0$, and this becomes

$$\rho \left(\frac{\partial \nabla \phi}{\partial t} + \frac{1}{2} \nabla |\nabla \phi|^2 \right) = -\nabla (P + \rho g z) \ . \tag{3.15}$$

But both sides are now total spatial derivatives, so we must have

$$\rho \frac{\partial \phi}{\partial t} + \frac{1}{2} \rho |\nabla \phi|^2 + P + \rho g z = f(t) \ . \tag{3.16}$$

Here the function $f(t)$ is an integration "constant" but, because we had only spatial derivatives in (3.15), this "constant" can be a function of time but must be genuinely constant in space. The result (3.16) is a version of Bernoulli's principle for irrotational flows. We then implement our final boundary condition (3.8) by imposing (3.16) on the surface $z = \eta(x, y, t)$, with the pressure replaced by the atmospheric pressure

$$\rho \left(\frac{\partial \phi}{\partial t} + \frac{1}{2} |\nabla \phi|^2 \right)_{z=\eta} + P_0 + \rho g \eta = f(t) \ . \tag{3.17}$$

This completes our setting up of the equations. We must solve the Laplace equation $\nabla^2 \phi = 0$ subject to the boundary conditions (3.12), (3.13), and (3.17). The Laplace equation is easy, but these boundary conditions look hard. In particular, even the domain on which $\phi(\mathbf{x}, t)$ is defined is weird and time-dependent, with $-H < z < \eta(x, y, t)$. As with so many other problems in fluid mechanics, we need to find an appropriate approximation scheme.

The Linearised Approximation

To make progress, we will assume that the waves are small and flat. The first condition is the statement that the amplitude is small, so

$$|\eta| \ll H \ . \tag{3.18}$$

The second condition is the statement that the derivatives of the amplitude are also small

$$\frac{\partial \eta}{\partial x}, \ \frac{\partial \eta}{\partial y} \ll 1 \ . \tag{3.19}$$

The boundary condition (3.13) is imposed at $z = \eta(x, y, t)$ but, since η is small, we can view this as "close" to a boundary condition at $z = 0$ by Taylor expanding as

$$\frac{\partial \phi}{\partial z} \bigg|_{z=\eta} = \frac{\partial \phi}{\partial z} \bigg|_{z=0} + \eta \frac{\partial^2 \phi}{\partial z^2} \bigg|_{z=0} + \dots \ . \tag{3.20}$$

The second order term is smaller than the first and can be dropped. Relatedly, the velocities u_x and u_y will be assumed to be small and we will drop all quadratic terms in the above boundary conditions. This means that we can ignore the $(\partial\phi/\partial x)(\partial\eta/\partial x)$ terms in (3.13) and the $|\nabla\phi|^2$ term in (3.17). The upshot is that our rather complicated set of boundary conditions reduce to the linear equations

$$\left.\frac{\partial\phi}{\partial z}\right|_{z=-H} = 0 \quad \text{and} \quad \left.\frac{\partial\phi}{\partial z}\right|_{z=0} = \frac{\partial\eta}{\partial t} \quad \text{and} \quad \left.\frac{\partial\phi}{\partial t}\right|_{z=0} + g\eta = \tilde{f}(t) \quad (3.21)$$

where, in the last of these, we have absorbed the pressure P_0 and density ρ into the redefined function $\tilde{f}(t)$. It is, it turns out, a significantly easier task to solve the Laplace equation subject to these conditions.

Finally, Some Waves

We will consider wave solutions that move in the x-direction and are independent of the y-direction. We make the ansatz

$$\phi(x,z,t) = \phi_0(z)\, e^{ikx - i\omega t} \quad \text{and} \quad \eta(x,t) = \eta_0\, e^{ikx - i\omega t} \ . \quad (3.22)$$

You may be surprised that the right-hand side is suddenly complex, while both quantities on the left-hand side are clearly real! There's nothing deep going on here, only laziness. Because the equations are linear, if we find a complex solution then the real and imaginary parts are also solutions. But it's often simpler to work with complex numbers $e^{i(\text{something})}$ rather than cos and sin functions. Moreover, this will be particularly useful in Chapter 4 when we come to study instabilities, since these manifest themselves as complex frequencies or wavenumbers for which the solution grows exponentially in time or space. For now, whenever you see equations like those above, you should implicitly think that we are taking the real (or imaginary) part.

The ansatz (3.22) depends on two numbers in the exponent, k and ω. Here k is called the *wavenumber*. From the equation, we see that the successive peaks, also known as *wavecrests*, are spaced a distance apart given by

$$\lambda = \frac{2\pi}{k} \ . \quad (3.23)$$

The distance λ is known as the *wavelength*. Meanwhile, ω is the *frequency* of the wave. Part of our goal is to determine the relationship between ω and k to get the function $\omega(k)$, and we'll see many examples of this relationship for different kinds of waves as this chapter proceeds. In particular, it's useful to trivially rewrite the exponent in the wave ansatz (3.22) as

$$kx - \omega(k)t = k(x - c(k)t) \ . \quad (3.24)$$

This highlights the fact that a mode with wavenumber k travels with velocity

$$c(k) = \frac{\omega(k)}{k} \; . \tag{3.25}$$

This is known as *phase velocity*. We'll meet a slightly different measure of velocity, known as *group velocity*, shortly.

The other part of our goal is to fix the function $\phi_0(z)$. If we substitute the ansatz above into the Laplace equation, we get

$$\frac{d^2\phi_0}{dz^2} = k^2\phi_0 \; . \tag{3.26}$$

This has the solution

$$\phi_0(z) = A\cosh(kz + kH) \tag{3.27}$$

where we've chosen one integration constant to ensure that the first boundary condition in (3.21) is satisfied, and the overall amplitude A is still to be fixed. The second boundary condition in (3.21) tells us

$$Ak\sinh(kH) = -i\omega\eta_0 \; . \tag{3.28}$$

Finally, the third, Bernoulliesque, condition in (3.21) tells us

$$\left(-iA\omega\cosh(kH) + g\eta_0\right)e^{ikx - i\omega t} = \tilde{f}(t) \; . \tag{3.29}$$

The left-hand side depends on the combination $kx - \omega t$, while the right-hand side is a function only of time t and not of space x. This is consistent only if both sides vanish, meaning that we must have

$$A\omega\cosh(kH) = -ig\eta_0 \; . \tag{3.30}$$

Dividing (3.28) by (3.30), we can eliminate A/η_0 to get the desired relationship between the frequency ω and wavenumber k,

$$\omega^2 = gk\tanh(kH) \; . \tag{3.31}$$

Equations of this kind are called *dispersion relations*. They are important in many different places in physics. In books that take place in the quantum world, we will see similar equations that relate energy (associated to frequency) and momentum (associated to wavenumber).

We find that, for surface waves, the frequency depends on the wavelength. This means that waves of different wavelength travel at different speeds (3.25), with

$$c = \sqrt{\frac{g}{k}\tanh(kH)} \; . \tag{3.32}$$

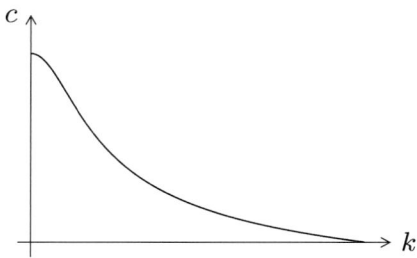 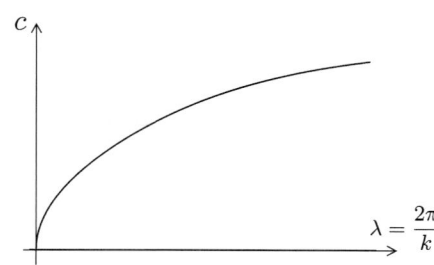

Fig. 3.3 The phase velocity of surface waves, plotted on the left as a function of wavenumber k, and plotted on the right as a function of wavelength $\lambda = 2\pi/k$.

For many kind of waves, including sound and light, the speed is independent of the wavelength. Not so for surface waves. For fixed height H, the speed is a monotonically decreasing function of k. In other words, long wavelength waves travel faster than short wavelength waves. In Figure 3.3, we plot the phase velocity c as a function of wavenumber k (on the left) and as a function of wavelength $\lambda = 2\pi/k$ (on the right).

There are two interesting limits that we can take:

- In deep water, $H \gg \lambda = 2\pi/k$ so we have the limit $kH \gg 1$. Here the speed becomes

$$c \approx \sqrt{\frac{g}{k}} \ . \tag{3.33}$$

  We could have anticipated this. When the ocean is very deep, we don't expect the speed of surface waves to depend on H simply because the floor is a long way from the surface. Then, on dimensional grounds, the only thing that we can write down is $c \sim \sqrt{g/k}$.

- For long wavelengths, we have $kH \ll 1$ and the speed becomes

$$c \approx \sqrt{gH} \ . \tag{3.34}$$

  Now the speed is independent of the wavelength of the wave. In this limit, the wave goes faster in deeper water than in shallow. There is a nice consequence of this. When waves come into the beach at an angle, the wave front that is further out travels faster and so the wave rotates until it is parallel to the beach.

  We can put some numbers on this. In a river, we might have $H \approx 2$ m. When it floods, the waves propagate at a speed $c \approx 4.5$ m s^{-1}. In the ocean, we have $H \approx 5$ km. But the requirement that $kH \ll 1$ means that this approximation is only valid for wavelengths $\lambda \gg 2\pi H$. This is easily satisfied by tsunamis that can have $\lambda \approx 200$ km. These waves travel at a speed $c \approx 220$ m s^{-1}. That's fast! It's close to 500 mph, the same ballpark as a commercial airliner.

The Velocity Field

It is a simple matter to compute the velocity field of the fluid. Substituting for the various integration constants, we have the potential

$$\phi = \text{Re}\left[-i\frac{\omega\eta_0}{k}\frac{\cosh(kz+kH)}{\sinh(kH)}e^{ikx-i\omega t}\right] \ . \tag{3.35}$$

which now just has a single undetermined integration constant η_0 that fixes the amplitude of the wave. Our approximations above mean that the solution should be trusted only when $\eta_0 k \ll 1$. For once we've explicitly reminded ourselves that we should take the real part of the potential when computing the velocity $\mathbf{u} = \nabla\phi$. We have

$$\begin{pmatrix} u_x \\ u_z \end{pmatrix} = \frac{\omega\eta_0}{\sinh(kH)} \begin{pmatrix} \cosh(kz+kH)\cos(kx-\omega t) \\ \sinh(kz+kH)\sin(kx-\omega t) \end{pmatrix} \ . \tag{3.36}$$

The velocity profile is plotted in Figure 3.4 for deep water waves (on the left) and for shallow water waves (on the right). In both cases, the velocity of the water is mostly up/down, despite the fact that the wave travels to the right. In the trough of the wave, the water is moving up on the left and down on the right. In the peak of the wave, this is reversed: the water moves down on the left and up on the right. The net effect is that the wave travels to the right.

There's something misleading about the figure for deep water waves. In this case, $e^{-kH} \approx 0$ and the velocity profile is well approximated by

$$\begin{pmatrix} u_x \\ u_z \end{pmatrix} \approx \omega\eta_0 e^{kz} \begin{pmatrix} \cos(kx-\omega t) \\ \sin(kx-\omega t) \end{pmatrix} \ . \tag{3.37}$$

We see that the magnitude of the velocity $|\mathbf{u}| \approx \omega\eta_0 e^{kz}$ decreases exponentially from its value at the surface $z = 0$. It means that all the action is really taking place within a depth of one wavelength or so from the surface. In contrast, for shallow water waves the speed does not vary greatly with height.

For deep water waves, the ratio of the fluid speed to the wave speed is $|\mathbf{u}|/c \approx k\eta_0 e^{kz}$. The condition (3.19) is tantamount to the requirement that

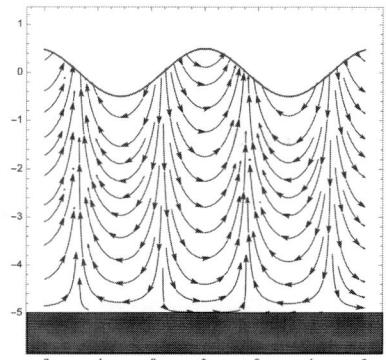 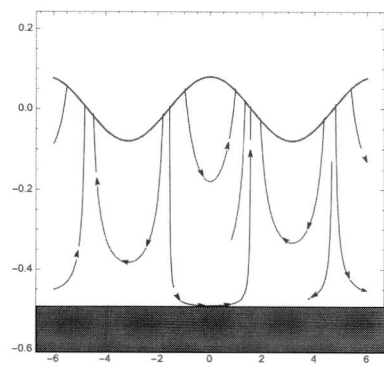

Fig. 3.4 The velocity field for deep water waves on the left, with $kH = 5$, and shallow water on the right, with $kH = 0.5$. (Note the different scales on the vertical axes!) The streamplot shows only the direction of the velocity field, not its size. In deep water, the velocity is exponentially smaller at the bottom than the top.

$k\eta_0 \ll 1$. (The small parameter $k\eta_0$ is known as the *steepness* of the wave.) We learn that, in deep water, the wave travels much faster than the fluid from which it's made.

Particle Paths

Suppose that you drop a small ball into the flow that follows an element of fluid on its travels. What path does it take? As we described in Section 1.1, the trajectory $\mathbf{x}(t)$ is called a pathline and is governed by the equation (1.1)

$$\frac{d\mathbf{x}}{dt} = \mathbf{u}(\mathbf{x}(t), t) \tag{3.38}$$

which we should solve given some initial starting point $\mathbf{x}(0) = \mathbf{x}_0$.

To solve this, we will assume that the particle doesn't get far from its original starting position and approximate the velocity field $\mathbf{u}(\mathbf{x}, t)$ by its Taylor expansion about $\mathbf{x}_0$

$$\mathbf{u}(\mathbf{x}, t) \approx \mathbf{u}(\mathbf{x}_0, t) + ((\mathbf{x} - \mathbf{x}_0) \cdot \nabla)\mathbf{u}(\mathbf{x}_0, t) + \ldots . \tag{3.39}$$

If we keep just the first term, the equation for the pathline becomes

$$\frac{d\mathbf{x}}{dt} = \frac{\omega\eta_0}{\sinh(kH)} \begin{pmatrix} \cosh(kz_0 + kH)\cos(kx_0 - \omega t) \\ \sinh(kz_0 + kH)\sin(kx_0 - \omega t) \end{pmatrix}$$

$$\implies \quad \mathbf{x}(t) = \mathbf{x}_0 + \frac{\eta_0}{\sinh(kH)} \begin{pmatrix} -\cosh(kz_0 + kH)\sin(kx_0 - \omega t) \\ \sinh(kz_0 + kH)\cos(kx_0 - \omega t) \end{pmatrix} . \tag{3.40}$$

This is telling us that the particle travels in ellipses, squashed in the vertical direction. For deep water waves, these ellipses become circles with

$$\mathbf{x}(t) = \mathbf{x}_0 + \eta_0 e^{k z_0} \begin{pmatrix} -\sin(k x_0 - \omega t) \\ \cos(k x_0 - \omega t) \end{pmatrix} . \tag{3.41}$$

The ellipses or circles become exponentially smaller as the depth increases. The vertical component of the velocity is in phase with the crests of the wave, $\eta \sim \cos(kx - \omega t)$. Meanwhile, the horizontal component ensures that the particle goes clockwise for waves that propagate to the right.

We can also look at the effect of the second term in (3.39). Things are simplest if we restrict attention to deep water waves, with velocity (3.37) and particle position (3.41). If we use our leading order expression (3.41) for $\mathbf{x}(t)$ we find, after a little algebra,

$$((\mathbf{x}(t) - \mathbf{x}_0) \cdot \nabla)\mathbf{u} = \omega k \eta_0^2 e^{2kz} \begin{pmatrix} 1 \\ 0 \end{pmatrix} . \tag{3.42}$$

When substituted into (3.38), this has the interpretation of a constant, horizontal drift velocity for the particles, given by

$$v_{\text{drift}} = \omega k \eta_0^2 e^{2kz} = c(k \eta_0 e^{kz})^2 . \tag{3.43}$$

This is known as *Stokes drift*. The ellipses traced by the particles don't quite close, but slowly inch their way in the direction in which the wave propagates. Note that there is a hierarchy of speeds,

$$v_{\text{drift}} \ll |\mathbf{u}| \ll c \tag{3.44}$$

with $k \eta_0 e^{kz} \ll 1$ the small, dimensionless number that governs successive ratios. The Stokes drift v_{drift} is the speed at which matter bobbing in the waves moves.

3.1.3 Group Velocity and Dispersion

The waves described above extend to infinity in the x-direction, varying over wavelength $\lambda = 2\pi/k$. We can create a localised wavepacket by summing over many different wavenumbers and, because the equations are linear (after linearisation!) this too is a solution. In this way, a bump in the fluid surface can be written as

$$\eta(x, t) = \int \frac{dk}{2\pi} a(k) e^{ikx - i\omega(k)t} \tag{3.45}$$

with some Fourier coefficients $a(k)$. As we've seen in (3.25), different Fourier modes will travel with different phase velocities

$$c(k) = \frac{\omega(k)}{k} \ . \tag{3.46}$$

This means that if we start off with a wavepacket that takes some particular shape at time $t = 0$, then this shape will be distorted over time. This phenomenon is known as *dispersion*.

While the phase velocity tells us the speed at which individual Fourier modes move, there is a different way to characterise the speed of the whole wavepacket. To see this, suppose that the Fourier modes $a(k)$ are peaked around some particular wavenumber $k = \bar{k}$. Then we can Taylor expand the frequency and write

$$\omega(k) = \omega(\bar{k}) + (k - \bar{k}) \left. \frac{\partial \omega}{\partial k} \right|_{k=\bar{k}} + \dots \ . \tag{3.47}$$

Substituting this into the expression (3.45) for the wavepacket, we have

$$\eta(x,t) \approx e^{-i(\omega(\bar{k}) - v_g \bar{k})t} \int \frac{dk}{2\pi} \ a(k) \, e^{ik(x - v_g t)} \tag{3.48}$$

where

$$v_g(\bar{k}) = \left. \frac{\partial \omega}{\partial k} \right|_{k=\bar{k}} \ . \tag{3.49}$$

This is called the *group velocity* of the wave. It's clear from the form (3.48) that v_g is the speed at which the wavepacket moves. It is also the speed at which energy (and, in other contexts, information) is transported by the wave. For the surface waves in deep water (3.33), we have $\omega \sim k^{1/2}$ and so the group velocity and phase velocity are related by $v_g(k) = \frac{1}{2}c(k)$. The wavepackets travel at half the speed of the individual Fourier modes.

From this discussion, it's clear that waves with a dispersion relation $\omega \sim k$ are special. In this case, the phase velocity (3.46) and group velocity (3.49) coincide and both measures of speed are independent of the wavelength $\lambda = 2\pi/k$ of the wave. This also means that the wavepacket retains its initial shape as it evolves. Such waves are called *non-dispersive*. Examples include the shallow water waves (3.34) and, as we saw in the book on Electromagnetism, light propagating in the vacuum.

Fig. 3.5   Even an AI generated image of a duck knows that the wake comes out at $39°$ (more or less).

3.1.4 Water Off a Duck's Back

There's a very cute application of the difference between the phase velocity and the group velocity. When a duck, or boat, moves through water, it leaves behind a familiar V-shaped wake, as shown in Figure 3.5. The angle of the wake is $39°$. This result doesn't depend on the speed of the duck, nor on the strength of gravity. It follows simply from the fact that the group velocity v_g is related to the phase velocity c by $v_g = \frac{1}{2}c$ for deep water waves.

Here we present a simple derivation of this fact. We take the duck to be moving in the positive x-direction with speed U. The duck creates waves of different frequencies ω and wavelength k, related by the deep water dispersion relation

$$\omega = \sqrt{gk} \ . \tag{3.50}$$

Correspondingly, the phase speed is

$$c(k) = \sqrt{\frac{g}{k}} \ . \tag{3.51}$$

The key point is that the wave-crests move along with the duck in the x-direction. To see what this means, we first focus on a particular wavenumber k. This wave will come out at an angle θ, as shown in the figure. This

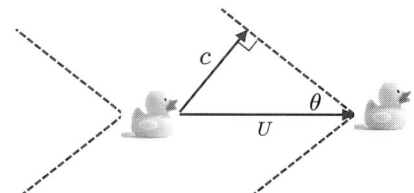

requires that we have

$$c = U \sin \theta \ . \tag{3.52}$$

Waves with different wavenumbers travel at different speeds, and hence will come out at different angles θ.

The coherent superposition of different waves carries energy, and this travels at the lesser speed $v_g = \frac{1}{2}c$. This means that, in some time, the wave profile of the superposition will have reached just half the distance of the peak. We denote the relevant angle as ϕ, as shown in the figure. Using the sine rule, we have

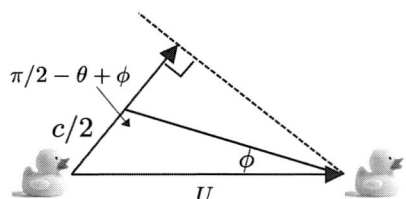

$$\frac{U}{\sin(\pi/2 + \theta - \phi)} = \frac{c}{2 \sin \phi} \quad \Longrightarrow \quad 2U \sin \phi = c \cos(\theta - \phi) \ . \tag{3.53}$$

Dividing (3.53) by (3.52), and playing a few $\sin^2 \theta + \cos^2 \theta = 1$ games, we get the relation between the two angles:

$$\tan \phi = \frac{\tan \theta}{2 + \tan^2 \theta} \ . \tag{3.54}$$

Both $\theta(k)$ and $\phi(k)$ are functions of the wavenumber k. However, the function $\phi(k)$ has the interesting feature that there is a maximum possible value it can take. To compute this, we just need to evaluate

$$\frac{d(\tan \phi)}{d(\tan \theta)} = \frac{2 - \tan^2 \theta}{(2 + \tan^2 \theta)^2} = 0 \ . \tag{3.55}$$

Clearly this holds when

$$\tan \theta = \sqrt{2} \quad \Longrightarrow \quad \tan \phi = \frac{1}{2\sqrt{2}} \quad \Longrightarrow \quad \phi \approx 19.5° \ . \tag{3.56}$$

This is the result that we want. There is a maximum opening angle $2\phi \approx 39°$ at which the waves emerge from behind a duck. This is known as a *Kelvin wake*

3.1.5 Surface Tension

If you're a molecule, a liquid is a nice, comfortable place to spend your time. You're attracted to all your neighbouring molecules, but are afforded enough freedom to wander off on your own.

Things get more precarious at the surface of the liquid. There are now fewer neighbours to keep you company. As each neighbour offers a welcoming, attractive potential, the fact that you now find yourself a little isolated means that you are sitting in a higher energy state. This, in turn, means that, collectively, the molecules in a liquid can lower their energy by keeping the area of the surface as small as possible. This results in a force called *surface tension*. This force is the reason that sufficiently small droplets of water, or soap bubbles, are round: the sphere has the minimal surface area.

The existence of surface tension means that pressure need no longer be continuous across the surface. Instead, the surface can tolerate a local pressure difference by bending slightly and letting the surface tension push back. Said another way, the surface tension provides another restoring force for the wave motion.

This physics is captured by a change to the boundary condition (3.8). For a surface with embedding $z = \eta(x, y, t)$, the pressure difference should now be

$$P(x, y, \eta(x, y)) - P_0 = -\gamma \nabla^2 \eta \qquad (3.57)$$

with $\gamma > 0$ the surface tension and $\nabla^2 \eta = \partial^2 \eta / \partial x^2 + \partial^2 \eta / \partial y^2$ the 2d Laplacian, which is the appropriate characterisation of the curvature of the surface. As a sanity check, if the atmospheric pressure P_0 is greater than the pressure of the fluid at the surface in some region, then this equation tells us that $\nabla^2 \eta > 0$, which means that the surface has a dip in that region, as expected.

We would like to understand how the existence of surface tension affects the dynamics of waves. If we follow through our derivation of the time-dependent Bernoulli principle, equation (3.17) is replaced by

$$\rho \left(\frac{\partial \phi}{\partial t} + \frac{1}{2} |\nabla \phi|^2 \right)_{z=\eta} + P_0 + \rho g \eta - \gamma \nabla^2 \eta = f(t) . \qquad (3.58)$$

After linearisation, the final condition in (3.21) becomes

$$\left. \frac{\partial \phi}{\partial t} \right|_{z=0} + g\eta - \frac{\gamma}{\rho} \nabla^2 \eta = \tilde{f}(t) \qquad (3.59)$$

with $\tilde{f}(t)$ a function that can depend on time but, crucially, must be independent of space. We now make our usual ansatz for waves propagating in the x-direction

$$\phi(x, z, t) = \phi_0(z) \, e^{ikx - i\omega t} \quad \text{and} \quad \eta(x, t) = \eta_0 \, e^{ikx - i\omega t} . \qquad (3.60)$$

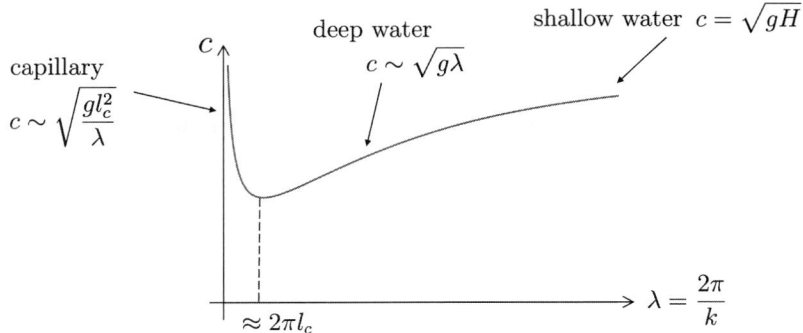

The phase velocity c for surface waves, including the effect of surface tension.

We can see how the surface tension affects our previous story just by staring at (3.59). With the wave ansatz, we replace $\nabla^2 \eta = -k^2 \eta$ and the surface tension always accompanies the gravitational acceleration g, replacing

$$g \longrightarrow g + \frac{\gamma k^2}{\rho} = g\left(1 + l_c^2 k^2\right) \tag{3.61}$$

in all our previous formulae. Here we've introduced the length scale

$$l_c = \sqrt{\frac{\gamma}{g\rho}} \ . \tag{3.62}$$

This is known as the *capillary length*. From (3.61), we see that long wavelength modes with $\lambda \gg l_c$, so $l_c k \ll 1$, are pretty much unaffected by surface tension. In contrast, surface tension effects dominate when the wavelength becomes short, $\lambda \ll l_c$, so $l_c k \gg 1$. Waves with $\lambda \lesssim l_c$ are referred to as *capillary waves*.

For water at room temperature, $l_c \approx 3$ mm. The capillary waves are little ripples on the water, up to a wavelength of 1 cm or so, with the factor of 2π in the definition of the wavelength raising us above l_c.

The general dispersion relation is

$$\omega^2 = gk(1 + l_c^2 k^2)\tanh(kH) \tag{3.63}$$

while the phase velocity is

$$c = \frac{\omega}{k} = \sqrt{\frac{g}{k}\left(1 + l_c^2 k^2\right)\tanh(kH)} \ . \tag{3.64}$$

For capillary waves, with $l_c k \gg 1$, in deep water, so $kH \gg 1$, we have

$$c \approx \sqrt{\frac{\gamma k}{\rho}} \, . \tag{3.65}$$

In contrast to surface waves driven by gravity (3.33), the short wavelength modes now travel faster. Furthermore, the group velocity (3.49) is $v_g(k) = \frac{3}{2} c$. The wavepackets now travel faster than the individual Fourier modes. The phase velocity is plotted in Figure 3.6.

3.2 Internal Gravity Waves

Gravitational waves are ripples of the spacetime continuum that emerge from violent events such as the collision of two black holes. That, sadly, is not the topic of discussion here. Instead, "gravity waves" describe the disappointingly mundane phenomenon of fluids bobbing up and down due to gravity. If you want to learn more about gravitational waves, you'll need to open the book on General Relativity. Otherwise, read on.

Gravity waves are simply waves in fluids where the restoring force is provided by gravity. The surface waves above are examples of gravity waves, at least those with wavelength longer than the capillary length where surface tension is negligible. In this section we study gravity waves in the bulk of the fluid, as opposed to on the surface. While these waves might not be quite as sexy as their spacetime counterparts, they play a prominent role in the physics of the climate which, arguably, makes them considerably more important. They are also, as we shall see, surprisingly subtle. Many of these subtleties can be traced to a phenomenon known as *stratification*.

Stratified Flows and Buoyancy Frequency

A flow is said to be *stratified* if the density ρ varies from place to place. Typically this happens because of gravity and the density is a function of the vertical direction: $\rho = \rho(z)$.

Consider a small ball immersed in a stratified flow. If the ball has density $\rho_0 = \rho(z_0)$ for some height z_0 then, by Archimedes' principle, it will naturally sit at height $z = z_0$. This is where the weight of fluid that the ball displaces is equal to its own weight.

Suppose now that we displace the ball upwards by some small amount δz.

The density of the fluid there is

$$\rho(z_0 + \delta z) \approx \rho(z_0) + \left.\frac{\partial \rho}{\partial z}\right|_{z_0} \delta z . \tag{3.66}$$

Now the weight of the displaced water differs from that of the ball, resulting in a net *upwards* force,

$$\text{upwards force} \approx g \left.\frac{\partial \rho}{\partial z}\right|_{z_0} \delta z . \tag{3.67}$$

If $\partial \rho/\partial z > 0$ then the ball's original position was unstable, and it flies upwards. But most stratified flows have density larger at the bottom than at the top, so $\partial \rho/\partial z < 0$ and the "upwards force" is negative and, hence, actually downwards. In this case, the ball oscillates about its equilibrium position, enacting simple harmonic motion with a frequency

$$N^2 = -\frac{g}{\rho_0}\frac{\partial \rho}{\partial z} . \tag{3.68}$$

This is called the *buoyancy frequency* or, sometimes, the *Brunt–Väisälä frequency*. In what follows, we'll look at similar motion but for the fluid itself.

Note that we haven't specified how $\rho(z)$ depends on the height z. Nor will we do this throughout the rest of this section. We'll address this in section 3.4 when we introduce an equation of state relating pressure and density. (We'll also meet a rather different way of fixing the density profile in section 4.4 known as the Boussinesq approximation.)

Equations for Gravity Waves

Until now, the incompressibility condition was forced upon us by the requirement that the density is constant. For stratified flows, this is no longer the case. Nonetheless, it is still physically sensible to insist on a divergence-free velocity field (at least for speeds smaller than the sound speed)

$$\nabla \cdot \mathbf{u} = 0 . \tag{3.69}$$

With this, mass conservation becomes the requirement,

$$\frac{\partial \rho}{\partial t} + \nabla \cdot (\rho \mathbf{u}) = 0 \quad \Longrightarrow \quad \frac{D\rho}{Dt} = \frac{\partial \rho}{\partial t} + \mathbf{u} \cdot \nabla \rho = 0 . \tag{3.70}$$

In addition, we will ignore viscosity and look at gravity waves in the Euler equation, now in the presence of gravity

$$\rho \left(\frac{\partial \mathbf{u}}{\partial t} + \mathbf{u} \cdot \nabla \mathbf{u} \right) = -\nabla P - \rho(z) g \hat{\mathbf{z}} . \tag{3.71}$$

We start by considering a boring background, with $\mathbf{u} = 0$ and the hydrostatic pressure $P_0(z)$ related to the density $\rho_0(z)$ through the Euler equation by

$$\frac{dP_0}{dz} = -g\rho_0(z) \ . \tag{3.72}$$

Now we look at small perturbations around this background. The gravity waves of interest travel in the horizontal x-direction, while bobbing up and down in the vertical z-direction. To this end, we look for solutions of the form

$$\mathbf{u}(\mathbf{x}, t) = (u_x, 0, u_z)e^{ik_x x + ik_z z - i\omega t} \tag{3.73}$$

with u_x and u_z constant. Both the density and pressure exhibit the same wave-like behaviour, with

$$\rho(x, z, t) = \rho_0(z) + \tilde{\rho}\, e^{ik_x x + ik_z z - i\omega t}$$
$$P(x, z, t) = P_0(z) + \tilde{P}\, e^{ik_x x + ik_z z - i\omega t} \ . \tag{3.74}$$

We've denoted the perturbations around the constant background with tildes on top, like $\tilde{\rho}$ and $\tilde{P}$. This is a convention that we'll use throughout this chapter and the next one. The divergence-free condition $\nabla \cdot \mathbf{u} = 0$ tells us that

$$k_x u_x + k_z u_z = 0 \ . \tag{3.75}$$

Even before we proceed, this equation is telling us that $\mathbf{k} \cdot \mathbf{u} = 0$. In other words, the waves are transverse. This is like light waves (which have $\mathbf{E} \cdot \mathbf{k} = \mathbf{B} \cdot \mathbf{k} = 0$ with $\mathbf{E}$ and $\mathbf{B}$ the electric and magnetic fields, respectively) but contrasts with the sound waves that we will meet in Section 3.4.

For the other equations, we linearise, throwing away any terms quadratic in perturbations. Mass conservation gives

$$-i\omega\tilde{\rho} + u_z \frac{d\rho_0}{dz} = 0 \tag{3.76}$$

and the two components of the Euler equation are

$$-i\rho_0 \omega u_x = -ik_x \tilde{P} \quad \text{and} \quad -i\rho_0 \omega u_z = -ik_z \tilde{P} - g\tilde{\rho} \ . \tag{3.77}$$

Solving these simultaneous equations gives us the dispersion relation for the frequency of gravity waves,

$$\omega = \pm N \frac{k_x}{\sqrt{k_x^2 + k_z^2}} \tag{3.78}$$

with N the buoyancy frequency (3.68). Note that we necessarily have $\omega \leq N$. Moreover, the frequency is non-vanishing only if $k_x \neq 0$. The maximum frequency arises when $k_z = 0$ so that $\omega = N$. In this case, the divergence-free condition (3.75) tells us that we must have $u_x = 0$. This, in turn, means that we have a wave propagating in the direction $\mathbf{k} = (k_x, 0, 0)$ with the fluid bobbing up and down with velocity $\mathbf{u} = (0, 0, u_z)$.

In general, the gravity wave propagates in the direction

$$\mathbf{k} = (k_x, 0, k_z) \ . \tag{3.79}$$

The slight surprise comes when we compute the group velocity. For a one-dimensional wave, this is $v_g = \partial \omega / \partial k$. For higher dimensional waves, like we have here, the relevant definition is

$$\mathbf{v}_g = \frac{\partial \omega}{\partial k_x} \hat{\mathbf{x}} + \frac{\partial \omega}{\partial k_z} \hat{\mathbf{z}} \ . \tag{3.80}$$

For the dispersion relation (3.78), this gives

$$\mathbf{v}_g = \frac{N k_z}{(k_x^2 + k_z^2)^{3/2}} (k_z, 0, -k_x) \ . \tag{3.81}$$

This reveals the most curious feature of internal gravity waves: the group velocity is perpendicular to the direction of the wave, $\mathbf{v}_g \cdot \mathbf{k} = 0$. Both wavepackets and energy propagate in the direction $\mathbf{v}_g$, but this is orthogonal to the direction $\mathbf{k}$ of the wave itself! In other words, the energy propagates parallel to the crests and troughs of the wave. It is somewhat less surprising when you realise that $\mathbf{v}_g$ is parallel to the velocity $\mathbf{u}$ of the fluid.

3.3 An Introduction to Geophysical Flows

In this section we take something of a diversion. We will explore some novel phenomena that arise when fluids rotate. The main motivation for this comes from the fact that Earth spins and this gives rise to some new types of waves with rather interesting properties.

Recall from Volume 1 on Classical Mechanics that, if we sit in a reference frame that rotates with constant angular velocity $\boldsymbol{\Omega}$, then we experience two fictitious forces. These are the centrifugal force, proportional to $\boldsymbol{\Omega} \times (\boldsymbol{\Omega} \times \mathbf{x})$, and the Coriolis force, proportional to $2\boldsymbol{\Omega} \times \mathbf{u}$. For fluids, these appear as forces on the right-hand side of the Navier–Stokes equation. Throughout this section, we will neglect viscosity and work with fluids with constant density. The Euler equation in a rotating frame is then

$$\frac{\partial \mathbf{u}}{\partial t} + \mathbf{u} \cdot \nabla \mathbf{u} = -\frac{1}{\rho} \nabla P + \mathbf{g} - 2\boldsymbol{\Omega} \times \mathbf{u} - \boldsymbol{\Omega} \times (\boldsymbol{\Omega} \times \mathbf{x}) \ . \tag{3.82}$$

The centrifugal force is not particularly interesting for our purposes. Locally, it simply redefines what we mean by "down" since, like gravity, it can be written as the gradient of a potential energy. We will simply ignore it. As we will see, all the interesting physics arises from the Coriolis force.

3.3.1 The Shallow Water Approximation

In what follows, we will make the so-called *shallow water* approximation. We will assume that the extent of the fluid in the horizontal directions, labelled by x and y, is much greater than the height of the fluid in the vertical z-direction. For our purposes, the Atlantic Ocean counts as "shallow" since it is, on average, around 3.5 km deep but several thousand kilometres wide. Similarly, the atmosphere also counts as "shallow" and the phenomena that we describe can be found in both.

Our choice of coordinates is shown in the figure below. Locally, "up" is in the z-direction, "north" is in the y-direction, and "east" is the x-direction.

There is some jargon here. The x-direction, corresponding to east/west, is referred to as the *zonal* direction. This coordinate parametrises a "zone" which sits at a fixed latitude and so might be expected to exhibit similar phenomena. Examples of such zones include the topical, equatorial, and polar regions. Meanwhile, the y-coordinate, corresponding to north/south, is called the *meridional* direction.

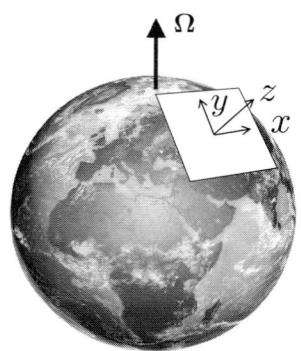

We define the *Coriolis parameter*

$$f = 2\mathbf{\Omega} \cdot \hat{\mathbf{z}} \ . \tag{3.83}$$

If we're considering flows where we can neglect the curvature of the Earth, then we restrict attention to a given tangent plane as shown in the figure and take f to be constant. In contrast, if we need to take into account the curvature of the Earth, then f will be a function $f = f(y)$, reflecting the fact that as we move along the surface the local "up" direction $\hat{\mathbf{z}}$ changes, while the spin $\mathbf{\Omega}$ remains fixed. In what follows, we will consider both situations in which f is taken to be constant and, in Section 3.3.5, situations in which f varies.

Our initial set-up will be similar to that of water waves described in Section 3.1. We'll take the average depth of the water to be H, with a flat, solid base at $z = -H$ and a varying surface at $z = \eta(x, y, t)$ with $|\eta| \ll H$ as shown in Figure 3.2.

Next, we assume that the velocities in the horizontal direction are independent of the depth, so

$$\mathbf{u} = (u, v, w) \quad \text{with} \quad \begin{cases} u = u(x, y, t) \\ v = v(x, y, t) \\ w = w(x, y, z, t) \end{cases} . \tag{3.84}$$

Note that this is where our set-up starts to differ from the water waves of Section 3.1.

The vertical velocity can be eliminated in favour of the height fluctuation $\eta(x, y, t)$ by using the incompressibility condition

$$\nabla \cdot \mathbf{u} = 0 \quad \Longrightarrow \quad \frac{\partial w}{\partial z} = -\frac{\partial u}{\partial x} - \frac{\partial v}{\partial y} . \tag{3.85}$$

We integrate over the vertical z-direction, and use the free boundary condition (3.6), which tells us that $w(z = \eta) = D\eta/Dt$ and $w(z = -H) = 0$. We then have

$$\frac{\partial \eta}{\partial t} + u\frac{\partial \eta}{\partial x} + v\frac{\partial \eta}{\partial y} = -(H + \eta)\left(\frac{\partial u}{\partial x} + \frac{\partial v}{\partial y}\right) . \tag{3.86}$$

Next, we assume that the pressure in the vertical direction adapts to balance the gravitational force. This *hydrostatic approximation* is what led us to Archimedes' principle in Section 1.2.3. We also need the boundary condition $P = P_0$ on the surface at $z = \eta$, meaning that we take the pressure to be

$$P = P_0 - \rho g(z - \eta) . \tag{3.87}$$

In the Navier–Stokes equation (3.82), we can then replace $-\frac{1}{\rho}\nabla P + \mathbf{g} = -g\nabla\eta$.

With these pieces in place, the remaining two Navier–Stokes equations read

$$\frac{\partial u}{\partial t} + u\frac{\partial u}{\partial x} + v\frac{\partial u}{\partial y} = fv - g\frac{\partial \eta}{\partial x} , \tag{3.88}$$

$$\frac{\partial v}{\partial t} + u\frac{\partial v}{\partial x} + v\frac{\partial v}{\partial y} = -fu - g\frac{\partial \eta}{\partial y} . \tag{3.89}$$

The triplet of equations, (3.86) for the height, and (3.88) and (3.89) for the velocity, are called the *shallow water equations*.

As usual, we want an excuse to drop the non-linear terms to make life easy. If a flow has characteristic velocity U, changing over some length scale L, then these non-linear terms scale as U^2/L. This should be compared with the Coriolis terms which scale as fU. We introduce a dimensionless ratio called the *Rossby number*

$$Ro = \frac{U}{fL} \ . \tag{3.90}$$

It's appropriate to drop the non-linear terms for flows with $Ro \ll 1$. The rotation of the Earth is $\Omega \approx 2\pi \times (\text{day})^{-1} \approx 10^{-4} \text{ s}^{-1}$, while typical atmospheric or oceanic speeds are around $U \sim 10 \text{ m s}^{-1}$. That means that

$$Ro \approx \frac{10^5 \text{ m}}{L} \ . \tag{3.91}$$

We see that we can think about dropping the non-linear terms only for very long wavelength perturbations. For $L \sim 10^3$ km, we have $Ro \approx 0.1$ which, while admittedly < 1, is barely $\ll 1$. Nonetheless, this is the approximation that we will make. We further linearise the first equation (3.86), leaving us with our three linear shallow water equations

$$\frac{\partial \eta}{\partial t} = -H\frac{\partial u}{\partial x} - H\frac{\partial v}{\partial y} \ , \tag{3.92}$$

$$\frac{\partial u}{\partial t} = fv - g\frac{\partial \eta}{\partial x} \ , \tag{3.93}$$

$$\frac{\partial v}{\partial t} = -fu - g\frac{\partial \eta}{\partial y} \ . \tag{3.94}$$

In the rest of this section, we will solve these equations in various scenarios for $u(x,y,t)$, $v(x,y,t)$, and $\eta(x,y,t)$.

3.3.2 Geostrophic Balance and Poincaré Waves

We're going to find a number of different solutions to the linearised shallow water equations (3.92), (3.93), and (3.94). Among these will be wave-like solutions. But, more surprisingly, we will also find some time-independent solutions that are more interesting than just an ocean with a flat surface $\eta = $ constant.

It's simple to see the existence of time-independent solutions by setting $\partial/\partial t = 0$ in (3.92), (3.93), and (3.94). Solutions can be built from any divergence-free flow, with $\nabla \cdot \mathbf{u} = 0$, that obeys

$$u = -\frac{g}{f}\frac{\partial \eta}{\partial y} \quad \text{and} \quad v = +\frac{g}{f}\frac{\partial \eta}{\partial x} \ . \tag{3.95}$$

Here the height η acts like a stream function of the kind we met in Section 1.1.4. Steady-state solutions of this form are said to be in *geostrophic balance*.

It's easy to understand the balance of forces underlying geostrophic balance. Suppose that there is some bump in the height of the fluid. Gravity, of course, wants to pull this down but, because the underlying fluid is incompressible, it results in a horizontal force in the direction $\nabla \eta$. The velocity in geostrophic balance is such that it gives rise to a Coriolis force that exactly counteracts the force of gravity.

Flows in geostrophic balance (3.95) obey $\mathbf{u} \cdot \nabla \eta = 0$. In other words, the flow is along lines of constant height η. But, from hydrostatic balance (3.87), we know that the pressure in the fluid is proportional to the height. In other words, the flow is along isobars. This is familiar from weather maps, where wind blows along lines of constant pressure, rather than from high to low pressure as one might naively expect. The large scale flow of both the ocean and atmosphere is largely in geostrophic balance.

Potential Vorticity

Our next task is to understand time-dependent solutions to the shallow water equations. To do this, it's best to first look more closely at the various conserved quantities.

In fact, it's best if we briefly return to the full non-linear equations (3.86), (3.88), and (3.89). These admit two conserved quantities. The first is simply the height, whose conservation follows from the underlying conservation of mass

$$\frac{\partial h}{\partial t} + \nabla \cdot (\mathbf{u}h) = 0 \quad \text{with} \quad h = H + \eta \,. \tag{3.96}$$

In this equation, both ∇ and $\mathbf{u}$ are now 2d vectors, rather than 3d vectors. The second is conservation of vorticity. It can be checked that

$$\frac{\partial W}{\partial t} + \nabla \cdot (\mathbf{u}W) = 0 \quad \text{with} \quad W = \frac{\partial v}{\partial x} - \frac{\partial u}{\partial y} + f \,. \tag{3.97}$$

Note that the vorticity includes the extra $+f$ contribution from the Coriolis force.

Both (3.96) and (3.97) are continuity equations, which is the usual conservation law that we know and love. Elsewhere in this book, we've been able to use the incompressibility condition $\nabla \cdot \mathbf{u} = 0$ to extract the velocity $\mathbf{u}$ from the clutches of the spatial derivative and write equations of this form as the vanishing of a material derivative. But we're not allowed to do this in the present context because the 2d velocity $\mathbf{u}$ does not necessarily obey $\nabla \cdot \mathbf{u} = 0$. The 3d fluid is still incompressible of course, but the 2d velocity $\mathbf{u}$ field can pile up at some point at the expense of increasing the height. Indeed, this is what our first equation (3.96) is telling us. Nonetheless, we can combine (3.96) and (3.97) to construct a quantity that has vanishing material derivative. This is

$$\mathcal{Q} = \frac{W}{h} = \frac{1}{H + \eta} \left(\frac{\partial v}{\partial x} - \frac{\partial u}{\partial y} + f \right) \tag{3.98}$$

which obeys

$$\frac{D\mathcal{Q}}{Dt} = \frac{\partial \mathcal{Q}}{\partial t} + \mathbf{u} \cdot \nabla \mathcal{Q} = 0 \ . \tag{3.99}$$

The quantity $\mathcal{Q}$ is called the *potential vorticity*. The equation $D\mathcal{Q}/Dt = 0$ is telling us that the value of the potential vorticity doesn't change as we follow the flow.

The discussion above is for the full non-linear equations. Something rather striking happens when we restrict to the linear equations. We linearise the conservations laws (3.96) and (3.97) about $h = H$ and $W = f$, to find

$$\frac{\partial h}{\partial t} + H \nabla \cdot \mathbf{u} = 0 \quad \text{and} \quad \frac{\partial W}{\partial t} + f \nabla \cdot \mathbf{u} = 0 \ . \tag{3.100}$$

The surprising fact is that these both have the same current: it is simply the velocity $\mathbf{u}$. This means that we can eliminate the current to find the linearised conservation law

$$\frac{\partial Q}{\partial t} = 0 \quad \text{with} \quad Q = \frac{\partial v}{\partial x} - \frac{\partial u}{\partial y} - \frac{f\eta}{H} \ . \tag{3.101}$$

The quantity Q is (up to constant term, and a scaling by H) the *linearised potential vorticity*. We see that Q is independent of time. That's a much stronger statement than our usual conservation laws. Usually when something is conserved, its value at a given point in space can change in time only if the conserved stuff moves to a neighbouring point. That's the physics of the continuity equation. But (3.101) is telling us something much stronger: the function $Q(\mathbf{x})$ is fixed for all time! This feature, which holds only in the linearised theory, adds a rigidity to the system that will be responsible for some of the features we'll see below.

Poincaré Waves

With this understanding of potential vorticity in hand, we'll now turn to some wave solutions of the linearised shallow water equations (3.92), (3.93), and (3.94).

If there were no rotation, it's clear what would happen. With $f = 0$, it's simple to check that the equations (3.92), (3.93), and (3.94) become the wave equation $\ddot{\eta} = c^2 \nabla^2 \eta$ with $c^2 = gH$. This describes shallow water surface waves propagating with speed c and reproduces our previous result (3.34) for long wavelength waves.

The Coriolis force changes this. If we assume that $f = $ constant (which means that we are neglecting the effects of the curvature of the Earth), then the wave equation that we derive from (3.92), (3.93), and (3.94) is

$$\frac{\partial^2 \eta}{\partial t^2} = c^2 \nabla^2 \eta - Hf \left(\frac{\partial v}{\partial x} - \frac{\partial u}{\partial y} \right) \quad \text{with} \quad c^2 = gH \ . \tag{3.102}$$

The additional terms can be rewritten in terms of the linearised potential vorticity (3.101) to get

$$\frac{\partial^2 \eta}{\partial t^2} - c^2 \nabla^2 \eta + f^2 \eta = -HfQ \tag{3.103}$$

where Q is the potential vorticity which, as we have seen above, is a constant function that doesn't change with time. For a given problem, one might have to solve (3.103) for some fixed Q. But, in addition, one can always add solutions to the complementary solution which solves the homogeneous equation

$$\frac{\partial^2 \eta}{\partial t^2} - c^2 \nabla^2 \eta + f^2 \eta = 0 \ . \tag{3.104}$$

This is a rather famous equation that, in the world of quantum field theory, is known as the *Klein–Gordon equation*. It is a simple matter to find solutions by writing

$$\eta(\mathbf{x}, t) = \tilde{\eta} e^{i\mathbf{k} \cdot \mathbf{x} - i\omega t} \tag{3.105}$$

with $\mathbf{x} = (x, y)$ and $\mathbf{k} = (k_x, k_y)$. This solves (3.104) provided that the frequency ω and wavevector $\mathbf{k}$ obey the dispersion relation

$$\omega^2 = c^2 k^2 + f^2 \ . \tag{3.106}$$

These are known as *Poincaré waves*. They are a form of gravity wave, since gravity acts as the restoring force, as seen in the speed $c = \sqrt{gH}$. But their properties are affected by the Coriolis force. They are sometimes referred to as *inertia-gravity waves*.

For long wavelengths, $k \to 0$, Poincaré waves have a finite frequency, set by the Coriolis parameter $\omega \to f$. In the language of quantum mechanics, we say that the spectrum is *gapped*, the "gap" being the smallest frequency at which the system oscillates.

The cross-over from "short" to "long" wavelengths happens at the length scale

$$R = \frac{c}{f} = \frac{\sqrt{gH}}{f} \ . \tag{3.107}$$

This is known as the *Rossby radius of deformation*. It is the characteristic length scale in the shallow water equations. For the ocean at mid-latitudes, one has $R \approx 1000$ km. Short wavelength modes, with $k \gg R^{-1}$, act just like usual surface waves, with $\omega \approx ck$. It's the long wavelength modes, with $k \ll R^{-1}$, that feel the effect of the Coriolis force. In this limit, we can neglect the η-terms in (3.92) and (3.93) to find that the velocities obey $\dot{u} = fv$ and $\dot{v} = -fu$. This tells us that the wave velocity in the x- and y-directions are $\pi/2$ out of phase.

In preparation for what follows, it's worth redoing the above calculation in a slightly different way. We write our three, linearised shallow water equations (3.92), (3.93), and (3.94) as a combined matrix eigenvalue equation

$$i\frac{\partial \Psi}{\partial t} = \begin{pmatrix} 0 & -ic\partial_x & -ic\partial_y \\ -ic\partial_x & 0 & if \\ -ic\partial_y & -if & 0 \end{pmatrix} \Psi \quad \text{with} \quad \Psi = \begin{pmatrix} \sqrt{g/H}\eta \\ u \\ v \end{pmatrix} . \tag{3.108}$$

We've done some cosmetic manipulations to get the equation in this form. In addition to rescaling the η variable, we've also multiplied everything by a factor of i. This makes the resulting equation look very much like a time-dependent Schrödinger equation. In particular, the matrix is Hermitian. (Recall that $-i\partial_x$ and $-i\partial_y$ are Hermitian operators for the same reason that the momentum operator is Hermitian in quantum mechanics.) With our wave ansatz $\Psi = \tilde{\Psi}e^{i\mathbf{k}\cdot\mathbf{x}-i\omega t}$, this becomes a standard eigenvalue problem

$$\begin{pmatrix} 0 & ck_x & ck_y \\ ck_x & 0 & if \\ ck_y & -if & 0 \end{pmatrix} \tilde{\Psi} = \omega\tilde{\Psi} \ . \tag{3.109}$$

Because this is a Hermitian matrix, the eigenvalues are guaranteed to be real. They are

$$\omega = \pm\sqrt{c^2 k^2 + f^2} \quad \text{and} \quad \omega = 0 \,. \tag{3.110}$$

We recognise the first of these as the dispersion relation for Poincaré waves (3.106). In addition, there are a collection of solutions with $\omega = 0$. In the context of condensed matter physics, this is known as a *flat band* (because if you plot ω vs k it is a flat plane.) The existence of the flat band follows from the functional conservation of the potential vorticity. It is telling us that there are additional, time-independent equilibrium solutions. These are solutions like (3.107) that exhibit geostrophic balance.

3.3.3 We Need to Talk About Kelvin Waves

Everyone likes a trip to the coast. Now it's our turn. For the purposes of this section, the coast is not going to be very exciting. It's simply a boundary of our fluid, which we will take to run north/south. The fluid exists only in the $x \geq 0$ direction. For $x < 0$, there is only land.

Obviously we must put a boundary condition $u = 0$ at $x = 0$, ensuring that no flow passes the boundary. In fact, we'll do something more extreme than this. We will search for solutions that have $u = 0$ everywhere. The linearised shallow water equation (3.93) then becomes

$$v = \frac{g}{f}\frac{\partial \eta}{\partial x} \,. \tag{3.111}$$

This is telling us that the fluid lives in geostrophic balance in the x-direction, with the pressure gradient from $\partial \eta / \partial x$ pushing against the Coriolis force that arises because the fluid has velocity v in the y-direction. Meanwhile, the other two shallow water equations (3.92) and (3.94) become

$$\frac{\partial \eta}{\partial t} = -H\frac{\partial v}{\partial y} \quad \text{and} \quad \frac{\partial v}{\partial t} = -g\frac{\partial \eta}{\partial y} \,. \tag{3.112}$$

These are standard wave-like equations. If we make the usual ansatz that $v = v_0(x)\, e^{iky - i\omega t}$ and $\eta = \eta_0(x)\, e^{iky - i\omega t}$, these become

$$\omega \eta_0 = kH v_0 \quad \text{and} \quad \omega v_0 = gk\eta_0 \quad \implies \quad \omega^2 = c^2 k^2 \tag{3.113}$$

with the speed given by $c = \sqrt{gH}$ as for our previous examples. So far, things look fairly standard. But there's a slight twist in the tail. This arises when we return to (3.111) which tells us the profile of the water near the boundary. We have

$$\frac{\partial \eta_0}{\partial x} = \frac{f\omega}{kc^2}\eta_0 \,. \tag{3.114}$$

Our dispersion relation $\omega^2 = c^2 k^2$ naively suggests that we have two options: $\omega = +ck$ or $\omega = -ck$. But that's not right. Suppose that we take $f > 0$, which is appropriate if we are in the Northern Hemisphere. Then if we pick $\omega = +ck$ we're in trouble, because the height of the water will grow exponentially away from the boundary: $\eta_0(x) \sim e^{+fx/c}$. And that's bad. It means that we should throw away this solution. The only physical solution has

$$\omega = -ck \qquad\qquad (3.115)$$

with the water profile decaying exponentially away from the boundary, $\eta_0(x) \sim e^{-fx/c}$. This means that the boundary waves propagate only in one direction which, in the current set-up, is the negative y-direction, also known as south. These are known as *Kelvin waves*.

Waves that propagate only in one direction are said to be *chiral*. In the Northern Hemisphere, with $f > 0$, Kelvin waves propagate so that the land always sits to their right. (In other words, if these waves are propagating around a land mass, then they move in a clockwise direction.) The direction of this wave is shown in the figure. In the Southern Hemisphere, where $f < 0$, the same argument tells us that we must have the $\omega = +ck$ solution, so a Kelvin wave propagates with the land to its left as it moves.

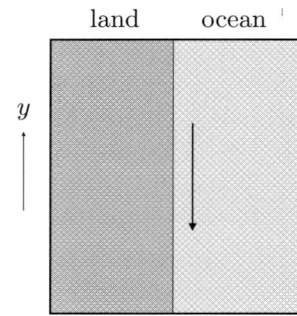

Chiral waves also make an appearance in various condensed matter systems where, as here, they typically live at the edge of some system. In that context, there is often some deep topological reason for the emergence of such chiral waves. The same is also true here and we will elaborate on this further in Section 3.3.6.

3.3.4 Rossby Waves

As we've seen, the linearised shallow water equations admit time-independent solutions in geostrophic balance, solving (3.95). But objects that are strictly unmoving are rare in nature. Which suggests that there may be something that can coax flows in geostrophic balance to move. And indeed there is. In this section, we will see that if we look at scales over which the Coriolis parameter f is no longer constant, then flows in geostrophic balance start to evolve in time. Such flows are said to be in *quasi-geostrophic balance*.

Crucially, the evolution of flows in quasi-geostrophic balance happens much more slowly than the dynamics of Poincaré waves that we saw above. That means that it is this quasi-geostrophic flow that governs the long-time dynamics of the ocean and atmosphere. The purpose of this section is to construct the equations that describe this flow.

At latitude θ, the Coriolis parameter is given by $f = 2\Omega \sin \theta$, where $\Omega = 2\pi \ \text{day}^{-1}$. To capture the variation of the Coriolis parameter, it will suffice to consider just the leading term in the Taylor expansion

$$f = f_0 + \beta y \tag{3.116}$$

where the meridional y-direction points north. Here the parameter β has dimension $[\beta] = L^{-1}T^{-1}$ and has order of magnitude $\beta \sim f_0/R$ with R the radius of the Earth.

We would like to understand what becomes of the flows in geostrophic balance now that f is no longer constant. Our strategy will be to turn to the conservation of potential vorticity (3.99)

$$\frac{D\mathcal{Q}}{Dt} = 0 \quad \text{with} \quad \mathcal{Q} = \frac{1}{H + \eta}\left(\frac{\partial v}{\partial x} - \frac{\partial u}{\partial y} + f\right) . \tag{3.117}$$

You can check that this equation remains valid even when $f = f(\mathbf{x})$. We will consider flows with Rossby number $Ro \ll 1$ that are very close to geostrophic balance (3.95). This means that we can replace the vorticity in the expression for $\mathcal{Q}$ with

$$\frac{\partial v}{\partial x} - \frac{\partial u}{\partial y} = \frac{g}{f}\nabla^2 \eta \approx \frac{g}{f_0}\nabla^2 \eta . \tag{3.118}$$

We further assume that variations in the height are small, so $\eta \ll H$, and the potential vorticity can be written solely in terms of the height fluctuations η. Ignoring an overall constant term, we have

$$\mathcal{Q} \approx \frac{f_0}{H^2}\left(\frac{c^2}{f_0^2}\nabla^2 \eta - \eta + \frac{\beta H}{f_0}y\right) . \tag{3.119}$$

As we've seen, potential vorticity is materially conserved and, using the geostrophic balance condition (3.95), this too becomes an equation that can be written solely in terms of the height

$$\frac{D\mathcal{Q}}{Dt} = 0 \quad \Longrightarrow \quad \dot{\mathcal{Q}} - \frac{g}{f_0}\frac{\partial \eta}{\partial y}\frac{\partial \mathcal{Q}}{\partial x} + \frac{g}{f_0}\frac{\partial \eta}{\partial x}\frac{\partial \mathcal{Q}}{\partial y} = 0 . \tag{3.120}$$

This is now a dynamical equation for the height η. It is known as the *shallow water quasi-geostrophic equation.*

The quasi-geostrophic equation looks a little daunting. But we can easily extract some simple physics. We linearise about a flat surface with $\eta = 0$ and drop any term quadratic in η. The equation then becomes

$$\frac{\partial}{\partial t}\left(c^2\nabla^2\eta - f_0^2\eta\right) + c^2\beta\frac{\partial\eta}{\partial x} = 0 \ . \tag{3.121}$$

We see clearly that the term with β, which captures the variation of the Coriolis parameter, is driving the dynamics. If we look for plane wave solutions with $\eta = \eta_0 e^{i\mathbf{k}\cdot\mathbf{x}-i\omega t}$, we find the dispersion relation

$$\omega = -\beta c^2 \frac{k_x}{c^2 k^2 + f_0^2} \ . \tag{3.122}$$

When $\beta = 0$, this gives us the flat band $\omega = 0$ that corresponds to steady-state geostrophically balanced flows. But once we take into account the variation of the Coriolis parameter, these flows start to move. The resulting waves are called *Rossby waves*. The minus sign in (3.122) is important. It is telling us that long wavelength (small k) waves travel in a westward direction. Moreover, the dispersion relation for the Rossby waves is much flatter than those of the Poincaré waves and, correspondingly, their group velocity is much slower. These Rossby waves are the dominant motion of the ocean seen in satellite images, which beautifully reveal Rossby waves that take months, or even years, to cross the Pacific Ocean.

It's useful to summarise what we've seen here. The shallow water equations admit two classes of solutions: fast-moving Poincaré waves and slow-moving quasi-geostrophic flows, including Rossby waves. The magic of the quasi-geostrophic equation (3.120) is that it has successfully filtered out the fast-moving Poincaré waves, leaving us just with the slow-moving modes. It is what is referred to in other areas of physics as the "low energy (or frequency) effective field theory". Historically, the development of the quasi-geostrophic equation was crucial in developing successful weather prediction.

3.3.5 Equatorial Waves

Next, we ask: What happens when we sit at the equator? Here the Coriolis parameter (3.83) vanishes

$$f = 2\mathbf{\Omega}\cdot\hat{\mathbf{z}} = 0 \tag{3.123}$$

and one might naively think that there can't be any interesting physics due to the Coriolis force. In fact, things are more subtle and more interesting.

To find the more interesting physics, we look a little away from the equator. If we Taylor expand, the Coriolis parameter again becomes position-dependent

$$f(y) = \beta y \tag{3.124}$$

where, as before, the y-direction is north, and $y = 0$ corresponds to the equator. We can form a distance scale

$$L_{\text{eq}} = \sqrt{\frac{c}{\beta}} \tag{3.125}$$

with $c = \sqrt{gH}$. For the Earth's oceans, this is around $L_{\text{eq}} \approx 250$ km. It is somewhat larger for the atmosphere.

We again arrange the height perturbation $\eta(x, y, t)$ and the velocities $u(x, y, t)$ and $v(x, y, t)$ as a vector $\Psi(x, y, t)$ as in (3.108). This time we will look for solutions that are localised near the equator but propagate as waves in the zonal x-direction (i.e. east/west)

$$\Psi(x, y, t) = \begin{pmatrix} \sqrt{g/H}\eta \\ u \\ v \end{pmatrix} = \tilde{\Psi}(y) e^{ikx - i\omega t} . \tag{3.126}$$

The shallow water equations now become

$$\begin{pmatrix} 0 & ck & -ic\partial_y \\ ck & 0 & i\beta y \\ -ic\partial_y & -i\beta y & 0 \end{pmatrix} \tilde{\Psi} = \omega \tilde{\Psi} . \tag{3.127}$$

Again, we're looking for eigenmodes of this equation. As in the case when f was constant, we expect different branches.

Equatorial Kelvin Waves

To kick us off, there is a special solution to (3.127). This occurs when $v = 0$, so there is no velocity in the y-direction. The equations coming from the first two components of (3.127) are simply algebraic. They relate $\tilde{u} = (\omega/kH)\tilde{\eta}$ and result in the dispersion relation

$$\omega^2 = c^2 k^2 \quad \Longrightarrow \quad \omega = \pm ck . \tag{3.128}$$

We're left just with the third component of (3.127), which governs the profile of $\tilde{\eta}(y)$ and $\tilde{u}(y)$ in the y-direction,

$$\frac{c^2}{H}\frac{\partial \tilde{\eta}}{\partial y} = -\beta y \tilde{u} \quad \Longrightarrow \quad \frac{\partial \tilde{\eta}}{\partial y} = -\frac{\omega}{ck}\frac{y}{L_{\text{eq}}^2}\tilde{\eta} . \tag{3.129}$$

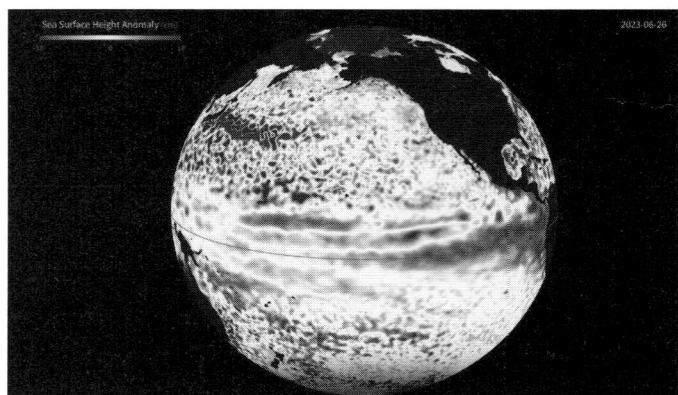

Fig. 3.7 A satellite image showing the height of the Pacific Ocean, measured to an accuracy of about 1 cm. The dark band in the Eastern Pacific is an equatorial Kelvin wave, hitting Ecuador. It then travels north, along the coast of Central America and (less visibly) south along the coastline of Peru. In the upper and lower parts of the ocean. Rossby waves drift slowly west. Credit: NASA Goddard Space Flight Center.

The key feature of the solution comes from that factor of ω/ck on the right-hand side. From the dispersion relation (3.128), this is either ± 1. However, the resulting solution is only normalisable if we take the positive sign

$$\omega = +ck \quad \Longrightarrow \quad \tilde{\eta} = \eta_0 e^{-y^2/2L_{\text{eq}}^2} \ . \tag{3.130}$$

The other choice of sign, with $\omega = -ck$, leads to a divergent solution $\tilde{\eta} \sim e^{+y^2}$, which is not physically permissible. The upshot is rather nice: we have waves at the equator that only travel in the positive x-direction. In other words, they only go east. In analogy with the coastal waves that we met in Section 3.3.3, these are known as *equatorial Kelvin waves*.

The equatorial Kelvin waves are close relatives of the coastal Kelvin waves. Suppose that the Earth had an ocean that fills the upper hemisphere and land in the lower hemisphere. Then there would be a coastal Kelvin wave, localised near the equator, propagating east to west, so the land is on its right. Now consider the inverted situation where the ocean fills the lower hemisphere, with land in the upper hemisphere. Again, there would be a coastal Kelvin wave but, because the Coriolis force has the opposite sign, it would again propagate east to west, this time with the land on its left. Now, finally, consider a world filled with water. Roughly, you can think of these two coastal waves combining to form the equatorial Kelvin wave. Both coastal and equatorial Kelvin waves can be seen in the satellite image shown in Figure 3.7.

Rossby, Poincaré, and Yanai Waves

Let's now return to the general problem of equatorial waves, given by the Schrödinger-like equation (3.127). The second component of (3.127) is algebraic and allows us to eliminate $\tilde{u}$ in favour of $\tilde{v}$ and $\tilde{\eta}$. This results in a pair of coupled, first order differential equations

$$-i\left(\frac{\partial}{\partial y} - \frac{\beta k y}{\omega}\right)\tilde{v} = \frac{1}{H}\left(\omega - \frac{c^2 k^2}{\omega}\right)\tilde{\eta}$$

$$-i\left(\frac{\partial}{\partial y} + \frac{\beta k y}{\omega}\right)\tilde{\eta} = \frac{H}{c^2}\left(\omega - \frac{\beta^2 y^2}{\omega}\right)\tilde{v} \ . \tag{3.131}$$

We can eliminate $\tilde{\eta}$ to manipulate this into a second order differential equation for $\tilde{v}$ alone. After a little bit of algebra, this is

$$\left(-c^2\frac{\partial^2}{\partial y^2} + \beta^2 y^2\right)\tilde{v} = \left(\omega^2 - c^2 k^2 - \frac{\beta c^2 k}{\omega}\right)\tilde{v} \ . \tag{3.132}$$

But this is a very famous equation: it is the Schrödinger equation for the harmonic oscillator. In that context, we write

$$\left(-\frac{\hbar^2}{2m}\frac{\partial^2}{\partial y^2} + \frac{1}{2}m\bar{\omega}^2 y^2\right)\tilde{v} = E_n\tilde{v} \tag{3.133}$$

where m is the mass of the particle and $\bar{\omega}$ is the frequency of the harmonic oscillator (not to be confused with ω, the frequency of our waves that we're trying to determine). We will not solve this equation here, but you can find the details (lots and lots of details!) in Volume 3 on Quantum Mechanics. In both the fluid and quantum contexts, we are interested in normalisable solutions which means that we can import the results directly. The velocity $\tilde{v}(y)$ is given by Hermite polynomials. More importantly, the energies of the harmonic oscillator are, famously,

$$E_n = \hbar\bar{\omega}\left(\frac{1}{2} + n\right) \quad \text{with} \quad n = 0, 1, 2, \dots \ . \tag{3.134}$$

Translating back into the variables of our equatorial waves, the dispersion relation is given by

$$\omega^3 - \omega\left(c^2 k^2 + \beta c(1 + 2n)\right) - \beta c^2 k = 0 \quad \text{with} \quad n = 0, 1, 2, \dots \ . \tag{3.135}$$

We'll now look at these for different n. We will find that the $n = 0$ waves are somewhat different from the $n \geq 1$ waves.

Let's start with the $n = 0$ waves. First note that, in this case, (3.135) has a root $\omega = -ck$. Naively, this looks like a wave moving in the opposite direction to the Kelvin wave. But it is a spurious solution. This is because, although $\tilde{v}(y)$ is normalisable, when we plug this solution into (3.131) we find

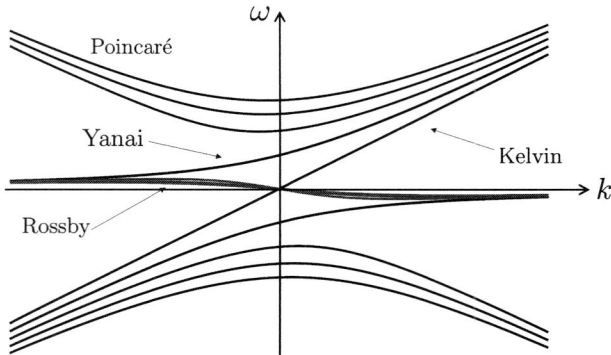

Fig. 3.8 The bestiary of equatorial waves includes the chiral Kelvin and Yanai waves, together with a discretum of Poincaré and Rossby waves.

that $\tilde{\eta}(y)$ is non-normalisable: it has a piece that diverges as $\tilde{\eta} \sim e^{+y^2/2L_{\mathrm{eq}}^2}$. So this solution should be thrown out. It turns out that it's the only spurious solution and all others are fine.

If we factor out the spurious $\omega = -ck$ solution, then we find a single $n = 0$ wave, with dispersion relation

$$\omega = \frac{ck}{2} \pm \frac{1}{2}\sqrt{c^2k^2 + 4\beta c} \,. \tag{3.136}$$

This too is a chiral wave. It is known as the *Yanai wave* and is generally accepted to be a little weird. To illustrate this, take the positive root in (3.136). In the limit $k \to \infty$, the Yanai wave has the same dispersion relation $\omega \sim +ck$ as the equatorial Kelvin wave, while in the opposite limit $k \to -\infty$, this wave has $\omega \to 0$. That's very odd! Because the Yanai wave interpolates between the Poincaré gravity waves and the Rossby waves, it is sometimes referred to as a *mixed Rossby-gravity wave*. The velocity profile is Gaussian around the equator, with $\tilde{v} \sim e^{-y^2/2L_{\mathrm{eq}}^2}$.

For $n \geq 1$, the general shape of the dispersion relation takes the same form. There are three branches of modes, which are modified versions of the dispersion relations (3.110) that we saw when f is constant. We again see the dispersion relations corresponding to Poincaré waves, with their characteristic gapped spectrum, asymptoting to $\omega \to \pm ck$.

In addition, we see that our flat band, which previously had $\omega = 0$, is again deformed. Now, it is no longer flat, but asymptotes to $\omega \to -\beta/k$ for large $|k|$. These are equatorial *Rossby waves*. The various modes for $n = 0, 1, 2, 3$, together with the Kelvin wave, are shown in Figure 3.8. As we noted previously, the dispersion relation for the Rossby waves is much flatter than those of the Poincaré waves and, correspondingly, the group velocity of the Rossby waves will be much slower.

3.3.6 Chiral Waves Are Topologically Protected

As we mentioned previously, chiral waves appear in various condensed matter systems. The most familiar example is the quantum Hall effect where a sample of electrons in a magnetic field has chiral modes propagating on its edge.

In the context of condensed matter, it turns out that the presence of chiral edge modes can often be traced to some interesting topological features of the system, an observation that led to many new developments in the field. The purpose of this section is to point out that, rather wonderfully, the same is true for chiral waves in fluids. I should warn you that this section is something of a departure from the rest of this book and the motivation is, in part, simply to illustrate the unity of physics. More details of the role that topology plays in electron systems can be found in the book on Condensed Matter.

We will describe the topology associated to equatorial chiral modes. (There is a similar, but more complicated, story for coastal Kelvin waves.) The idea is that the existence of the two chiral modes – Kelvin and Yanai – is a direct consequence of topology. But this topology isn't to be found in real space – instead, it's topology in the space of wavevectors. In quantum mechanics, this is usually said to be topology in momentum space.

To set the scene, we will return to the case of constant Coriolis parameter f. As we've seen in (3.110), there are three bands with dispersion

$$\omega = \pm\sqrt{c^2 k^2 + f^2} \quad \text{and} \quad \omega = 0 \ . \tag{3.137}$$

The resulting bands are shown in Figure 3.9 for three cases: $f > 0$, $f = 0$, and $f < 0$. For $f \neq 0$, there is a gap between the geostrophic flat band and the Poincaré waves. This gap closes when $f = 0$ and this is closely related to the existence of the chiral equatorial waves.

The question that the topological approach addresses is: How robust is this situation? Could we, for example, add some further parameters to the problem so that, as we vary f from positive to negative, the gap never closes? Topology tells us that the answer to this is no. There must always be some point that looks like the $f = 0$ figure where the gap closes.

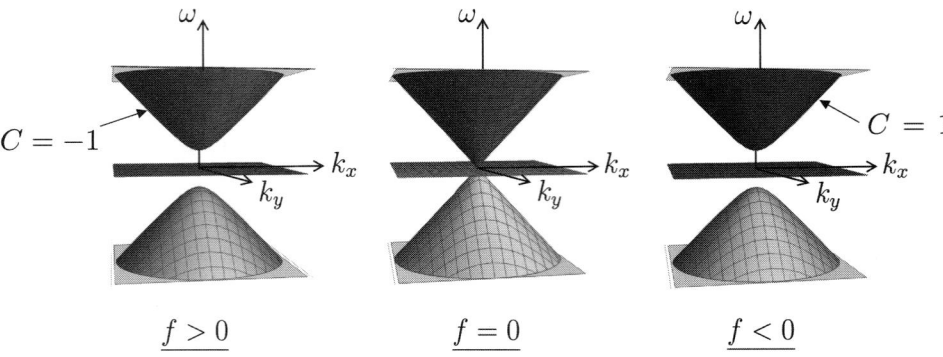

The band structure as a function of constant Coriolis parameter f. As f jumps sign, so too does the Chern number C.

The reason for this is that there is a subtle difference between the $f > 0$ and $f < 0$ situations. This difference doesn't show up in dispersion relations (3.137) which are clearly symmetric under $f \to -f$. Instead, we have to look more closely at what's going on in each band.

Recall from (3.109) that the frequencies arise as the solution to the following eigenvalue problem

$$\begin{pmatrix} 0 & ck_x & ck_y \\ ck_x & 0 & if \\ ck_y & -if & 0 \end{pmatrix} \tilde{\Psi} = \omega \tilde{\Psi} \ . \tag{3.138}$$

We will focus on the positive frequency band of Poincaré waves, with

$$\omega(\mathbf{k}) = +\sqrt{c^2 k^2 + f^2} \ . \tag{3.139}$$

As we've already mentioned, the eigenvalues are clearly invariant under $f \to -f$. To see the difference between $+f$ and $-f$ we need to look at the eigenvector. The normalised eigenvector is given by

$$\tilde{\Psi}_+(\mathbf{k}, f) = \frac{1}{\sqrt{2\omega^2 \mathbf{k}^2}} \begin{pmatrix} ck^2 \\ k_x \omega + if k_y \\ k_y \omega - if k_x \end{pmatrix} \ . \tag{3.140}$$

The eigenvector depends on the wavevector $\mathbf{k}$. This means that as we move around momentum space, labelled by $\mathbf{k} \in \mathbb{R}^2$, the eigenvector Ψ_+ evolves in $\mathbb{C}^3$. The key idea is that as we explore all of momentum space, the eigenvector may twist within the larger space $\mathbb{C}^3$. This twist is where topology enters the story.

The fact that eigenvectors twist and turn in a larger space is more familiar in the context of quantum mechanics where it goes by the name of *Berry phase*. (This topic was described in Volume 3 on Quantum Mechanics.) We will not review this in detail, but simply state how to characterise the topology of the eigenvector. First, given an eigenvector Ψ_+ we define the *Berry connection*

$$A_a(\mathbf{k}) = -i\tilde{\Psi}_+^\dagger \frac{\partial \tilde{\Psi}_+}{\partial k^a} \quad \text{with} \quad a = 1, 2 \ . \tag{3.141}$$

A short calculation shows that, for the eigenvector (3.140), the Berry connection is

$$A_a = -\frac{f}{\omega \mathbf{k}^2} (k_y, -k_x) \ . \tag{3.142}$$

The Berry connection has the same mathematical structure as the gauge potential in electromagnetism. In particular, as the next step we compute something akin to the magnetic field,

$$B = \partial_1 A_2 - \partial_2 A_1 = -\frac{c^2 f}{(f^2 + c^2 \mathbf{k}^2)^{3/2}} \ . \tag{3.143}$$

This is known as the *Berry curvature*. Finally, we integrate this curvature over momentum space to get an object known as the *Chern number*, which we calculate to be

$$C = \frac{1}{2\pi} \int_{\mathbb{R}^2} d^2k \ B = -\text{sign}[f] \ . \tag{3.144}$$

Note that, as promised, the Chern number distinguishes between f positive and f negative: we have $C = -1$ for $f > 0$ and $C = +1$ for $f < 0$.

At this stage the argument becomes slightly delicate. When the Chern number is computed by integrating over a compact space (i.e. one which doesn't stretch to infinity), then there is a mathematical result that says

$$C \in \mathbb{Z} \ . \tag{3.145}$$

In physics, this is usually referred to as Dirac quantisation. The fact that C is integer valued is important. It is telling us that we have some discrete way of characterising the system, even though the underlying fluids are continuous. This is the essence of topology.

However, things are not so straightforward for our fluids because the integral (3.144) is not over a compact space but instead over $\mathbb{R}^2$. (This is not a problem in condensed matter systems because the underlying spatial lattice means that momentum lives in a compact Brillouin zone.) And there's no mathematical theorem that says such an integral should be integer valued. Indeed, if you integrate the magnetic flux through a solenoid then you can get anything at all.

There are a couple of ways around this and we will take the cheapest. Note that asymptotically, as $|\mathbf{k}| \to \infty$, we have $A_a \to 0$. In fact, more importantly, we have $\oint A_a \, dk^a \to 0$ as the integration curve is taken out to infinity. This is a property of short wavelength modes and so should hold regardless of any deformation of the system which doesn't affect arbitrarily short wavelengths. So we insist that A_a is trivial asymptotically and this allows us to effectively compactify the problem, by adding a point at infinity and viewing $\mathbb{R}^2 + \{\infty\} = \mathbf{S}^2$. Correspondingly, we learn that the Chern number C – which is clearly an integer in (3.144) – should remain an integer no matter how we deform the system.

Now we're in business. For $f > 0$, we have $C = -1$. This is shown on the left-hand side of Figure 3.9. But because C is restricted to be an integer, it can't change as we vary parameters. The only exception to this is if the gap to some other band closes, because then the eigenvector Ψ_+ becomes degenerate with another eigenvector and the calculation above breaks down. But the Chern number for the upper band when $f < 0$ is $C = +1$, depicted on the right-hand side of Figure 3.9. So we learn that there is no path in any enlarged parameter space that takes us from $f < 0$ to $f > 0$ without closing the gap.

In fact, there's more to learn from this. The number of chiral edge modes that appear as we vary from $f < 0$ to $f > 0$ is given by the difference in the Chern numbers. In other words, the number of chiral waves is necessarily

$$\left| C[f > 0] - C[f < 0] \right| = 2 \, . \tag{3.146}$$

And this is indeed what we find, with two modes, the Kelvin and Yanai waves, appearing at the equator.

As we mentioned previously, the kind of calculation that we've performed above underlies various properties of materials, notably quantum Hall states and topological insulators. We will develop this story further in the book on Condensed Matter.

3.4 Sound Waves

Throughout this book, we've focussed on incompressible fluids obeying $\nabla \cdot \mathbf{u} = 0$. It is now time to abandon this assumption. Instead, we want to ask: What new physics arises when the density of a fluid changes? The answer, as we shall see, is sound.

Apart from its inherent interest, this question forces us to re-examine the fundamental equations of fluid mechanics. So far, we have mass conservation

$$\frac{\partial \rho}{\partial t} + \nabla \cdot (\rho \mathbf{u}) = 0 \tag{3.147}$$

and the Navier–Stokes equation

$$\rho \left(\frac{\partial \mathbf{u}}{\partial t} + \mathbf{u} \cdot \nabla \mathbf{u} \right) = -\nabla P + \mu \nabla^2 \mathbf{u} + \left(\frac{\mu}{3} + \zeta \right) \nabla (\nabla \cdot \mathbf{u}) \tag{3.148}$$

where, because the fluid is now compressible, there's a second viscosity term that can appear on the right-hand side. This comes with a new coefficient ζ, the *bulk viscosity*, in addition to μ, the dynamic shear viscosity. (In fact, we'll largely ignore the effects of both viscosities in this section, but it's worth keeping them in play while we return to the fundamentals.)

When the density was constant, these equations were all we needed. They are four equations that govern four independent, dynamical fields, $\mathbf{u}(\mathbf{x}, t)$ and $P(\mathbf{x}, t)$. However, when $\rho = \rho(\mathbf{x}, t)$ is also a dynamical field we need a further equation before we can get going. As we now explain, this additional equation is dictated by thermodynamics and forces us to think about the temperature of the fluid.

3.4.1 Compressible Fluids and the Equation of State

The *equation of state* is a relationship between the pressure P, the volume V, and the temperature T of a fluid. The simplest such example is very familiar: it is the ideal gas law

$$PV = Nk_B T \ . \tag{3.149}$$

Here N is the number of particles in the gas and k_B is Boltzmann's constant. The ideal gas equation describes a gas of non-interacting particles. Note that the "ideal" in "ideal gas" means something different from the "ideal" in "ideal fluid"! The latter just refers to something that obeys the Euler equation with no viscosity. Fluid dynamicists sometimes refer to the ideal gas as a "perfect gas" to avoid this confusion. The relation between these two ideals will be clarified in Chapter 6.

Other equations of state can be more complicated, capturing some internal interactions between the constituent molecules. For example, a simple generalisation of the ideal gas law, which holds for dilute gases, is the van der Waals equation

$$\frac{PV}{Nk_BT} = 1 - \frac{N}{V}\left(\frac{a}{k_BT} - b\right) \tag{3.150}$$

where a and b are two constants that characterise the interactions of the gas. You can find derivations of both these equations of state in the book on Statistical Physics.

For liquids, equations of state become rather complicated. There are no simple formulae that can be derived from first principles; instead, the equations of state are empirical formulae, typically with many variables chosen to match results from experiment. For that reason, much of what we say in this section will apply most straightforwardly to gases, rather than liquids.

We can make contact with the variables of fluid mechanics if we replace the volume with the density ρ of the fluid,

$$\rho = \frac{Nm}{V} \tag{3.151}$$

where m is the mass of each individual particle in the fluid. Then the ideal gas law becomes

$$P = \frac{\rho k_BT}{m} . \tag{3.152}$$

When we first meet the equation of state, we think of P, $\rho \sim 1/V$, and T as numbers that describe the global, equilibrium properties of the system. However, the whole point of fluid mechanics is that we can understand what happens as we move away from equilibrium. To achieve this, we assume that locally the system is still described by P, ρ, and T but these are now dynamical fields whose values can vary in space and in time. The equation of state now gives a *local* relationship between these quantities. So, for example, an ideal gas obeys

$$P(\mathbf{x}, t) = \frac{\rho(\mathbf{x}, t)k_BT(\mathbf{x}, t)}{m} . \tag{3.153}$$

The existence of the equation of state tells us why we need to start thinking about temperature. If the pressure P and density ρ are changing, then so too is the temperature T. Indeed, this is true even when ρ is constant, but throughout this book we have implicitly assumed that $T(\mathbf{x}, t)$ simply tracks the pressure $P(\mathbf{x}, t)$. Now, however, we need to think more carefully about how T changes.

3.4.2 Some Thermodynamics

The correct way to proceed is to derive an equation of motion for the temperature $T(\mathbf{x}, t)$. For now, however, we will take something of a shortcut which requires a brief foray into the laws of thermodynamics. (A much fuller discussion can be found in the book on Statistical Physics.) We will then describe the better approach in Section 3.4.4 and, more fully, in Chapter 6.

The shortcut that we have in mind is called the *adiabatic approximation.* Heuristically, this means that we assume that the time scale over which the fluid moves is much shorter than the times scale of heat diffusion within the fluid. Mathematically, it means that we assume a quantity called *entropy* is conserved. The purpose of this section is to lead us to the following, simple result: under the adiabatic approximation

$$\frac{P}{\rho^\gamma} = \text{constant} \tag{3.154}$$

where γ is the ratio of heat capacities $\gamma = C_P/C_V$ and will be defined below. For air, $\gamma \approx 1.4$. Starting in Section 3.4.3 we'll then make use of this result to study the properties of sound waves.

For now, we'll revert to the older setting where P, V, and T are just numbers that characterise the global property of an equilibrium system. Given any two of these, the third is fixed by the equation of state. In addition, there are a number of interesting functions of these variables that capture certain aspects of the gas. Here we'll need two such functions. The first is familiar: the energy of the gas can be viewed as a function $E = E(V, T)$. The second function is more subtle. It is called *entropy* and is denoted $S(V, T)$. As we proceed, we'll give expressions for both the energy and the entropy of an ideal gas. But, first, we describe some of their properties.

The *first law of thermodynamics* says that the energy E of a gas can change in one of two ways: either you can do work on the gas by squeezing it, or you can add heat to the gas by putting it in contact with something at a different temperature. The infinitesimal change in energy is then

$$dE = \delta W + \delta Q \tag{3.155}$$

where δW is the work added, and δQ is the heat added. As we mentioned above, the energy is a function of the system, meaning that if you know, say, V and T, then the energy $E(V, T)$ is fixed. In contrast, both heat and work are things that you do *to* the gas. There's no sense in which we can talk about the "work" contained in a gas or the "heat" contained a gas and, correspondingly there's no function "$W(V, T)$" or "$Q(V, T)$". Instead, work and heat are means of transporting energy. Roughly speaking, this is the reason that we write the terms on the right-hand side as δQ and δW instead of dQ and dW.

However, it should be possible to describe the effect of both the work done and the heat added in terms of changes to the state of the system. For the work done, this is straightforward. If the fluid has pressure P and we squeeze it by changing its volume, then the infinitesimal work done is

$$\delta W = -PdV \ . \tag{3.156}$$

But what is the analogous statement for the heat added to a gas? This is where the idea of entropy comes into play. There is a function of the gas, $S(V, T)$, known as *entropy*, so that the heat added can be written as

$$\delta Q = TdS \ . \tag{3.157}$$

This should be viewed as a definition of entropy, originally due to Clausius. This definition is really the essence of the second law of thermodynamics which is more commonly phrased as the statement that entropy can never decrease. Subsequently, Boltzmann understood entropy in terms of counting microscopic arrangements of atoms. A large part of the book on Statistical Physics is devoted to understanding why these two definitions are actually equivalent. For our purposes, we'll only need the definition (3.157). With this definition of entropy, the first law can be written as

$$dE = -PdV + TdS \ . \tag{3.158}$$

This suggests that the most natural way to think of energy is as a function $E(V, S)$.

An *adiabatic process* is one in which the system changes without adding any heat. This means that $\delta Q = 0$ and so entropy remains constant. It turns out that sound waves are an example of an adiabatic process, at least if we ignore viscosity, which we can do to a good approximation. You might think that this means we can simply ignore the heat term in the first law. Sadly, that's not quite true! We need to understand a little better what heat actually is before we can discard it.

Next, we need the idea of heat capacity. This is, as the name suggests, the capacity of the gas to absorb heat, measured by how much the temperature changes. The slightly subtle point is that you must specify what you are holding fixed when you do this experiment. You could, for example, hold the volume fixed. The corresponding heat capacity C_V is defined by

$$C_V = T \left.\frac{\partial S}{\partial T}\right|_V = \left.\frac{\partial E}{\partial T}\right|_V \tag{3.159}$$

where, in the second equality, we've used the first law of thermodynamics (3.158) where the $-PdV$ term doesn't contribute precisely because we're holding the volume fixed. Alternatively, you could add heat keeping the pressure fixed, rather than the volume. Again, using the first law, we have

$$C_P = T \left.\frac{\partial S}{\partial T}\right|_P = \left.\frac{\partial E}{\partial T}\right|_P + P \left.\frac{\partial V}{\partial T}\right|_P . \tag{3.160}$$

In this case, the temperature is expected to rise less because the energy from the heat must now also do work expanding the volume of the gas. Correspondingly, we expect $C_P > C_V$. We often talk about the *specific heats*, which is the heat capacity per unit volume: $c_V = C_V/V$ and $c_P = C_P/V$.

The Ideal Gas

So far, our discussion has been general. To make progress, we now focus on a specific system: the ideal gas, with the familiar equation of state

$$PV = Nk_BT . \tag{3.161}$$

This is a good approximation for dilute gases, like the air in the room you're sitting in. It's not a good approximation for liquids.

The final fact that we need is known as *equipartition*. It is the statement that, at temperature T, the energy of each microscopic degree of freedom is given by $\frac{1}{2}k_BT$. This means that if we have a gas of N *monatomic* particles, meaning that each particle is itself a structureless object, then

$$E = \frac{3}{2}Nk_BT \quad \text{for monatomic gases} \tag{3.162}$$

where the $\frac{3}{2}$ comes because each particle can move in three dimensions, each direction contributing $\frac{1}{2}$.

However, if the particles comprising the gas have additional internal degrees of freedom then equipartition ensures that these too contribute to the energy. For example, a diatomic molecule can be viewed as a dumbbell-like object. It has three translational degrees of freedom, but also two rotational degrees of freedom. (The rotation of the dumbell about its axis of symmetry doesn't count, rather wonderfully because of a quantum mechanical effect that will be described in the book on Statistical Physics.) This means that the energy is

$$E = \frac{5}{2} N k_B T \quad \text{for diatomic gases .} \tag{3.163}$$

Air is mostly N_2 and O_2, both of which are diatomic molecules, so this is the energy of air.

We can now compute the heat capacities for the different ideal gases. We have

$$C_V = \frac{3}{2} N k_B \quad \text{and} \quad C_P = \frac{5}{2} N k_B \quad \text{for monatomic gases ,} \tag{3.164}$$

and

$$C_V = \frac{5}{2} N k_B \quad \text{and} \quad C_P = \frac{7}{2} N k_B \quad \text{for diatomic gases .} \tag{3.165}$$

Note that, in both cases, $C_P - C_V = N k_B$, which follows from (3.160), together with the equation of state. It will be useful to define the ratio of the heat capacities

$$\gamma = \frac{C_P}{C_V} = \begin{cases} 5/3 & \text{for monatomic gases} \\ 7/5 & \text{for diatomic gases} \end{cases} . \tag{3.166}$$

This is where we get the statement that $\gamma \approx 1.4$ for air.

Finally, we can use the technology developed above to compute the entropy of an ideal gas. We start from the first law (3.158), now written as

$$dS = \frac{1}{T} dE + \frac{P}{T} dV = \frac{C_V}{T} dT + \frac{N k_B}{V} dV . \tag{3.167}$$

We now replace $N k_B = C_P - C_V$ and integrate to get

$$S = C_V \log \left(\frac{T}{T_0} \right) + (C_P - C_V) \log \left(\frac{V}{V_0} \right)$$

$$= C_V \log \left(\frac{T}{T_0} \left(\frac{V}{V_0} \right)^{\gamma - 1} \right) \tag{3.168}$$

$$= C_V \log \left(\frac{P}{P_0} \left(\frac{V}{V_0} \right)^{\gamma} \right) . \tag{3.169}$$

This means that if you want to change the gas in some way so that the entropy remains constant, then the pressure and volume must themselves change so that PV^γ is constant. Or, written in terms of the density $\rho \sim 1/V$,

$$\frac{P}{\rho^\gamma} = \text{constant} . \tag{3.170}$$

This is the result (3.154) that we advertised at the beginning of this section. When we study sound waves, this is the equation we will need. We can also use the equation of state for an ideal gas to write this expression in terms of other variables. In particular, from the line (3.168), we see that we can equivalently require

$$TV^{\gamma-1} = \text{constant} . \tag{3.171}$$

This form will be useful in Section 3.4.4 where we look more closely at the physics of heat transport in a fluid.

3.4.3 The Equations for Sound Waves

Finally, after that long preamble, we can turn to the subject of interest: sound waves. We will initially ignore viscosity (remedying this in Section 3.4.5) and work with the Euler equation

$$\rho \left(\frac{\partial \mathbf{u}}{\partial t} + \mathbf{u} \cdot \nabla \mathbf{u} \right) = -\nabla P . \tag{3.172}$$

Our starting point is the simplest possible solution to the Euler equation: a stationary fluid, with constant density and pressure

$$\mathbf{u} = 0 , \quad \rho = \rho_0 , \quad P = P_0 . \tag{3.173}$$

We then study small perturbations about this background. We write

$$\rho = \rho_0 + \tilde{\rho} \quad \text{and} \quad P = P_0 + \tilde{P} \tag{3.174}$$

with the perturbations small, meaning $\tilde{\rho} \ll \rho_0$ and $\tilde{P} \ll P_0$. We will also take $\mathbf{u}$ to be small, in the sense that we keep terms only linear in $\mathbf{u}$, $\tilde{\rho}$, and $\tilde{P}$. The linearised Euler equation then becomes

$$\rho_0 \frac{\partial \mathbf{u}}{\partial t} = -\nabla \tilde{P} . \tag{3.175}$$

We augment this with the equation of mass conservation which, after linearisation, becomes

$$\frac{\partial \tilde{\rho}}{\partial t} + \rho_0 \nabla \cdot \mathbf{u} = 0 . \tag{3.176}$$

We can combine these by taking the gradient ∇ of the first and the time derivative of the second. This gives

$$\frac{\partial^2 \tilde{\rho}}{\partial t^2} - \nabla^2 \tilde{P} = 0 \ . \tag{3.177}$$

At this point, we need to invoke the adiabatic approximation for an ideal gas (3.170) which, after linearising, becomes

$$\frac{(P_0 + \tilde{P})}{(\rho_0 + \tilde{\rho})^\gamma} = \text{constant} \quad \Longrightarrow \quad \tilde{P} - \frac{P_0}{\rho_0}\gamma\tilde{\rho} = 0 \ . \tag{3.178}$$

The equation (3.177) then becomes

$$\frac{\partial^2 \tilde{\rho}}{\partial t^2} - c_s^2 \nabla^2 \tilde{\rho} = 0 \ . \tag{3.179}$$

This is the *wave equation.* As we will soon see, it describes waves propagating with speed c_s which, for us, is the speed of sound given by

$$c_s = \sqrt{\frac{\gamma P_0}{\rho_0}} \ . \tag{3.180}$$

For an ideal gas, the equation of state (3.152) relates this to the temperature T_0 of the background fluid, and the mass m of the constituent particles

$$c_s = \sqrt{\frac{\gamma k_B T_0}{m}} \ . \tag{3.181}$$

We see that the speed of sound depends on the temperature. For the air at $20°$ the speed is $c_s \approx 340 \text{ m s}^{-1}$. This was first measured by Newton by clapping his hands in Nevile's Court, Trinity College, Cambridge. (He got a value around 300 m s^{-1}.)

A General Fluid

The equation (3.177) holds for any fluid while, for the subsequent derivation of the wave equation, we restricted to the ideal gas. But we get the same wave equation for any equation of state; it's just the speed of sound that changes. This allows us to get a general expression for the sound speed for any fluid, including liquids.

It's useful to think of the pressure as a function

$$P = P(\rho, S) \tag{3.182}$$

rather than the more familiar $P = P(V, T)$. For an adiabatic process, like sound waves, the entropy is constant so if we Taylor expand the pressure about $\rho = \rho_0 + \tilde{\rho}$, we have

$$P(\rho_0 + \tilde{\rho}) = P(\rho_0) + \left.\frac{\partial P}{\partial \rho}\right|_S \tilde{\rho} = P_0 + \tilde{P} \ . \tag{3.183}$$

The steps from (3.177) to (3.179) then proceed, but with the sound speed given by

$$c_s = \sqrt{\left.\frac{\partial P}{\partial \rho}\right|_S} \ . \tag{3.184}$$

Both left- and right-hand sides can be measured experimentally. The derivative on the right-hand side is usually expressed in terms of the *bulk modulus*, defined to be $K = \rho\,\partial P/\partial\rho$. For water at $20°$ this is $K \approx 200 \ \mathrm{N\,m^{-2}}$. It's much higher than the corresponding value for gases, reflecting the fact that it is more difficult to squeeze water than air. The density of water is $\rho_0 \approx 10^3 \ \mathrm{kg\,m^{-3}}$. The speed of sound in water is then much higher than in air, with $c_s \approx 1500 \ \mathrm{m\,s^{-1}}$.

Sound Waves Are Longitudinal

The wave equation (3.179) is solved by any Fourier mode

$$\tilde{\rho}(\mathbf{x}, t) = \hat{\rho}\, e^{i\mathbf{k}\cdot\mathbf{x} - i\omega t} \ . \tag{3.185}$$

Here $\hat{\rho}$ is the constant amplitude of the wave. In the exponent, ω is the frequency and $\mathbf{k}$ is the wavevector which points in the direction of propagation. The two are related by the dispersion relation

$$\omega = c_s|\mathbf{k}| \ . \tag{3.186}$$

This is now a dispersion relation that doesn't disperse, in the sense that all wavelengths propagate with the same speed. As we've seen, this contrasts with the surface waves of Section 3.1. Because the wave equation is linear, we can combine many Fourier modes to make a wavepacket. If this is made from wavevectors $\mathbf{k}$ that all point in the same direction, then the wavepacket will keep its shape as it moves. We can also see this directly from the wave equation. If the wave is moving in the x-direction, then the wave equation is solved by any function of the form

$$\tilde{\rho} = F(t - x/c_s) + G(t + x/c_s) \ . \tag{3.187}$$

Here F and G are the profiles of two wavepackets, moving to the right and left respectively.

We can reconstruct the pressure and velocity oscillations from our original, first order equations. The pressure perturbations are simply given by $\tilde{P} = c_s^2 \tilde{\rho}$. From (3.176) we have

$$\mathbf{u}(\mathbf{x}, t) = \frac{\hat{\mathbf{k}}}{\rho_0 c_s} \tilde{P}(\mathbf{x}, t) = \frac{c_s \hat{\mathbf{k}}}{\rho_0} \tilde{\rho}(\mathbf{x}, t) \ . \tag{3.188}$$

The oscillations of the fluid velocity and the pressure are all in phase with the density. The velocity oscillations are also parallel to the direction $\mathbf{k}$ in which the wave travels. Such waves are called *longitudinal*.

Spherically Symmetric Waves

Although we can construct any solution from the Fourier modes (3.185), that's often not the best way to proceed. For example, if we have some localised source, which, for convenience, we will assume is spherically symmetric, then it's clear that we are best served by working in spherical polar coordinates. Ignoring the angular directions, the wave equation becomes

$$\frac{\partial^2 \tilde{\rho}}{\partial t^2} - c_s^2 \nabla^2 \tilde{\rho} = 0 \quad \Longrightarrow \quad \frac{\partial^2 (r\tilde{\rho})}{\partial t^2} - c_s^2 \frac{\partial^2 (r\tilde{\rho})}{\partial r^2} = 0 \ . \tag{3.189}$$

This is now a 1d wave equation. It is solved, analogously to (3.187), by any two functions

$$\tilde{\rho}(r, t) = \frac{1}{4\pi r} \left[F(t - r/c_s) + G(t + r/c_s) \right] \ . \tag{3.190}$$

The factor of 4π is there for the same reason that it sits in the Coulomb force law: it's a convenience that reflects the fact that subsequent equations will be multiplied by the area of a unit sphere. The function F describes the outgoing wave, while G describes the incoming wave. In many situations, there's no wave coming in from infinity so we set $G = 0$. This is the choice we make here.

The associated velocity field is most simply computed from (3.175) using $\tilde{P} = c_s^2 \tilde{\rho}$. To write down the solution, we need to integrate the wave profile. We write

$$F(t - r/c_s) = \frac{\dot{Q}(t - r/c_s)}{c_s^2} \ . \tag{3.191}$$

In spherical polars, we then have

$$\nabla \tilde{P} = -\frac{1}{4\pi} \left[\frac{\dot{Q}(t - r/c_s)}{r^2} + \frac{\ddot{Q}(t - r/c_s)}{c_s r} \right] \hat{\mathbf{r}} \tag{3.192}$$

and, comparing to (3.175), the velocity field is radial, with

$$\mathbf{u}(r,t) = \frac{1}{4\pi\rho_0}\left[\frac{Q(t-r/c_s)}{r^2} + \frac{\dot{Q}(t-r/c_s)}{c_s r}\right]\hat{\mathbf{r}} . \tag{3.193}$$

Close to the source, the first term dominates; far away, the second term dominates. The quantity Q has the interpretation as the mass flux emerging from the origin.

As an example, consider the sound waves generated by a pulsating sphere of radius a. We'll take this sphere to beat in and out with frequency ω and amplitude ϵ, so the radius changes with time as

$$R(t) = a + \epsilon e^{i\omega t} \quad\Longrightarrow\quad \dot{R} = i\omega\epsilon e^{i\omega t} . \tag{3.194}$$

The solution must take the form (3.193) for some $Q(t) = Ae^{i\omega t}$. This means that

$$\mathbf{u}(r,t) = \frac{A}{4\pi\rho_0}\left[\frac{1}{r^2} + \frac{i\omega}{c_s r}\right]e^{i\omega(t-r/c_s)}\hat{\mathbf{r}}. \tag{3.195}$$

This is subject to the requirement that the fluid velocity matches that of the sphere on its surface, so that

$$\mathbf{u}(R(t),t) = \dot{R}\,\hat{\mathbf{r}} \quad\Longrightarrow\quad \mathbf{u}(a,t) + \frac{\partial\mathbf{u}}{\partial r}\epsilon e^{i\omega t} + \ldots = i\omega\epsilon e^{i\omega t}\hat{\mathbf{r}} . \tag{3.196}$$

Since $\mathbf{u} \sim \mathcal{O}(\omega\epsilon)$, the second term in the above expression is lower order and it will suffice to set

$$\mathbf{u}(a,t) = i\omega\epsilon e^{i\omega t}\hat{\mathbf{r}} \quad\Longrightarrow\quad \frac{A}{4\pi\rho_0 a^2}\left[1 + \frac{i\omega a}{c_s}\right]e^{-i\omega a/c_s} = i\omega\epsilon \tag{3.197}$$

which fixes the overall coefficient A.

3.4.4 Briefly, Heat Transport

All the results on sound waves that we derived above relied on the adiabatic approximation which, for an ideal gas, is given by (3.170) or, equivalently, (3.171)

$$\frac{P}{\rho^\gamma} \sim TV^{\gamma-1} = \text{constant} . \tag{3.198}$$

There is a more sophisticated way of viewing things which involves working directly with the temperature field $T(\mathbf{x},t)$ and understanding its dynamics. This has the advantage that it allows us to go beyond the adiabatic approximation.

First, we'll see how to reproduce the results above. For an ideal gas, the adiabatic dynamics of the temperature field $T(\mathbf{x}, t)$ is governed by the transport equation

$$\left(\frac{\partial}{\partial t} + \mathbf{u} \cdot \nabla\right) T + (\gamma - 1)T \nabla \cdot \mathbf{u} = 0 \ . \qquad (3.199)$$

We will derive this equation in Chapter 6 where we discuss kinetic theory. (The result is derived in equation (6.167).) For now, we merely show that this heat equation reproduces our previous expression (3.198) for adiabatic changes.

To do this, we need to think afresh about what we mean by an adiabatic process. The novelty is that we're now once again dealing with fields, with variables like $T(\mathbf{x}, t)$ and $S(\mathbf{x}, t)$ functions of space and time. The statement that entropy is constant in (3.198) is $TV^{\gamma-1} \sim T\rho^{1-\gamma} = $ constant, and this should now be interpreted as a local statement, meaning

$$\frac{D}{Dt}(T\rho^{1-\gamma}) = 0 \ . \qquad (3.200)$$

This is the local statement of adiabatic evolution: as you follow the local entropy with the flow, it remains unchanged. We then have the following result.

Claim: The heat transport equation (3.199) implies adiabatic evolution in the form (3.200).

Proof: Both T and ρ are now fields, and we can see how these evolve within the fluid. We have

$$\frac{D}{Dt}(T\rho^{1-\gamma}) = \left(\frac{\partial}{\partial t} + \mathbf{u} \cdot \nabla\right)(T\rho^{1-\gamma})$$

$$= (1 - \gamma)\rho^{1-\gamma}T\nabla \cdot \mathbf{u} + T\left(\frac{\partial}{\partial t} + \mathbf{u} \cdot \nabla\right)\rho^{1-\gamma} \ . \qquad (3.201)$$

Here the first term follows from the heat equation (3.199). We can evaluate the second term using the conservation of mass (3.147). This immediately gives us the desired result (3.200). $\qquad \square$

In the next section, we will see how we can recover the equations of sound waves using the transport equation (3.199). But the real advantage of this new approach is that it allows us to go beyond ideal fluids. In fact, the heat transport equation (3.199) should be viewed as analogous to the Euler equation for the velocity: both are missing the effect of dissipation. For the velocity field, this is captured by viscosity. For the temperature field, it is captured by the *thermal diffusivity*, α. This appears as an overall coefficient in an additional term in the transport equation which, more generally, reads

$$\left(\frac{\partial}{\partial t} + \mathbf{u} \cdot \nabla\right) T + (\gamma - 1)T \nabla \cdot \mathbf{u} = \alpha \nabla^2 T \ . \tag{3.202}$$

In the absence of any flow, so $\mathbf{u} = 0$, this reduces to the heat equation

$$\frac{\partial T}{\partial t} = \alpha \nabla^2 T \ . \tag{3.203}$$

The adiabatic approximation is essentially the statement that the diffusion of heat can be neglected in the problem of interest. But now we're in a position to understand how the physics changes when we include diffusion.

3.4.5 Viscosity and Damping

It is natural to ask: How does viscosity affect the propagation of sound? Because viscosity is dissipative, we might anticipate that any process will necessarily increase the entropy and so is no longer adiabatic. This means that we can't just use the simple relation $P\rho^{-\gamma}$ and must instead turn to the more sophisticated description in terms of the heat transport equation (3.202). This should be augmented with the Navier–Stokes equation

$$\rho\left(\frac{\partial \mathbf{u}}{\partial t} + \mathbf{u} \cdot \nabla \mathbf{u}\right) = -\nabla P + \mu \nabla^2 \mathbf{u} + \left(\frac{\mu}{3} + \zeta\right) \nabla(\nabla \cdot \mathbf{u}) \tag{3.204}$$

together with mass conservation and an appropriate equation of state that relates P, ρ, and T. We'll stick with the ideal gas equation of state, so

$$P = \frac{k_B T \rho}{m} \tag{3.205}$$

and we substitute this into the Navier–Stokes equation. For dilute gases, it turns out that $\zeta \approx 0$ so we choose to set it to zero. (It doesn't qualitatively change the physics because, as you can see, the shear viscosity μ already appears in the relevant term.) Our goal is to reproduce our previous results about sound waves in this framework, and then to understand how these results are affected by the viscosity μ and the thermal diffusivity α.

As before, we start with a stationary fluid but now also include the fact that it has constant temperature

$$\mathbf{u} = 0 \ , \quad \rho = \rho_0 \ , \quad T = T_0 \ . \tag{3.206}$$

We then consider time-dependent perturbations,

$$\mathbf{u} = \tilde{u}\hat{\mathbf{k}}\, e^{i\mathbf{k}\cdot\mathbf{x}-i\omega t}$$
$$\rho = \rho_0 + \tilde{\rho}\, e^{i\mathbf{k}\cdot\mathbf{x}-i\omega t} \tag{3.207}$$
$$T = T_0 + \tilde{T}\, e^{i\mathbf{k}\cdot\mathbf{x}-i\omega t}\ .$$

Note that we're looking for longitudinal waves, with $\mathbf{u}$ parallel to $\mathbf{k}$. Linearising, the mass conservation equation tells us that

$$\omega\tilde{\rho} = \rho_0 k \tilde{u}\ . \tag{3.208}$$

The linearised heat transport equation (3.202) is

$$-i\omega\tilde{T} + i(\gamma - 1)T_0 k\tilde{u} = -\alpha k^2 \tilde{T}\ . \tag{3.209}$$

Last, the linearised Navier–Stokes equation is

$$-i\rho_0\omega\tilde{u} = -\frac{ik_B k}{m}\left(T_0\tilde{\rho} + \rho_0\tilde{T}\right) - \frac{4}{3}\mu k^2 \tilde{u}\ . \tag{3.210}$$

We can write these simultaneous equations as a matrix,

$$M\begin{pmatrix} \tilde{\rho} \\ \tilde{u} \\ \tilde{T} \end{pmatrix} = \omega \begin{pmatrix} \tilde{\rho} \\ \tilde{u} \\ \tilde{T} \end{pmatrix} \tag{3.211}$$

with

$$M = \begin{pmatrix} 0 & \rho_0 k & 0 \\ k_B k T_0/m\rho_0 & -\frac{4}{3}i\mu k^2/\rho_0 & k_B k/m \\ 0 & (\gamma-1)T_0 k & -i\alpha k^2 \end{pmatrix}\ . \tag{3.212}$$

The frequencies of the perturbations ω are given by the eigenvalues of the matrix M. As we will see, this will give the dispersion relation between ω and k. Note, moreover, that the elements of the matrix are real except for those that multiply the dissipative coefficients μ and α. We'll see what this means for the physics shortly.

First let's look at what happens when $\mu = \alpha = 0$. There are solutions

$$\begin{pmatrix} \tilde{\rho} \\ \tilde{u} \\ \tilde{T} \end{pmatrix} = \epsilon \begin{pmatrix} \rho_0 \\ \omega/k \\ (\gamma-1)T_0 \end{pmatrix} \tag{3.213}$$

with ϵ some small, dimensionless parameter needed for the linearised approximation to be valid. This immediately solves the first and third equations, while the second requires

$$mw^2 = \gamma k_B T_0 k^2 \quad \Longrightarrow \quad \omega = \pm\sqrt{\frac{\gamma k_B T_0}{m}}\, k = \pm c_s k \ . \qquad (3.214)$$

But this is just our previous result (3.181) for the speed of sound. Moreover, we see that this perturbation has $(\gamma - 1)T_0\tilde{\rho} - \rho_0\tilde{T} = 0$, which means that $T/\rho^{\gamma-1}$ is constant to linear order. But this is the expected behaviour (3.168) for an adiabatic deformation of the fluid. So, in the limit that the dissipative effects vanish, we do indeed recover the adiabatic sound waves of the previous section.

There is also a novel solution to the equation (3.211) with $\mu = \alpha = 0$ that we haven't seen previously. This has $\tilde{u} = 0$ and

$$T_0\tilde{\rho} + \rho_0\tilde{T} = 0 \qquad (3.215)$$

a combination that ensures that $P \sim \rho T$ is constant in this perturbation. It solves the matrix equation with eigenvalue $\omega = 0$. This is telling us that there are additional static solutions to the equations of motion that do not evolve in time. These arise because the temperature variation balances the density variation, ensuring that pressure is constant so there is no restoring force.

Having made contact with our previous result, we can now see how things change when we turn on viscosity μ and thermal diffusivity α. Rather than directly finding the eigenvectors, we can take a bit of a shortcut to extract just the eigenvalues ω. First note that the determinant and trace of M are given by

$$\det M = i\frac{\alpha c_s^2}{\gamma} k^4 \quad \text{and} \quad \text{Tr}\, M = -i\left(\alpha + \frac{4}{3}\frac{\mu}{\rho_0}\right) k^2 \ . \qquad (3.216)$$

The product of the three eigenvalues must be equal to $\det M$. When $\mu = \alpha = 0$, we know that one of the eigenvalues vanishes and the other two were $\pm c_s k$. But now we see that the three must multiply to give something proportional to α. This means that, to leading order in α, the zero eigenvalue that arose from perturbations of constant pressure must change to

$$-c_s^2 k^2 \omega = \det M \quad \Longrightarrow \quad \omega = -\frac{i\alpha k^2}{\gamma} \ . \qquad (3.217)$$

The frequency is imaginary and negative. This is telling us that the modes decay exponentially quickly. To see this, write $\omega = -i\Gamma$. Then the behaviour of all modes goes as $e^{-i\omega t} = e^{-\Gamma t}$. The behaviour that we find above scales as $\omega \sim -ik^2$. This is characteristic of diffusion. It is the kind of behaviour that we get from the heat equation.

The two remaining modes are what becomes of sound waves. These too are expected to get a dissipative contribution. If we anticipate that they take the form

$$\omega = \pm c_s k - i\tilde{\Gamma} \tag{3.218}$$

possibly with some change to the sound speed, then we can compute $\tilde{\Gamma}$ by noting that the trace must equal the sum of all three eigenvalues, so

$$-2i\tilde{\Gamma} - i\frac{\alpha}{\gamma}k^2 = \operatorname{Tr} M \implies \tilde{\Gamma} = \frac{1}{2}\left(\frac{4}{3}\frac{\mu}{m\rho_0} + \frac{\gamma-1}{\gamma}\alpha\right)k^2 . \tag{3.219}$$

We see that the effects of viscosity and of heat conduction are similar: the sound waves diffuse and decay over time, with their lifetime set by $1/\tilde{\Gamma}$.

In addition, we can ask about velocity perturbations that are transverse to the wave, so that $\mathbf{k} \cdot \mathbf{u} = 0$. These are known as *shear perturbations*. It's straightforward to see that mass conservation and heat transport require $\tilde{\rho} = \tilde{T} = 0$, while the linearised Navier–Stokes equation gives the dispersion relation

$$\omega = -i\frac{\mu}{\rho_0}k^2 . \tag{3.220}$$

We see that these modes also behave diffusively.

3.5 Non-Linear Sound Waves

So far, throughout this chapter, we've only considered linear wave equations. For surface waves we went to some lengths to pick an approximation which made our equations linear and for sound waves we dropped the $\mathbf{u} \cdot \nabla \mathbf{u}$ term in the Navier–Stokes equation. This is a good first step since linear equations are significantly easier to solve than non-linear equations. But it's natural to wonder: Under what circumstances are the non-linearities important? And what effect do they have? Here we start to address such questions, albeit in the somewhat restricted context of waves propagating in one dimension.

We'll revisit our analysis of sound waves, but now in 1d. Our defining equations are the continuity equation

$$\frac{\partial \rho}{\partial t} + \frac{\partial(\rho u)}{\partial x} = 0 \tag{3.221}$$

and the Euler equation

$$\frac{\partial u}{\partial t} + u\frac{\partial u}{\partial x} = -\frac{1}{\rho}\frac{\partial P}{\partial x} \ . \tag{3.222}$$

Previously we dropped the $u\partial u/\partial x$ term. Our goal now is to understand what role it plays.

So far we have two equations for three variables: u, P, and ρ. As we stressed previously in Section 3.4, we must add one further equation. Rather than getting all hot and bothered by introducing temperature, we will instead work directly with an adiabatic equation of state that relates the pressure to the density

$$P = P(\rho) \ . \tag{3.223}$$

For example, for the ideal gas undergoing adiabatic deformations, we showed that the relevant equation is $P\rho^{-\gamma} = $ constant with $\gamma = c_P/c_V$ the ratio of specific heats. (See equation (3.170).) We'll turn to this example later but for now we keep things general with the function $P(\rho)$. We also saw in the previous section that, in the linearised approximation, the speed of sound is given by (3.184)

$$c_s^2(\rho) = \frac{dP}{d\rho} \ . \tag{3.224}$$

(Previously we wrote this as a partial derivative, keeping entropy S fixed. In this section we implicitly assume that entropy is fixed and view P only as a function of ρ.) One of the things we would like to learn is the sense in which c_s retains its interpretation as the speed of sound waves beyond the linearised approximation.

3.5.1 The Method of Characteristics

From the definition of c_s^2, together with (3.221), we have

$$\frac{\partial P}{\partial t} + u\frac{\partial P}{\partial x} = c_s^2\left(\frac{\partial \rho}{\partial t} + u\frac{\partial \rho}{\partial x}\right) = -\rho c_s^2\frac{\partial u}{\partial x} \ . \tag{3.225}$$

To make progress, we're going to rewrite the Euler equation (3.222) and our equation for pressure (3.225) in a clever way. Starting from (3.222), we have

$$\left(\frac{\partial}{\partial t} + (u - c_s)\frac{\partial}{\partial x}\right)u = -\frac{1}{\rho}\frac{\partial P}{\partial x} - c_s\frac{\partial u}{\partial x}$$

$$= \frac{1}{\rho c_s}\left(\frac{\partial}{\partial t} + (u - c_s)\frac{\partial}{\partial x}\right)P \tag{3.226}$$

where, to get to the second line, we've used (3.225). There's a nice symmetry between the left- and right-hand sides of this equation, with the same differential operator appearing in both. The only difference between them is that extra function $1/\rho c_s$ sitting on the right-hand side. To make things look even more symmetric, we define the new variable,

$$Q(\rho) = \int_{\rho_0}^{\rho} d\rho' \; \frac{c_s(\rho')}{\rho'} \tag{3.227}$$

with ρ_0 some useful fiducial, constant density such as the asymptotic value of the density if such a thing exists. This has the property that

$$\frac{\partial Q}{\partial t} = \frac{c_s}{\rho} \frac{\partial \rho}{\partial t} = \frac{1}{\rho c_s} \frac{\partial P}{\partial t} \quad \text{and} \quad \frac{\partial Q}{\partial x} = \frac{1}{\rho c_s} \frac{\partial P}{\partial x} \; . \tag{3.228}$$

This means that we can write (3.226) as

$$\left(\frac{\partial}{\partial t} + (u - c_s) \frac{\partial}{\partial x} \right) (u - Q) = 0 \; . \tag{3.229}$$

The same argument, with some minus signs flipped, also gives

$$\left(\frac{\partial}{\partial t} + (u + c_s) \frac{\partial}{\partial x} \right) (u + Q) = 0 \; . \tag{3.230}$$

This motivates the introduction of the *Riemann invariants*

$$R_{\pm} = u \pm Q \; . \tag{3.231}$$

These obey the *Riemann wave equation*

$$\left(\frac{\partial}{\partial t} + (u \pm c_s) \frac{\partial}{\partial x} \right) R_{\pm} = 0 \; . \tag{3.232}$$

We next want to understand what this equation is telling us. To this end, for a given flow $u(x,t)$ with density $\rho(x,t)$, we construct two collections of *characteristic curves*, $\mathcal{C}_{\pm}$, as shown in Figure 3.10. These are worldlines in the spacetime parametrised by (x,t), defined by

$$\mathcal{C}_{\pm} : \quad \frac{dx}{dt} = u(x,t) \pm c_s(x,t) \; . \tag{3.233}$$

We introduce two new coordinates in spacetime: ξ_+ and ξ_-. These have the property that $\xi_{\pm}$ are constant on the characteristic curves $\mathcal{C}_{\pm}$ respectively. Then the meaning of (3.232) is as follows.

Claim: $R_{\pm}$ is constant on characteristic curves $\mathcal{C}_{\pm}$.

Proof: To show this, we just need to think carefully about what depends on

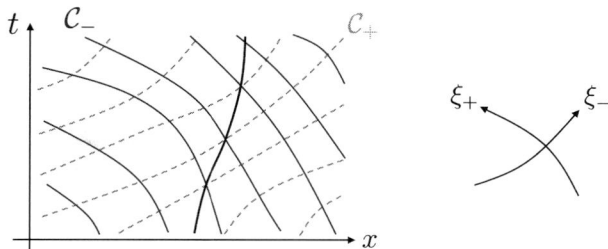

Fig. 3.10 The (dotted) characteristic curves $\mathcal{C}_+$ run from bottom left to top right, and the (solid) curves $\mathcal{C}_-$ run from bottom right to top left. The curves depend locally on both the flow $u(x,t)$ and, through $c_s(\rho)$, the density $\rho(x,t)$. The thick (verticalish) line depicts one integral curve of the fluid flow $u(x,t)$. The coordinates $\xi_\pm$ are constant on $\mathcal{C}_\pm$ respectively. This means that, as shown in the axes on the right, ξ_+ increases as we move in the up-left direction, while ξ_- increases as we move in the up-right direction.

what. Suppose that we vary both ξ_+ and ξ_- a tiny tiny bit. Then we move in the t-direction an infinitesimal amount

$$ dt = \left.\frac{\partial t}{\partial \xi_+}\right|_{\xi_-} d\xi_+ + \left.\frac{\partial t}{\partial \xi_-}\right|_{\xi_+} d\xi_- \tag{3.234}$$

and we move in the x-direction an infinitesimal amount

$$ dx = \left.\frac{\partial x}{\partial \xi_+}\right|_{\xi_-} d\xi_+ + \left.\frac{\partial x}{\partial \xi_-}\right|_{\xi_+} d\xi_- . \tag{3.235}$$

On the characteristic curves $\mathcal{C}_+$ we know that ξ_+ is constant, so we have

$$ \mathcal{C}_+ : \quad d\xi_+ = 0 \quad \Longrightarrow \quad \left.\frac{\partial x}{\partial \xi_-}\right|_{\xi_+} = \frac{dx}{dt}\left.\frac{\partial t}{\partial \xi_-}\right|_{\xi_+} = (u + c_s)\left.\frac{\partial t}{\partial \xi_-}\right|_{\xi_+} . \tag{3.236}$$

Now, if we view $R_+(x,t)$ as a function $R_+(\xi_+,\xi_-)$, then

$$
\begin{aligned}
\left.\frac{\partial R_+}{\partial \xi_-}\right|_{\xi_+} &= \frac{\partial R_+}{\partial t}\left.\frac{\partial t}{\partial \xi_-}\right|_{\xi_+} + \frac{\partial R_+}{\partial x}\left.\frac{\partial x}{\partial \xi_-}\right|_{\xi_+} \\
&= \left(\frac{\partial R_+}{\partial t} + (u + c_s)\frac{\partial R_+}{\partial x}\right)\left.\frac{\partial t}{\partial \xi_-}\right|_{\xi_+} = 0 .
\end{aligned} \tag{3.237}
$$

In other words, $R_+(\xi_-,\xi_+)$ is really just a function of a single variable, $R_+(\xi_+)$. The same argument also tells us that $R_- = R_-(\xi_-)$. So if we move along a characteristic curve $\mathcal{C}_+$, where ξ_+ is constant, then R_+ doesn't change. Similarly, R_- doesn't change if we move along a characteristic curve $\mathcal{C}_-$. $\qquad\square$

It's worth taking stock of what we've achieved. Our goal is to solve for the flow $u(x,t)$ and the density $\rho(x,t)$. We haven't done this yet! However, we have showed that, *if* we can solve for the flow, then we can construct characteristic curves $\mathcal{C}_{\pm}$ on which the variables $R_{\mp}$ are constant. And $R_{\pm}$, in turn, depend on u and ρ that we are trying to figure out. All of which means that the Riemann invariants don't immediately solve our problem, but they should contain some information that we can exploit.

Furthermore, if it's possible to somehow figure out $R_{\pm}(x,t)$ then it's straightforward to reconstruct the velocity field which, from (3.231), is given by

$$u(x,t) = \frac{1}{2}(R_+(\xi_+) + R_-(\xi_-)) \,. \tag{3.238}$$

This is the generalisation of the more familiar solution to the linearised wave equation (3.187)

$$u(x,t) = F(x - c_s t) + G(x + c_s t) \tag{3.239}$$

which describes wavepackets moving left and right at a constant speed c_s.

3.5.2 Soundcones

The equations (3.232) are telling us that the something is propagating in the fluid with speed c_s relative to the flow.

To see this more clearly, consider some initial disturbance with $u(x,0) \neq 0$ for $|x| < L$ as shown in Figure 3.11. We'll also assume that the density $\rho(x,t)$ differs from some asymptotic value ρ_0 only within this same region. From (3.227), this ensures that $Q = 0$ outside of this region so $R_{\pm}(x,0) = 0$ for $|x| > L$.

We can draw this on a spacetime diagram, with the vertical axis labelled by $c_0 t$ where $c_0 = c_s(\rho_0)$ is the asymptotic sound speed. This ensures that linearised sound waves travel at $\pm 45°$ in the diagram, rather like light rays in Minkowski space. In analogy with special relativity, we will say that the pair of characteristic curves $\mathcal{C}_{\pm}$ emerging from any point form a *soundcone*. (In fact, the analogy works better with general relativity where the lightcones depend on the curvature of spacetime, just like the soundcones depend on the background flow $u(x,t)$.)

Consider the soundcones emerging from the points $x = \pm L$ at time $t = 0$. These are shown in Figure 3.11. They divide spacetime for $t > 0$ into six distinct regions with the following properties:

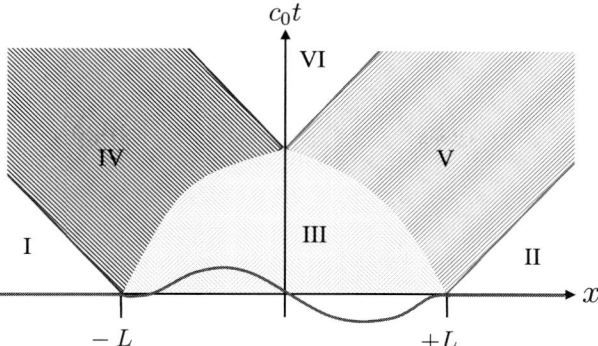

An initial disturbance in a region $|x| < L$ propagates to the left and to the right. Regions I,II and VI are undisturbed; regions IV and V have only left-moving and right-moving waves respectively, and anything can happen in region III.

- Regions I and II have the property that both $\mathcal{C}_+$ and $\mathcal{C}_-$ characteristic curves pass through the x-axis in the region with $R_\pm(x, 0) = 0$. Because $R_\pm$ are constant on $\mathcal{C}_\pm$ respectively, this means that we must have $R_+ = R_- = 0$ throughout the regions I and II. This makes sense: the initial disturbance takes time to propagate and, just as signals can't travel faster than the speed of light in special relativity, here they can't travel faster than the (local) speed of sound. Hence the regions I and II know nothing about the disturbance. Both characteristic curves are straight lines at $\pm 45°$ in these regions.

- Region III has a complicated flow, with both left- and right-travelling waves. Here we would expect both R_+ and R_- to be non-vanishing and there is no reason to think that the characteristic curves $\mathcal{C}_\pm$ will be straight.

- In Region IV, we can trace the $\mathcal{C}_+$ curves back to the region in which $R_+ = 0$. So we know that

$$R_+ = u + Q = 0 \quad \Longrightarrow \quad u = -Q \ . \tag{3.240}$$

Correspondingly, $R_- = u - Q = 2u$. We know that R_- is constant on characteristic curves $\mathcal{C}_-$. So this tells us that both u and Q, and hence ρ and $c_s(\rho)$, must also be constant on $\mathcal{C}_-$. In other words, all of these are functions only of ξ_-,

$$u(x, t) = u(\xi_-) \ , \quad \rho(x, t) = \rho(\xi_-) \ , \quad c_s(x, t) = c_s(\xi_-) \ . \tag{3.241}$$

These are purely left-moving waves. Now, the defining equation for the characteristic curve $\mathcal{C}_-$ is

$$\frac{dx}{dt} = u(\xi_-) - c_s(\xi_-). \tag{3.242}$$

But for a given $\mathcal{C}_-$ curve, defined by some fixed value of ξ_-, the right-hand side is constant. This means that the $\mathcal{C}_-$ characteristic curves are straight lines in region IV. Although these curves are all straight lines in region IV, they need not necessarily be parallel: the gradient $u(\xi_-) - c_s(\xi_-)$ can depend, as advertised, on ξ_- and typically will. Of course, if they're not parallel then there is the possibility that they will converge and cross at some point. That's somewhat confusing because $R_-(\xi_-)$ is expected to have different values on different curves and if two curves collide then it looks like, say, the velocity $u(x,t)$ will have two different values at that point! We will learn how to think about this shortly.

• Region V has similar properties to Region IV, now with the characteristic curves $\mathcal{C}_+$ straight lines. This is a purely right-moving wave. In general, situations where one of the Riemann invariants vanish (or, indeed, is constant) over a region of space are referred to as *simple waves* and are associated to waves that are either purely left-moving or purely right-moving.

• In Region VI, the same arguments apply as for Region I and II: you can trace back both $\mathcal{C}_+$ (to the left) and $\mathcal{C}_-$ (to the right) to regions where $R_+ = 0$ and $R_- = 0$ respectively. This means that this is once again a region of calm, with $u = 0$ and $\rho = \rho_0$. The disturbance has passed. This is because there is no option to dawdle in this system: all modes must travel at the speed of sound. The only question is how that speed of sound changes.

3.5.3 Wave Steepening and a Hint of Shock

We can illustrate these ideas further. We'll first give a heuristic discussion of the physics and then fill in some of the details. To do this, it's useful to pick a concrete example and we'll choose to look at an ideal gas obeying $P\rho^{-\gamma} = \text{constant}$.

The speed of sound for the ideal gas is (3.180)

$$c_s^2(\rho) = \frac{\gamma P}{\rho} . \tag{3.243}$$

The function $Q(\rho)$ defined in (3.227) is

$$Q(\rho) = \int_{\rho_0}^{\rho} d\rho' \, \frac{c_s(\rho')}{\rho'} = \frac{2}{\gamma - 1} \left(c_s(\rho) - c_0 \right) \tag{3.244}$$

where $c_0 = c_s(\rho_0)$. The Riemann invariants are then

$$R_\pm = u \pm \frac{2}{\gamma - 1}(c_s - c_0) \ . \tag{3.245}$$

These are normalised so that, in a boring, static flow with $u = 0$ and $\rho = \rho_0$, both Riemann invariants vanish: $R_\pm = 0$.

Now consider a simple, right-moving flow in which $R_- = 0$ everywhere. This means that

$$R_- = u - \frac{2}{\gamma - 1}(c_s - c_0) = 0 \quad \Longrightarrow \quad c_s = c_0 + \frac{1}{2}(\gamma - 1)u \ . \tag{3.246}$$

This gives us a relation between the velocity field and the speed of sound. In fact, this contains the key bit of physics. Any region of fluid with $u > 0$ has a speed of sound $c_s > c_0$. And any region of fluid with $u < 0$ has a speed of sound $c_s < c_0$ (assuming $\gamma > 1$).

The wave propagates along the characteristic curves $\mathcal{C}_+$, on which u and c_s are both constant. These curves are given by (3.233)

$$\mathcal{C}_+ : \quad \frac{dx}{dt} = u(\xi_+) + c_s(\xi_+) = c_0 + \frac{1}{2}(\gamma + 1)u(\xi_+) \ . \tag{3.247}$$

Suppose that we're given some initial data at time $t = 0$. We're told that

$$u(x, 0) = U(x) \ . \tag{3.248}$$

The requirement that we've got a purely right-moving wave then fixes the density (or, equivalently, the speed of sound) along this slice using (3.246). We know that ξ_+ labels the different $\mathcal{C}_+$ curves, but we haven't yet got a natural way to fix the normalisation of this coordinate. We'll resolve this by choosing ξ_+ to be the value of x where a given curve $\mathcal{C}_+$ intersects the $t = 0$ axis. Then the characteristic curves (3.247) are simply the straight lines

$$\mathcal{C}_+ : \quad x(t; \xi_+) = \xi_+ + \left[c_0 + \frac{1}{2}(\gamma + 1)U(\xi_+) \right] t \ . \tag{3.249}$$

We can see the slope of the characteristic curves in the square bracket. Those parts of the fluid that had an initial velocity $U > 0$ travel along characteristic curves that have an angle greater than $45°$ (as measured from the vertical axis). And those parts of the fluid with $U < 0$ travel along lines that sit at less than $45°$.

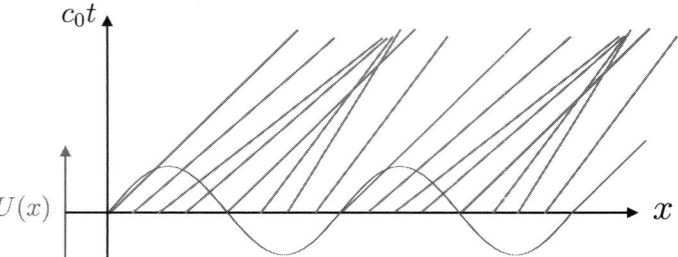

An initial, purely right-moving, sine wave has characteristic curves $\mathcal{C}_+$ that intersect.

The resulting characteristic curves are shown in Figure 3.12 for the simple initial data where $U(x)$ is a sine wave. Importantly, we see that, as we anticipated previously, the characteristic curves meet. But this is very confusing! On a given characteristic curve, the velocity of the fluid is fixed as $U(x)$. Wherever two curves intersect, the velocity must be multi-valued. In other words, the non-linearities have pushed our nice, simple initial sine wave into a solution that is discontinuous in the velocity field!

In fact, there's a straightforward interpretation of this. The non-linearities cause the peak of the wave, where $U > 0$, to move faster than the trough of the wave, where $U < 0$. This means that the wave will become skewed, and increasingly sawtooth-like. This is known as *wave steepening* and is shown in Figure 3.13 where we show the steepening of an initial Gaussian wavepacket. (We choose a Gaussian wavepacket, rather than the sine wave we considered earlier, simply because it's easier to draw!)

Eventually, the peak will catch up with the trough, at which point the velocity field is no longer single-valued. This is the reason that the characteristic curves cross. It turns out that this is telling us that a *shock* forms. We'll understand more about what this means in Section 3.6. For now, we will simply adopt the terminology and say that a shock is tantamount to two characteristic curves intersecting.

We can ask: How long does it take for the shock to form? Two curves intersect whenever a shift to an adjacent curve doesn't require a shift in space. Or, in equations,

$$\frac{\partial x}{\partial \xi_+}\bigg|_t = 0 \ . \tag{3.250}$$

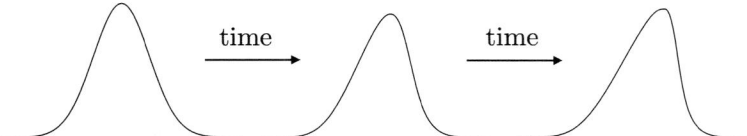

Fig. 3.13 The steepening of an initial Gaussian wavepacket over time.

From (3.249), this gives the requirement

$$t = -\frac{2}{\gamma+1}\frac{1}{U'(\xi_+)} \ . \tag{3.251}$$

We want the value of ξ_+ that minimises $t > 0$ since this is when two of the characteristic curves first cross. We take, as our initial condition, a sine wave as shown in Figure 3.12

$$u(x,0) = U(x) = U\sin kx \tag{3.252}$$

with U the overall amplitude of the wave. The first shock then forms when

$$t_{\text{shock}} = \frac{2}{\gamma+1}\frac{1}{kU} \ . \tag{3.253}$$

It's useful to compare this time scale with the period T of the wave itself. Recall from Section 3.4 that linear sound waves have the simple dispersion relation $\omega = c_s|k|$ and the period is $T = 2\pi/\omega \approx 2\pi/c_0|k|$. This then gives

$$\frac{t_{\text{shock}}}{T} = \frac{1}{\pi(\gamma+1)}\frac{c_0}{U} \ . \tag{3.254}$$

This is our first sign of an important dimensionless quantity called the *Mach number M*, the ratio of the velocity of the fluid flow to the speed of sound

$$M = \frac{U}{c_0} \ . \tag{3.255}$$

We can put some numbers into this. The decibel scale is the familiar scale used to measure how loud a sound is. It's a $\log_{10}$ scale such that if the amplitude of the wave U changes by a factor of 10 then the decibels increase additively by 10. A quiet chat down the pub with a friend will involve sound waves with frequency around $\omega \sim 1000$ s^{-1} and a corresponding period of $T \sim 3 \times 10^{-3}$ s. The volume is around 60 dB and this corresponds to a Mach number $M \sim 10^{-13}$. Apparently you would need to wait about 1000 years for a shock wave to form! We can instead crank up the volume. If you stand next to a rocket at take off, you will suffer around 180 dB. (And get permanent ear damage, although this may well be the least of your problems.) Now $M \sim 10^{-1}$. Perhaps unsurprisingly, you can expect a shock wave to form in a very short time.

3.5.4 Burgers' Equation

We can elucidate the analysis above a little further with a simple change of variables. We continue to study a purely right-moving wave for which $R_+ = 2u$ and, from (3.246), the local speed of sound c_s is related to the fluid velocity by

$$c_s(u) = c_0 + \frac{1}{2}(\gamma - 1)u \ . \tag{3.256}$$

The Riemann wave equation (3.232) is now a non-linear equation for u

$$\left(\frac{\partial}{\partial t} + \left[c_0 + \frac{1}{2}(\gamma + 1)u \right] \frac{\partial}{\partial x} \right) u = 0 \ . \tag{3.257}$$

We introduce the co-moving coordinate

$$X = x - c_0 t \ . \tag{3.258}$$

This would travel along with the wave if the wave were travelling at a constant speed c_0. Of course, the wave is not travelling at constant speed, which is where much of the fun lies. We also introduce the rescaled velocity field

$$v = \frac{1}{2}(\gamma + 1)u \tag{3.259}$$

and we think of $v = v(X, t)$. Then the Riemann wave equation becomes

$$\frac{\partial v}{\partial t} + v \frac{\partial v}{\partial X} = 0 \tag{3.260}$$

where the partial time derivative is now taken with X held fixed, rather than x held fixed as before. This is the *inviscid Burgers' equation*. (Inviscid because throughout this section we've neglected viscosity.)

The Burgers' equation (3.260) takes a particularly simple form. We'll again take the initial data to be

$$v(X, 0) = V(X) \tag{3.261}$$

and identify X with ξ_+ when $t = 0$. (The function $V(X)$ differs from our previous $U(x)$ defined in (3.248) only by the constant factor $2/(\gamma+1)$.) The characteristic curves $\mathcal{C}_+$ given in (3.249) are then just

$$\mathcal{C}_+ : \quad X(t; \xi_+) = \xi_+ + V(\xi_+)t \tag{3.262}$$

along which v is constant. This means that $v(X, t) = v(\xi_+, 0)$. Or, substituting the expression for X above, we have the solution

$$v(X, t) = v(X - V(\xi_+)t, 0) \ . \tag{3.263}$$

It's worth pausing to parse what this solution means. First, for a given X and t, we need to use (3.262) to figure out what ξ_+ is. In other words, what characteristic curve $\mathcal{C}_+$ the point (X, t) lies on. Clearly this depends on the initial data $v(\xi_+, 0)$ that we're given. It's an algebraic computation and, for a given initial condition, the answer may not be available in closed form. Nonetheless, it's doable numerically. With this in hand, the solution (3.263) then tells us how the initial data evolves. Indeed, it's just a rewriting of what we saw previously: the points with higher initial velocity propagate at a faster speed.

Steepening Again

The coordinate X has the advantage that it keeps up with the propagating wave, at least on average. The slope of the wave is

$$\frac{\partial v}{\partial X}\bigg|_t = \frac{\partial v(\xi_+, 0)}{\partial \xi_+}\frac{\partial \xi_+}{\partial X}\bigg|_t = \frac{V'(\xi_+)}{1 + V'(\xi_+)t} \tag{3.264}$$

where, in the second equality, we've used (3.262). This now gives us a better handle on the phenomenon of wave steepening. Those parts of the wave with $V'(\xi_+) < 0$ get steeper over time; those parts of the wave with $V'(\xi_+) > 0$ become flatter. The shock occurs when the wave becomes infinitely steep, so

$$\frac{\partial v}{\partial X}\bigg|_t = \infty \implies t = -\frac{1}{V'(\xi_+)} \,. \tag{3.265}$$

This agrees with our earlier condition (3.251).

Very Briefly, the Effect of Viscosity

So far our discussion has neglected viscosity. But we can see at the level of equations what it changes. If we go back to the 1d Euler equation (3.222) and trace through various changes of variables, we find that the Burgers' equation (3.260) is replaced by

$$\frac{\partial v}{\partial t} + v\frac{\partial v}{\partial X} = \nu\frac{\partial^2 v}{\partial X^2}$$

where ν is, as always, the kinematic viscosity and we've inadvertently stumbled upon the rather unfortunate situation of having both ν and v in the same equation. (It won't be for long and we can live with it.) This is the full *Burgers' equation*.

We know that viscosity causes the velocity to diffuse, and this mitigates large velocity gradients. This can be shown to remove the formation of the discontinuous shock.

3.5.5 Traffic

We can illustrate some of these ideas in a much simpler situation that is, I'm sure, close to all our hearts: the flow of traffic. The ideas described below were first developed by Lighthill and Whitham, and subsequently extended by Richards to include shock waves. It is sometimes called the LWR model.

Our variables are the density of cars $\rho(x, t)$ along a road, and their velocity $u(x, t)$. We define the flow of traffic to be $q(x, t) = \rho u(x, t)$. The conservation of cars (a law of physics that follows almost immediately from the Standard Model) tells us that

$$\frac{\partial \rho}{\partial t} + \frac{\partial q}{\partial x} = 0 \ . \tag{3.266}$$

In contrast to the fluids that we described above, we don't have a second differential equation for the velocity $u(x, t)$. Instead, we just jump straight in with an "equation of state" which, for us, relates the velocity directly to the density, so there is some given function

$$u(x, t) = u(\rho(x, t)) \ . \tag{3.267}$$

We will assume that $u(\rho)$ is a monotonically decreasing function of ρ. On an empty road, we have $u(0) = u_{\max}$ where $u_{\max}$ is the speed limit, or perhaps something not very much greater than the speed limit. Importantly, we also have $u(\rho_\star) = 0$ for some value of $\rho_\star$. This is sensible: if the density of cars is so great that they're bumper to bumper, then no one is getting home any time soon. Reasonably, $\rho_\star$ is known as the *jam density*.

It is more useful to think of the flow $q(\rho) = \rho u(\rho)$ as a function of the density. With the assumptions above, $q(\rho)$ is a convex function, vanishing at both $\rho = 0$ and $\rho = \rho_\star$, with a maximum value at $\rho = \rho_{\mathrm{crit}}$. This is shown in Figure 3.14. In traffic circles (or roundabouts to give them their correct name) the (ρ, q)-graph is called the *fundamental diagram*.

We can rewrite the continuity equation (3.266) as a non-linear wave equation

$$\frac{\partial \rho}{\partial t} + c(\rho)\frac{\partial \rho}{\partial x} = 0 \quad \text{with} \quad c(\rho) = \frac{\partial q}{\partial \rho} \ . \tag{3.268}$$

Here $c(\rho)$ should be viewed as the speed of density waves. One way to make this manifest is to compare to our previous discussion of characteristics.

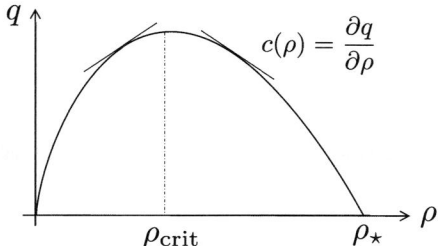

Fig. 3.14 The grandly named "fundamental diagram" for traffic flow. The speed of propagation of density waves is given by the slope of the curve.

Indeed, the wave equation already takes the form of the Riemann wave equation (3.232) with the characteristic $\mathcal{C}$ given by the straight line

$$\mathcal{C}: \quad \frac{dx}{dt} = \frac{dq}{d\rho} \ . \tag{3.269}$$

In this case, the Riemann invariant is simply the density $\rho(x,t)$. This means that along any straight line $\mathcal{C}$ the density is constant, at least provided that nothing singular happens like a shock wave. It's straightforward to check this: if we move a little bit along the line $\mathcal{C}$ then the density changes as

$$d\rho = \frac{\partial \rho}{\partial t} dt + \frac{\partial \rho}{\partial x} dx = \left(\frac{\partial \rho}{\partial t} + \frac{\partial \rho}{\partial x} \frac{dx}{dt} \right) dt = 0 \ . \tag{3.270}$$

We see that the speed at which density waves propagate is $c(\rho)$, the slope in the fundamental diagram. We have $dq/d\rho = u + \rho \, du/d\rho$ and, because $du/d\rho < 0$, we necessarily have $c < u$. This means that the density waves propagate at a speed that is less than the traffic speed. That makes sense: you react if someone suddenly brakes in front of you, but not so much if they brake behind you.

We can also ask about the speed of these density waves relative to the road. If the traffic is reasonably light, with $\rho < \rho_{\text{crit}}$, then $c > 0$ and the density wave propagates forwards along the road. But if the traffic is congested, with $\rho > \rho_{\text{crit}}$, then $c < 0$ and the density wave propagates backwards.

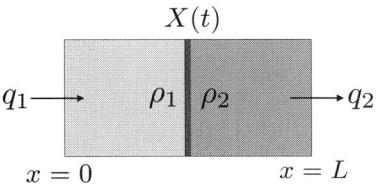

In the context of traffic flow, a shock is a discontinuity in the density. This is what happens in a traffic jam, where light traffic with density ρ_1 catches up with heavier traffic with density ρ_2. We model this by considering a length of road, between $x = 0$ and $x = L$ as shown in the figure. The traffic coming in at $x = 0$ has flow $q_1 = q(\rho_1)$ and that exiting at $x = L$ has flow $q_2 = q(\rho_2)$. Somewhere in between is the point $X(t)$ of the shock, with $0 < X < L$. We want to understand how the position of the shock – the back of the traffic jam – evolves with time.

The total number of cars is $N = \rho_1 X + (L - X)\rho_2$, so

$$\frac{dN}{dt} = \frac{dX}{dt}\left(\rho_1 - \rho_2\right) . \tag{3.271}$$

This change in the number of cars is accounted for by those entering and exiting the region, so

$$\frac{dN}{dt} = q_1 - q_2 \quad \Longrightarrow \quad \frac{dX}{dt} = \frac{q_1 - q_2}{\rho_1 - \rho_2} . \tag{3.272}$$

This is telling us that the shock can travel either forwards (with $dX/dt > 0$) or backwards (with $dX/dt < 0$) along the road, depending in a simple way on the traffic conditions.

3.6 Shocks

> This is a manifest absurdity. No step, however, of the reasoning by which this result has been obtained can be controverted. What then is the meaning of it?
>
> *The Rev. James Challis, in 1848, expressing shock on first discovering that initial data in the Euler equation gives rise to discontinuities in the velocity*

We got our first hints of shock waves in Section 3.5 where a discontinuity in the velocity field (or in the density of traffic) arises as the flow evolves over time. Here we'll explore some properties of these shocks in the context of fluid dynamics. We'll see that the discontinuities have a remarkably simple and constrained structure, all of which follows from conservation laws, together with a little bit of thermodynamics.

We're going to upgrade from one dimension to two. We'll consider flows of a compressible fluid in the (x, y)-plane, with a shock that sits at some specific point in the x-direction and extends along the y-direction. The flows themselves will be invariant under translations in the y-direction. This means that we will restrict our attention to flows of the form

$$\mathbf{u}(\mathbf{x}, t) = (u(x, t), v(x, t)) . \tag{3.273}$$

We have mass conservation

$$\frac{\partial \rho}{\partial t} + \frac{\partial (\rho u)}{\partial x} = 0 \tag{3.274}$$

and the non-linear Euler equation

$$\rho \frac{\partial u}{\partial t} + \rho u \frac{\partial u}{\partial x} + \frac{\partial P}{\partial x} = 0 \implies \frac{\partial (\rho u)}{\partial t} + \frac{\partial (\rho u^2)}{\partial x} + \frac{\partial P}{\partial x} = 0 \tag{3.275}$$

$$\rho \frac{\partial v}{\partial t} + \rho u \frac{\partial v}{\partial x} = 0 \implies \frac{\partial (\rho v)}{\partial t} + \frac{\partial (\rho u v)}{\partial x} = 0 \tag{3.276}$$

where to get the second set of equations we use the continuity equation (3.274) associated to mass conservation. We've rewritten the Euler equation in this way to stress that they too are continuity equations, describing the conservation of momentum.

We've got three equations for four variables, ρ, u, v, and P. We need a fourth. This is a road that we've been down before: for adiabatic variations, the fourth condition is a relation

$$P = P(\rho) \tag{3.277}$$

and our standard example is the ideal gas with $P/\rho^\gamma = \text{constant}$.

As we will see, the physics of the shock is all about understanding conserved quantities. The final conservation law that we need is for energy. But here there's a subtlety because there are additional contributions to the energy for a compressible fluid. As we will now show, these additional contributions are fully determined by the relation $P = P(\rho)$.

We start by introducing the energy density of the fluid

$$\text{energy} = \frac{1}{2}\rho(u^2 + v^2) + \rho e(\rho) \tag{3.278}$$

where $e(\rho)$ is some internal energy (per unit mass) that we will determine below. In addition, we introduce the energy current

$$\text{energy current} = \frac{1}{2}\rho u(u^2 + v^2) + \rho u h(\rho) \tag{3.279}$$

where the additional quantity $h(\rho)$ is known as the *enthalpy*. Again, we'll figure out what this is shortly. Then using the same kind of manipulations that we saw when first deriving Bernoulli's principle in Section 1.2.4, we can write the conservation of energy as

$$\frac{\partial}{\partial t}\left(\frac{1}{2}\rho(u^2 + v^2) + \rho e(\rho)\right) + \frac{\partial}{\partial x}\left(\frac{1}{2}\rho u(u^2 + v^2) + \rho u h(\rho)\right) = 0 \ . \tag{3.280}$$

A little algebra shows that this equation holds if the internal energy $e(\rho)$ and enthalpy $h(\rho)$ obey the equations

$$\frac{d(\rho e)}{d\rho} = h \quad \text{and} \quad \frac{d(\rho h)}{d\rho} = h + \frac{dP}{d\rho} \;. \tag{3.281}$$

These equations can then be solved given the relation $P = P(\rho)$. For example, for the ideal gas with $P = A\rho^\gamma$ for some constant A, we can solve these to get

$$e = \frac{A}{\gamma - 1}\rho^{\gamma-1} = \frac{1}{\gamma - 1}\frac{P}{\rho} \quad \text{and} \quad h = \frac{A\gamma}{\gamma - 1}\rho^{\gamma-1} = \frac{\gamma}{\gamma - 1}\frac{P}{\rho} \;. \tag{3.282}$$

Note that $h = e + P/\rho$.

Equations (3.274), (3.275), (3.276), and (3.280) are our starting point. Our goal now is to search for discontinuous solutions to these equations describing shock waves.

3.6.1 Jump Conditions

A shock is a discontinuous flow. But we don't allow any old discontinuity. Instead, the discontinuity itself has certain properties. And these are derived from the conservation laws described above.

The discontinuity splits the flow into two, as shown in the figure. On the left, the flow has values $\mathbf{u}_1$, ρ_1, and P_1; on the right values $\mathbf{u}_2$, ρ_2, and P_2. To make life particularly simple, we'll assume that each of these flows is constant in space and time. All of the physics arises from the discontinuity.

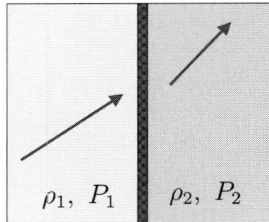

We'll assume that the shock itself is not propagating, but is fixed at some position, say $x = 0$. We model the discontinuity as an infinitely thin surface and, as such, it can't carry any mass density. Any mass that enters from one side must exit through the other. The same holds for momentum and energy. This means that each of the currents in (3.274) through to (3.280) must coincide on the left and right. Mass conservation (3.274) tells us

$$\rho_1 u_1 = \rho_2 u_2 \;. \tag{3.283}$$

Momentum conservation tells us

$$\rho_1 u_1^2 + P_1 = \rho_2 u_2^2 + P_2 \qquad (3.284)$$

and

$$\rho_1 u_1 v_1 = \rho_2 u_2 v_2 \quad \Longrightarrow \quad v_1 = v_2 \qquad (3.285)$$

where the second equation follows from (3.283). We learn that the velocity tangent to the shock remains constant. Finally, energy conservation gives

$$\rho_1 u_1 \left(\frac{1}{2}(u_1^2 + v_1^2) + h_1 \right) = \rho_2 u_2 \left(\frac{1}{2}(u_2^2 + v_2^2) + h_2 \right)$$

$$\Longrightarrow \quad \frac{1}{2} u_1^2 + h_1 = \frac{1}{2} u_2^2 + h_2 \qquad (3.286)$$

where now the second equation follows from (3.283) and (3.285). This is Bernoulli's theorem applied to the shock. Equations (3.283), (3.284), (3.285), and (3.286) are called the *Rankine–Hugoniot jump conditions.*

It's simplest to transform to a frame in which $v_1 = v_2 = 0$, so that all the action is taking place transverse to the shock wave. We're then left with three conditions which fix u_2, ρ_2, and P_2 in terms of the initial flow data. To see this in more detail, we first use (3.283) and (3.284) to derive the relation

$$\rho_1 u_1^2 = \left(1 - \frac{\rho_1}{\rho_2} \right)^{-1} (P_2 - P_1) . \qquad (3.287)$$

Since the left-hand side is positive, the right-hand side must also be positive. That gives us two possibilities: either pressure and density both increase across the shock

$$P_2 > P_1 \quad \text{and} \quad \rho_2 > \rho_1 \qquad (3.288)$$

or the opposite happens. Clearly these two options are related by a parity flip, so we'll assume that the above occurs and the pressure is greater on the right of the shock. Then, from (3.283), we have

$$u_2 = \frac{\rho_1}{\rho_2} u_1 . \qquad (3.289)$$

This tells us that $|u_2| < |u_1|$, so the *speed* of the flow is smaller on the right of the shock. Note, however, that we haven't yet said anything about the sign of u_1 and u_2, i.e. is the flow left-to-right or right-to-left? We'll come to this shortly.

The Size of the Shock

The *shock compression ratio* is defined to be

$$r = \frac{\rho_2}{\rho_1} = \frac{u_1}{u_2} \ . \tag{3.290}$$

It's a measure of how big the discontinuity is. We can also get an expression for r in terms of the pressure difference on each side but, for this, we need our final matching condition associated to conservation of energy (3.286). For an ideal gas, this reads

$$\frac{1}{2}u_1^2 + \frac{\gamma}{\gamma - 1}\frac{P_1}{\rho_1} = \frac{1}{2}u_2^2 + \frac{\gamma}{\gamma - 1}\frac{P_2}{\rho_2} \ . \tag{3.291}$$

We can use (3.287) and (3.289) to write this as

$$\frac{1}{2}\frac{\rho_1 + \rho_2}{\rho_1\rho_2}(P_2 - P_1) = \frac{\gamma}{\gamma - 1}\left(\frac{P_2}{\rho_2} - \frac{P_1}{\rho_1}\right) \ . \tag{3.292}$$

Now substituting $\rho_2 = r\rho_1$, we find

$$r = \frac{(\gamma - 1)P_1 + (\gamma + 1)P_2}{(\gamma + 1)P_1 + (\gamma - 1)P_2} \ . \tag{3.293}$$

This form of r puts some bounds on the strength of the shock. For a *strong shock*, $P_2 \gg P_1$. In the limit $P_2/P_1 \to \infty$, we have

$$r \to r_{\max} = \frac{\gamma + 1}{\gamma - 1} \ . \tag{3.294}$$

This is the largest compression factor that we can have. For a monatomic gas, with $\gamma = 5/3$, we have $r_{\max} = 4$. Note that the discontinuity is very different for pressure and speed: if the pressure changes by an infinite amount, the speed changes only by a factor of 4.

There's also something familiar hiding in this unfamiliar setting. Suppose that we have a *weak shock*, meaning $P_2 = P_1 + \Delta P$ with $\Delta P \ll P_1$. Then we have $r \approx 1 + \Delta P/\gamma P_1$. We also write $\rho_2 = \rho_1 + \Delta\rho$ and this gives $r \approx 1 + \Delta\rho/\rho_1$. Equating these, we have

$$\frac{\Delta P}{\Delta\rho} = \frac{\gamma P_1}{\rho_1} \ . \tag{3.295}$$

But this is the equation for the speed of sound in an ideal gas (see, for example, (3.180) and (3.184))

$$\frac{dP}{d\rho} = c_s^2 = \frac{\gamma P}{\rho} \ . \tag{3.296}$$

We previously derived this result for linearised (i.e. small) sound waves.

Here we make contact with the shock waves. A very weak shock wave can be viewed as the limit of a very strong sound wave.

The Entropy Jump

There's a very basic question that we haven't yet addressed. Which way is the flow going? Is the fluid moving left-to-right, so $u_1, u_2 > 0$ as shown in the earlier figure? Or is it moving right-to-left, with $u_1, u_2 < 0$. In other words, does the pressure increase in the direction of the flow or decrease in the direction of the flow? It turns out that the answer to this question lies in the second law of thermodynamics.

The entropy density $s = S/V$ for an ideal gas was computed in (3.168). It is given by

$$s = c_V \log \left(\frac{P}{P_0} \left(\frac{\rho_0}{\rho} \right)^{\gamma} \right) \tag{3.297}$$

with P_0 and ρ_0 some fiducial values. Indeed, this is where we got the now-familiar equation of state $P\rho^{-\gamma} = $ constant for adiabatic processes which have constant entropy.

We'd like to understand how the entropy changes across the shock,

$$\Delta s = s_2 - s_1 = c_V \log \left(\frac{P_2}{P_1} \left(\frac{\rho_1}{\rho_2} \right)^{\gamma} \right) . \tag{3.298}$$

The second law of thermodynamics means that entropy must increase. But is the change of entropy Δs or is it $-\Delta s$? In other words, does the fluid flow from region 1 to region 2, in which case the change in entropy is Δs. Or does it go from region 2 to region 1, in which case it's $-\Delta s$?

This is straightforward to answer using the expression for the compression ratio (3.290) and (3.293). The entropy jump Δs is then

$$\Delta s = c_V \log \left(\frac{r(\gamma + 1) - (\gamma - 1)}{(\gamma + 1) - r(\gamma - 1)} r^{\gamma} \right) . \tag{3.299}$$

For $r < r_{\max}$, we have $\Delta s > 0$. But the second law of thermodynamics then gives us a direction for the shock: the flow must propagate from left to right, as anticipated in the figure, so that Δs is the change of entropy. Correspondingly, the speed of the flow decreases and the density increases. We say that the shock is *compressive*.

Although we've studied the shock only for an ideal gas, it turns out that the result above is general: shocks are always compressive for any equation of state $P(\rho)$, with the velocity decreasing after the shock.

The fact that the entropy is not constant across the discontinuity means that shocks are necessarily dissipative. There's something a little surprising about this. We've worked with the Euler equation which, as mentioned previously, enjoys the symmetry of time reversal. Moreover, we've also used the adiabatic condition $P\rho^{-\gamma} = \text{constant}$ for an ideal gas. Nonetheless, the discontinuity is a violent event and allows dissipative behaviour to be hidden in the singularity, even though the underlying equations did not themselves have dissipation.

Physically, we've captured the dissipation by allowing the internal, heat energy $e(\rho)$ to increase downstream. A fuller understanding of the dissipation mechanism would need us to look more closely at the shock wave by understanding the role that viscosity plays in thickening the discontinuity. But the results above tell us that, ultimately, these microscopic details don't affect the amount of dissipation: that's fully determined by the properties of the initial flow and some basic conservation laws.

3.6.2 Shocks Start Supersonic

There is more physics to extract from our expressions for the compression ratio. We define the *normal Mach number*

$$\mathcal{M} = \frac{u}{c_s} \qquad\qquad (3.300)$$

where the speed of sound is (3.296)

$$c_s^2 = \frac{\gamma P}{\rho} \ . \qquad\qquad (3.301)$$

This is not quite the same thing as the Mach number (3.255) because we've ignored the tangential velocity v. Each side of the flow has a normal Mach number, $\mathcal{M}_1$ and $\mathcal{M}_2$. Note, in particular, that the speed of sound also differs on either side of the shock. We'll now show that we can express the normal Mach numbers $\mathcal{M}_1$ and $\mathcal{M}_2$ on either side of the flow directly in terms of the compression factor r.

Lemma: The algebra is a little fiddly so we'll tread carefully. To begin, we'll need:

$$\rho_1 u_1^2 = \frac{1}{2}\left[(\gamma - 1)P_1 + (\gamma + 1)P_2\right]$$

$$\rho_2 u_2^2 = \frac{1}{2}\left[(\gamma + 1)P_1 + (\gamma - 1)P_2\right] . \tag{3.302}$$

Proof: We start with two expressions for the compression factor r, the first following from (3.290) and the second (3.293),

$$r = \frac{\rho_1 u_1^2}{\rho_2 u_2^2} = \frac{(\gamma - 1)P_1 + (\gamma + 1)P_2}{(\gamma + 1)P_1 + (\gamma - 1)P_2} . \tag{3.303}$$

Note that if the first equation in (3.302) is true, then (3.303) immediately implies that the second is also true. So we just need to prove the first. This follows from (3.284) which reads $\rho_1 u_1^2 + P_1 = \rho_2 u_2^2 + P_2$. If we divide through by $\rho_1 u_1^2$ then, after a little rearranging, we find

$$\rho_1 u_1^2 = \frac{r}{r - 1}(P_2 - P_1) . \tag{3.304}$$

Now compute $r/(r - 1)$ using the expression in (3.303) involving pressure. This will give the result (3.302) that we want. $\qquad\square$

Now we've done the hard work. We use the expression for the speed of sound $c_s^2 = \gamma P/\rho$ to write the two equations in (3.302) as

$$\gamma \mathcal{M}_1^2 = \frac{1}{2}\left[(\gamma - 1) + (\gamma + 1)\frac{P_2}{P_1}\right]$$

$$\gamma \mathcal{M}_2^2 = \frac{1}{2}\left[(\gamma + 1)\frac{P_1}{P_2} + (\gamma - 1)\right] . \tag{3.305}$$

To finish, we just need an expression for the pressure ratios. We can easily get this from (3.303). It is

$$\frac{P_2}{P_1} = \frac{r(\gamma + 1) - (\gamma - 1)}{(\gamma + 1) - r(\gamma - 1)} . \tag{3.306}$$

Finally we get the results that we wanted: the normal Mach numbers before and after the shock are

$$\mathcal{M}_1^2 = \frac{2r}{(\gamma + 1) - r(\gamma - 1)}$$

$$\mathcal{M}_2^2 = \frac{2}{r(\gamma + 1) - (\gamma - 1)} . \tag{3.307}$$

The key takeaway from these equations is that, for $1 < r \leq r_{\max}$, we always have $\mathcal{M}_1 > 1$ and $\mathcal{M}_2 < 1$, as shown in Figure 3.15. This means that shocks only form in supersonic flows, where the speed of the fluid exceeds the speed of sound. After the shock, the speed of the fluid is reduced below the sound

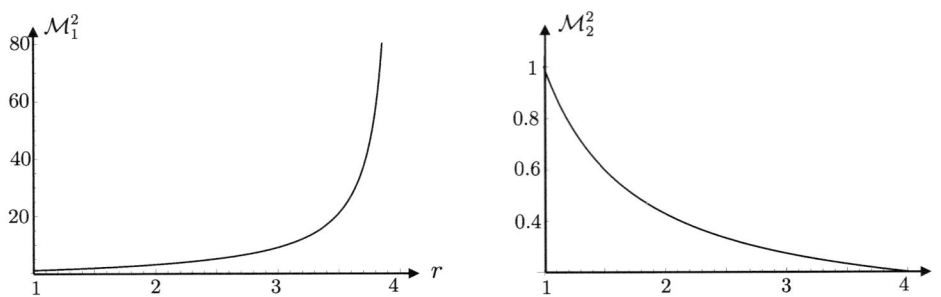

Fig. 3.15 The initial and final Mach numbers as a function of the compression ratio r, plotted for $\gamma = 5/3$. We have $\mathcal{M}_1 \geq 1$ and $\mathcal{M}_2 \leq 1$ for all values of r.

speed. Although, as the reduction of the fluid speed is limited by a factor of $r < r_{\mathrm{max}}$, for very fast flows this is achieved by increasing the pressure, and hence increasing the sound speed, rather than by reducing the flow speed.

From a physical perspective, the equations (3.307) are kind of backwards: the compression factor r doesn't determine the initial speed $\mathcal{M}_1$. It's the other way round! We can easily invert these equations to get the compression factor in terms of the initial Mach number,

$$r = \frac{(\gamma+1)\mathcal{M}_1^2}{2 + (\gamma-1)\mathcal{M}_1^2} \, . \tag{3.308}$$

Similarly, the jump in pressure, given in (3.306), is also determined by the initial speed

$$\frac{P_2}{P_1} = 1 + \frac{2\gamma}{\gamma+1}(\mathcal{M}_1^2 - 1) \, . \tag{3.309}$$

These equations are known as the *Rankine–Hugoniot relations*.

3.6.3 On Singularities and Physics

There is a general expectation that non-linear, partial differential equations will develop singularities in a finite time. When those non-linear equations describe something physical, these singularities are particularly interesting. A singularity is telling us that the equations are no longer sufficient to capture the underlying physics and must be replaced by something more fundamental. This suggests that singularities may offer a window into the microscopic realm.

Within classical physics, there are two pre-eminent sets of non-linear equations. These are the Navier–Stokes equation (or its baby brother, the Euler equation) for fluids, and the Einstein equations for gravity. Both these equations are rather special and the way in which singularities form, or fail to form, is surprising and poorly understood.

For fluids, it's useful to distinguish between the compressible and non-compressible cases. As we've seen above, the compressible Euler equation readily develops singularities in finite time. These are the shock waves that we've explored in this section, characterised by a discontinuity in the density ρ and other dynamical variables. As we anticipated above, the presence of the shock does mean that we have to introduce new physics. But the surprise of this section is that this new physics is the most minimal imaginable: it is just the second law of thermodynamics. Once we accept that entropy must increase when the shock develops, we have all that we need to tell us what happens to the subsequent evolution. We certainly don't need to resort to any detailed microscopic description involving atoms and the quantum world. This is rather remarkable. Shocks may be singular but, from a physical perspective, the singularity is very mild.

It is natural to ask: Is this same property shared by all singularities of the compressible Euler equation? Or, indeed singularities of the compressible Navier–Stokes equation? The answer is: we don't know. For example, what happens when many shocks collide and start to interact with each other? Is it still the case that we can track the singular evolution of the Euler equations using only the second law as our guide? This situation is complicated and we don't know the answer. Moreover, one may worry that there are singularities worse than shocks that can arise in the compressible Euler equation. For example, it may be possible that $\rho(\mathbf{x}, t) \to \infty$ in some finite time. This kind of singularity would surely need some detailed understanding of the underlying atoms to resolve. But does such a singularity actually occur? Again, the answer is: we don't know. It can be shown that such singularities occur for very special initial data, but to be physically relevant they should form for generic initial conditions, meaning initial conditions that lie within some open ball rather than at specific points. And it remains an open problem to show whether or not this happens.

The situation for the incompressible Euler and Navier–Stokes equations is somewhat simpler to state, but still not well understood. Here there is a conjecture that no singularities occur in a finite time. No counterexample is known, but a mathematical proof appears challenging to say the least. Indeed, proving the existence and smoothness of solutions to the Navier–Stokes equation is one of the Millennium Prize problems with a one million dollar prize attached. (If you're genuinely motivated by the money then I would suggest that mathematics may not be your true calling. There are easier ways to be both happy and rich.)

Finally, that leaves us with the Einstein equations of general telativity. Here the situation is most intriguing of all. It is straightforward to show that singularities do develop in finite time (at least with a suitable definition of "time"!). This arises when matter collapses to form a black hole, with a singularity forming in the centre where the curvature of spacetime becomes infinite. The presence of such a singularity is telling us that the laws of classical gravity are breaking down and must be replaced by something quantum. This means that black holes, with their singularities, provide a wonderful opportunity to teach us something new about the "atoms of spacetime", whatever that phrase means. Sadly, however, nature has made these singularities very difficult to access experimentally. It appears that they are generically shielded by an event horizon, so that they can't be seen by anyone sensible who chooses not to jump into the black hole. The idea that singularities necessarily sit behind an event horizon goes by the name of the *cosmic censorship conjecture*. From a mathematical perspective, it appears utterly miraculous and a proof is generally thought to be even more challenging than the Navier–Stokes existence and smoothness conjecture.

The upshot is that the laws of physics appear to be surprisingly robust against the formation of singularities. Even when singularities do arise – as in the compressible Euler equation and the Einstein equations – some poorly understood feature of the equations means that they are more innocuous than we would have naively thought. They are either hidden behind horizons, or neatly resolved by the second law. In both cases, we can largely carry on with our lives without worrying too much about what microscopic physics lurks inside the singularity.

It feels like there is an important lesson hiding within this story. The refusal of both the Navier–Stokes and the Einstein equations to develop readily accessible singularities, that require something atomic or quantum to fully understand, is a striking mathematical fact. It should have a striking physical reason behind it. But I don't know what it is.

3.7 Biological Waves

In this section, we take a short break from our obsession with the equations of fluid mechanics. Instead, we will look at waves in a class of equations known as *reaction-diffusion equations*. These describe the dynamics of some scalar quantity $\phi(\mathbf{x}, t)$, evolving as

$$\frac{\partial \phi}{\partial t} = D\nabla^2 \phi + F(\phi) \ . \tag{3.310}$$

Here D is a constant and $F(\phi)$ is a function that captures the relevant dynamics. Equations of this form are common when describing biological or chemical processes, from the spread of disease to the way wounds heal.

At first glance, it's surprising that reaction-diffusion equations admit wave-like solutions at all. If we set $F(\phi) = 0$, then we're left with the diffusion equation. We will devote Chapter 7 to exploring various properties of this equation, but the key feature follows on dimensional grounds alone: if there is some lump of the ϕ stuff, with size L, then it spreads out as $L \sim \sqrt{Dt}$. This is the characteristic feature of diffusion, and it stands in contrast to wave motion which travels as $L \sim t$.

Nonetheless, for suitably chosen $F(\phi)$, the equation does admit wave-like behaviour. And this is where our interest lies in this section. We will illustrate this through a particular example.

3.7.1 The Fisher–KPP Equation

We will consider a one-dimensional reaction-diffusion equation with the dynamical variable $p(x, t)$, described by

$$\frac{\partial p}{\partial t} = \frac{\partial^2 p}{\partial x^2} + p(1 - p) \ . \tag{3.311}$$

This is the *Fisher–KPP equation* (sometimes just the *Fisher equation*) with the additional initials reflecting the important work done by Kolmogorov, Petrovsky and Piskunov. Fisher originally introduced this equation in 1937 to describe the spread of advantageous genes, with $p(x, t)$ the percentage of the population that carries the gene.

To understand the meaning of this equation, we first ignore the spatial direction x and focus on the change in time. If we set $\partial p/\partial x = 0$, then we're

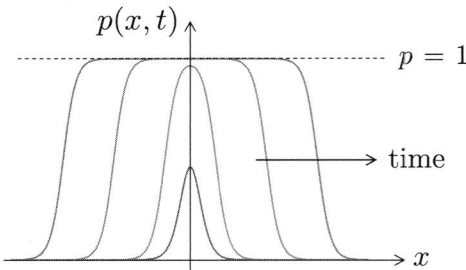

Fig. 3.16 A cartoon of the evolution of a small perturbation in the Fisher equation, spreading out over time as a wavefront.

left with the simple differential equation

$$\frac{\partial p}{\partial t} = p(1 - p) \ . \tag{3.312}$$

This is the *logistic equation*. It is one of the simplest equations used to describe how (suitably rescaled) populations evolve. For $0 < p \leq 1$, it has the solution

$$p(t) = \frac{1}{1 + \alpha e^t} \tag{3.313}$$

with $\alpha > 1$ an integration constant that can be thought of as setting the initial value $p(0) = 1/(1 + \alpha)$. As $t \to \infty$, we have $p(t) \to 1$. This equation does a pretty good job of modelling the fraction of a population that adapts in some advantageous way. For example, the fraction of farmers in various US states who adopted a new variation of hybrid corn is described pretty well by the logistic equation. Eventually, as $t \to \infty$, we have $p(t) \to 1$, which is telling us that everyone eventually succumbs. Viewed as a dynamical system, the logistic equation has two fixed points: $p = 0$ and $p = 1$. The $p = 0$ state is unstable, while the $p = 1$ state is stable.

Now we want to know what happens when we reintroduce the diffusion term in the Fisher–KPP equation (3.311). Suppose that we start close to the $p = 0$ state, but introduce a spatially localised disturbance, meaning that there's some small region where $p \neq 0$. How does this subsequently propagate?

Here is a guess. Suppose that we start with a small disturbance, localised in a region of size L around $x = 0$ at time $t = 0$. We might expect that this perturbation will grow in both height and width, with the top plateauing at the fixed point $p = 1$. If we're sitting at some distance $x \ll L$ from the initial perturbation, we have to wait some time until this perturbation hits us. A cartoon of this dynamics is shown in Figure 3.16.

This suggests that we might look for wave-like solutions to the Fisher–KPP equation. It's worth pointing out that we're not guaranteed that such solutions exist. Indeed, as we already mentioned, the basic diffusion equation does not support wave-like solutions. But the addition of a "reaction term" – meaning the $p(1-p)$ term in (3.311) – changes the story and, as we'll now show, such waves do exist.

We don't know how fast such a wave travels so we'll leave this as arbitrary for now and call it c. We will then look for solutions of the form

$$p(x,t) = f(\xi) \quad \text{with} \quad \xi = x - ct \tag{3.314}$$

with $c > 0$ the as yet unknown wave speed. If we substitute this into the Fisher–KPP equation, then we get an ordinary differential equation

$$-cf' = f'' + f(1-f) \tag{3.315}$$

where $f' = df/d\xi$. Our task is to analyse solutions to this equation. Here we offer a number of ways to do this.

Phase Plane Analysis

To start, we can turn our second order differential equation into a pair of first order differential equations,

$$f' = g \quad \text{and} \quad g' = -cg - f(1-f) . \tag{3.316}$$

This is a simple 2d dynamical system and there's a standard way to go about solving it. First, we look for fixed points such that $f' = g' = 0$. There are two: $(f,g) = (0,0)$ and $(f,g) = (1,0)$.

Next we look at stability of these fixed points. This boils down to determining the eigenvalues of the 2×2 Jacobian matrix

$$J = \begin{pmatrix} 0 & 1 \\ -1 + 2f & -c \end{pmatrix} \tag{3.317}$$

evaluated at the fixed point. If the real part of both eigenvalues are negative, then the fixed point is stable. If both are positive, then the fixed point is unstable. While if one is positive and one negative, then the fixed point is a saddle.

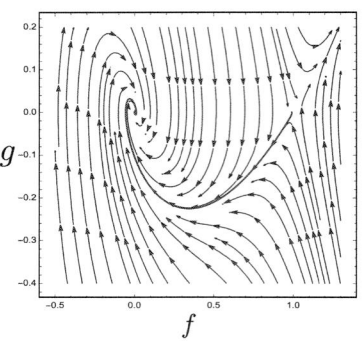 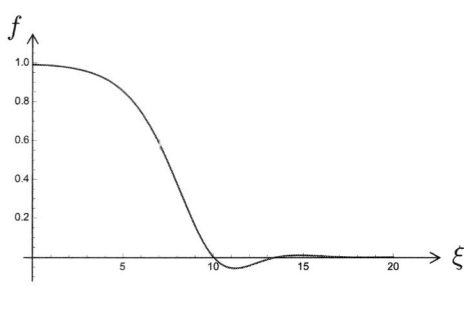

Numerical solutions with $c = 1$. Dynamics in the (f, g) phase plane is shown on the left, with the thick solid line depicting a trajectory that starts close to the fixed point $(f, g) = (1, 0)$. This same trajectory is shown as a function $f(\xi)$ on the right, where it's clear that f dips below zero, rendering the solution invalid.

For the fixed point $(f, g) = (1, 0)$, it's simple to see that we have one positive and one negative eigenvalue. (For example, the product of eigenvalues is $\det J = -1$.) This means that the fixed point is a saddle.

The other fixed point at $(f, g) = (0, 0)$ is more interesting. The eigenvalues λ of the Jacobian are

$$\lambda^2 + c\lambda + 1 = 0 \quad \Longrightarrow \quad \lambda = -\frac{c}{2} \pm \frac{1}{2}\sqrt{c^2 - 4} \,. \tag{3.318}$$

For $c < 2$, the eigenvalues are complex, with negative real part, whuch means that the fixed point is a stable focus, with trajectories spiralling in. For $c > 2$, the eigenvalues are real and negative and so the fixed point is stable.

The fact that the flows in the phase plane have qualitatively different behaviour for $c < 2$ and $c > 2$ is important. In particular, we can look at the kind of solutions we get with $c < 2$. These are plotted numerically in Figure 3.17. While these are fine formal solutions to the Fisher equation, because they spiral into the origin they necessarily have a region of ξ for which $f < 0$. But if we're thinking of $f(\xi)$ as the fraction of a population then we want $f \geq 0$. This means that, for our present purpose, we discard the solutions with $c < 2$.

That leaves us with $c \geq 2$. Here there is no such concern. A numerical plot of this solution (shown with $c = 2$) is depicted in Figure 3.18. Now there is a solution that starts near the unstable fixed point and heads directly

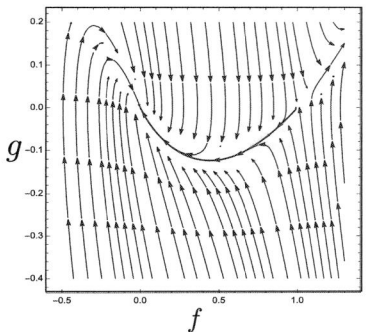 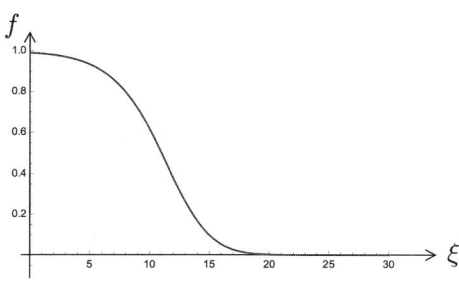

Fig. 3.18 Numerical solutions with $c = 2$. Dynamics in the (f, g) phase plane is shown in the left, again with a highlighted trajectory that starts close to $(f, g) = (1, 0)$. The evolution of $f(\xi)$ is shown on the right, now remaining positive for all ξ.

towards the stable fixed point. These are the class of solutions that we will be interested in.

It's worth pointing out that something a little strange has happened here. We wanted to find solutions where we start at $f = 0$ and then perturb slightly to see what happens. Instead, our phase space analysis has resulted in solutions that seem to go the opposite way, with

$$f(\xi) \to 1 \quad \text{and} \quad \xi \to -\infty \quad \text{and} \quad f(\xi) \to 0 \quad \text{as} \quad \xi \to +\infty \; . \quad (3.319)$$

In fact, this is just because the ξ coordinate is defined as $\xi = x - ct$ and that minus sign is the cause for the strange behaviour. For fixed x, these same solutions obey $f(t) \to 0$ as $t \to -\infty$ and $f(t) \to 1$ as $t \to +\infty$.

A Mechanical Analogy

There's a way to translate the story above into something familiar. The equation (3.315) is the kind of thing that we studied in Volume 1 on Classical Mechanics. We can write (3.315) as

$$f'' = f(f - 1) - cf' = -\frac{dV}{df} - cf' \quad (3.320)$$

with

$$V(f) = \frac{1}{2}f^2 - \frac{1}{3}f^3 \; . \quad (3.321)$$

It then looks like the equation of motion for a particle moving in a potential V with a friction term $-cf'$.

The potential is plotted in the figure to the right. It has two critical points, at $f = 0$ and at $f = 1$. In this analogy, the point $f = 1$ is again unstable and the point $f = 0$ is stable. That's compatible with what we saw above: in the $\xi = x - ct$ coordinate, we flow from $f = 1$ to $f = 0$ rather than the other way around.

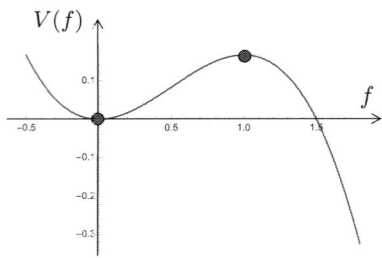

We've seen above that we get qualitatively different behaviour for $c < 2$ and for $c > 2$. It's simple to see why in this mechanical analogy, where c dictates the strength of the friction force. For $c < 2$, the system is underdamped, meaning that, as it rolls down the hill, it overshoots the minimum at $f = 0$, oscillating back and forth before settling down. This is the behaviour seen in Figure 3.17.

In contrast, for $c > 2$ the system is overdamped, slowing enough so that it stops when it ultimately reaches the minimum at $f = 0$. The phase plane analysis tells us that the crossover between these two behaviours happens at $c = 2$ when the system has critical damping.

Linearised Analysis

We can learn more about the travelling wave solution by looking at the leading edge of the wave, where $f \approx 0$. This means that we're looking at the region of the graph in Figure 3.18 where $f(\xi)$ is approaching the ξ-axis. Here it's appropriate to linearise the equation (3.315) and work with

$$-cf' = f'' + f \ . \tag{3.322}$$

We make the obvious ansatz $f(\xi) = e^{-\lambda\xi}$, with $\lambda > 0$ so that this solution decays towards $f \to 0$ as ξ increases. We see that this solves the equation if

$$\lambda^2 - c\lambda + 1 = 0$$
$$\implies \quad c = \lambda + \frac{1}{\lambda} \ . \tag{3.323}$$

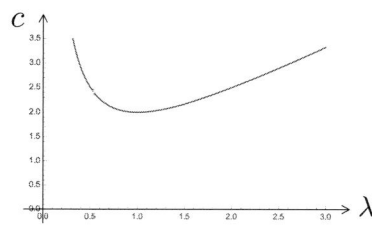

This is plotted in the figure to the right. We again see that the speed is bounded below by $c \geq 2$. Moreover, because we have $f(\xi) = e^{-\lambda\xi}$, we can think of $1/\lambda$ as the width of the wavefront. We learn that the speed and shape of the wave are related.

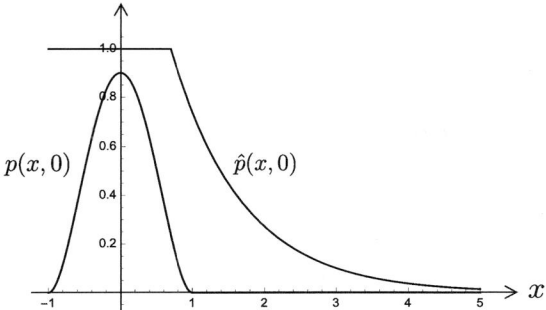

Fig. 3.19 The initial profile $p(x,0)$ that evolves through the full non-linear equation is bounded above by the exponentially decaying profile $\hat{p}(x,0)$ that will evolve through the linearised Fisher equation.

The Speed of the Non-Linear Wave

So far, nothing has told us what speed c the wave travels at if we start with a given initial, localised perturbation $p(x,0)$. We only know that this speed must be $c \geq 2$.

The full non-linear analysis is complicated but the final result, proven by Kolomgorov, is beautifully simple. If we start with some initial conditions that vanish outside of some interval, i.e. $p(x,0) = 0$ for all $|x| > x_0$, then the system will ultimately settle down to a wave that travels with speed $c = 2$. In other words, the non-linear system travels at the *slowest* possible speed of the linearised system.

We won't prove this result here, but we can motivate it with the following argument. First note that the non-linear speed must be *one* of the allowed linear speeds $c \geq 2$ just because the linearised analysis is valid at the wave-front. But, with suitably localised initial conditions, we can show that the non-linear speed must be less than (or equal to) that of a linear wave.

To see this, let's take initial conditions that are strictly localised in some region

$$p(x,0) = 0 \quad \text{for all } |x| \geq x_0 . \tag{3.324}$$

We'll evolve this with the full non-linear Fisher–KPP equation (3.311).

Our strategy is to set this profile in a race against a wavefront with initial profile

$$\hat{p}(x,0) = Ae^{-\lambda x} . \tag{3.325}$$

We will pick A so that $p(x,0) < \hat{p}(x,0)$. It's simple to check that it is always possible to construct such a bounding profile for any choice of $\lambda > 0$ simply by picking a suitable A. An example is shown in Figure 3.19. Crucially, we evolve the profile $\hat{p}(x,t)$ through the *linearised* Fisher–KPP equation

$$\frac{\partial \hat{p}}{\partial t} = \frac{\partial^2 \hat{p}}{\partial x^2} + \hat{p} \ . \tag{3.326}$$

Now we set these two profiles off. We will show that it's not possible for the non-linear $p(x,t)$ to overtake the linearised $\hat{p}(x,t)$. At heart, this follows because the missing term in the linearised equation is $-p^2$ and, with the minus sign, this only serves to delay the non-linear evolution.

We can put some meat on this argument by defining $\beta(x,t) = \hat{p}(x,t) - p(x,t)$. By construction, we have $\beta(x,0) > 0$. Watching this function evolve in time, we have

$$\frac{\partial \beta}{\partial t} = \frac{\partial \hat{p}}{\partial t} - \frac{\partial p}{\partial t} = \frac{\partial^2 \beta}{\partial x^2} + \beta + p^2 \geq \frac{\partial^2 \beta}{\partial x^2} + \beta \ . \tag{3.327}$$

We learn that the evolution of $\beta(x,t)$ is at least as fast as diffusion with linear growth. And with $\beta(x,0)$ positive, the function $\beta(x,t)$ can never go negative. This is telling us that $p(x,t)$ is bounded above by $\hat{p}(x,t)$ for all time. In other words, the non-linear wave can never overtake the linear wave.

But the analysis above holds for any choice of λ and, in particular, for $\lambda = 1$ which travels at the slowest speed $c = 2$. It tells us that the non-linear wave can travel no faster than $c = 2$. But, as we've seen previously, the wave ansatz only makes sense for $c \geq 2$. Hence the non-linear wave must travel at the slowest possible speed $c = 2$.

3.7.2 Front Propagation in Bistable Systems

Here is a simple modification of the Fisher–KPP equation

$$\frac{\partial p}{\partial t} = \frac{\partial^2 p}{\partial x^2} - p(p - r)(p - 1) \ . \tag{3.328}$$

We'll take $0 < r < 1$.

Now, in the homogeneous system, where $\partial p/\partial x = 0$, both $p = 0$ and $p = 1$ are stable fixed points. The intermediate value of $p = r$ is unstable. For this reason, systems of this type are referred to as *bistable*.

Suppose that we start in a system with $p = 0$ to the left, as $x \to -\infty$, and $p = 1$ to the right, as $x \to +\infty$. In between there has to be a transition, which we will call a "front". It looks something like the plot shown in the figure. The question that we would like to ask is: What happens next? Does the front advance to the left, or to the right? Since both $p = 0$ and $p = 1$ are stable fixed points, it's not immediately obvious which will win. We could think of this as a model for how diseases, mutations, or chemicals spread or die out.

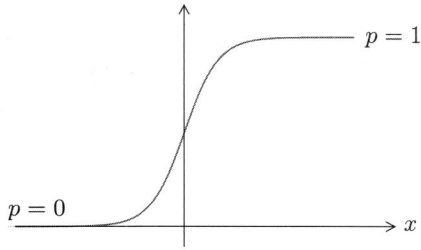

We can use the kind of analysis that we developed for the Fisher equation to answer this. We will again look for a travelling wave solution, with

$$p(x,t) = f(\xi) \quad \text{with} \quad \xi = x - ct \qquad (3.329)$$

for some velocity c. With this ansatz, the reaction-diffusion equation (3.328) becomes

$$-cf' = f'' - f(f - r)(f - 1) . \qquad (3.330)$$

This time we can get all the information we need from the mechanical analogy. We write

$$f'' = -\frac{dV}{df} - cf' \quad \text{with} \quad V(f) = -\frac{1}{4}f^4 + \frac{1}{3}(1 + r)f^3 - \frac{1}{2}rf^2 . \qquad (3.331)$$

This potential has two maxima, at $f = 0$ and $f = 1$. But, crucially, the shape of the potential depends on the value of r. For $r < 0.5$, the maximum at $f = 1$ is higher; for $r > 0.5$, the maximum at $f = 0$ is higher. Two representative examples are shown in Figure 3.20.

When $r < 0.5$, the ball rolls down from $f = 1$. If the friction term, captured by $-cf'$, is large, then the ball will ultimately come to rest at the local minimum at $f = r$. If the friction is low, then the ball will sail past the local maximum at $f = 0$ and into oblivion. In both cases, there is a formal solution to the reaction-diffusion equation but not one with the boundary conditions that we want. However, there is a special value of c such that the friction is just right and the ball rolls down from $f = 1$ where it sat at $\xi \to -\infty$, coming to rest at $f = 0$ at $\xi \to +\infty$. This is the velocity c that we want.

When $r > 0.5$, the heights of the two maxima are inverted, and now there

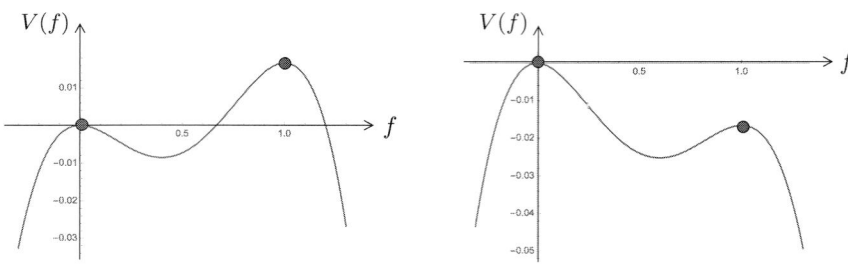

Fig. 3.20 On the left: the potential $V(f)$ plotted with $r = 0.4$. On the right: the potential plotted with $r = 0.6$.

is a critical value of the friction where the ball rolls the other way: from $f = 0$ to $f = 1$.

As in the analysis of the Fisher equation, we have to remember that the mechanical analogy reverses what actually happens because $\xi = x - ct$. This means that:

- For $r < 0.5$ we have, for fixed x, $p(x, t) \to 1$ as $t \to \infty$ and so the front moves to the left.

- For $r > 0.5$ we have, for fixed x, $p(x, t) \to 0$ as $t \to \infty$ and so the front moves to the right.

In contrast to the Fisher equation, the mechanical analogy tells us that the front must travel at a very specific velocity c. But what is it? We can make progress by computing the analog of the work done in our mechanical system. We can take the equation of motion (3.331), multiply by f', and integrate. We have

$$\int_{-\infty}^{+\infty} d\xi \; f' f'' = \int_{-\infty}^{+\infty} d\xi \; \left(-f' \frac{dV}{df} - c f'^2 \right)$$
$$= \int_{-\infty}^{+\infty} d\xi \; \left(-\frac{dV}{d\xi} - c f'^2 \right) . \tag{3.332}$$

The left-hand side is a total derivative of $\frac{1}{2} f'^2$ but our boundary conditions mean that $f' \to 0$ as $\xi \to \pm\infty$. This leaves us with an expression for the velocity

$$c = -\frac{\Delta V}{\int d\xi \; f'^2} . \tag{3.333}$$

Here ΔV is the difference in energy between the two maxima,

$$\Delta V = V(f = 1) - V(f = 0) = \frac{1}{6}\left(\frac{1}{2} - r\right) . \qquad (3.334)$$

Note that if $r < 0.5$, then $\Delta V > 0$ and so $c < 0$ and the front travels to the left. Meanwhile, if $r < 0.5$ then $c > 0$ and the front travels to the right. This agrees with what we saw above.

To determine the actual speed c, we still need to do the integral $\int d\xi \, f'^2$. And, for this, we typically need to first figure out the solution $f(\xi)$. That's not so easy, but there is one value of r for which we can solve the equation of motion exactly. This is the value $r = 0.5$, when the local maxima of the potential have the same height and, correspondingly, the front is static with $c = 0$ and doesn't move. In this case, you can check that the function $f(\xi)$ is given by

$$f = \frac{1}{2}\left(1 - \tanh\left(\frac{\xi}{2\sqrt{2}}\right)\right) . \qquad (3.335)$$

If we are cheeky and use this as a proxy for the function $f(\xi)$ in the integral, then we get a speed

$$c = 2\left(r - \frac{1}{2}\right) . \qquad (3.336)$$

Strictly, this calculation holds only when $r = 0.5$ where the speed vanishes! But we can view this as a good approximation to the speed for $r \approx 0.5$. Indeed, if we were more careful we could set up a perturbation expansion in $r - 0.5$, with this the leading order term.

Localised Perturbations

There's a closely related question that we can ask of this system? Suppose that we start in the $p = 0$ state but then introduce a small region of the $p = 1$ state, localised around the origin. What then happens? Does this $p = 1$ state expand to take over the world? Or does the $p = 0$ state fight back and shrink it?

The analysis that we've done above goes part way towards answering this. If $r > 0.5$ then the $p = 0$ state will win and the $p = 1$ insurgents will be crushed. Meanwhile, when $r < 0.5$ then there's an opportunity for the $p = 1$ state to expand. But it's not guaranteed. That's because our previous analysis assumed that the system settled down to some wavelike behaviour with the fronts propagating at some constant speed. But the question of

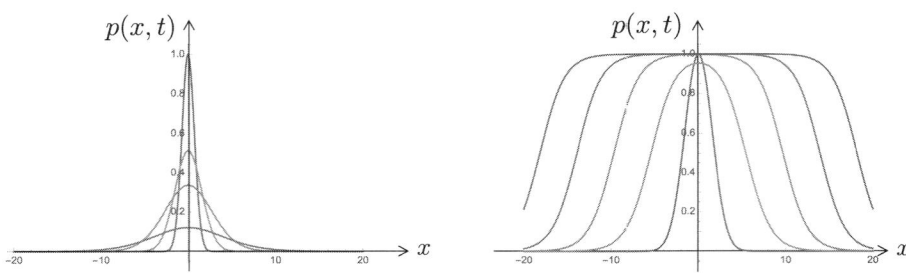

Fig. 3.21 Numerical solutions for $r = 0.2$. The plot on the left starts with the initial condition $p(x, 0) = e^{-x^2}$. The one on the right starts with the wider initial condition $p(x, 0) = e^{-x^2/5}$. We see that the first flounders while the second flourishes.

when this happens depends on how wide the initial localised perturbation is.

We won't give any detailed analysis of this behaviour here, but instead exhibit some simple numerical solutions. Figure 3.21 shows the time evolution of two different initial perturbations, the one on the left narrower than the one on the right. You can see clearly that the one on the left shrinks over time, while the one on the right thrives, ultimately forming wavefronts that are described by the analysis above.

We started this book by studying some simple laminar flows in which the fluid moves smoothly through space. One important question that we haven't yet addressed is: Are these flows stable? If we perturb them in some way, does the flow persist or does it get driven to something more complicated?

The answer is that many flows, even very simple ones, can be unstable. This statement is especially true at high Reynolds number, where viscosity fails to dampen the perturbations. You can see an example in the photographs of Poiseuille flow, describing fluid flowing along a pipe, shown in Figure 4.1. The fluid in the pictures is transparent and the flow is moving from left to right, but some black dye is released from the same point on the left-hand edge and traces out what is known as a *streakline*. (For time-independent flows, the streakline coincides with both the pathline and the streamline that we introduced previously. For time-dependent flows, it differs.)

The picture on the left of Figure 4.1 shows the flow when it is slow, so that $Re \lesssim 1$ and the fluid is well described by the solution that we already met in Section 2.2.3. The flow is laminar, with the velocity field lying parallel to the pipe and the picture looks boring.

But as the flow speeds up, so that $Re \approx 100$ or so, things become more interesting. This is shown in the middle picture where the streakline starts to wobble in places, with the wobbles dragged along by the flow.

Finally, by the time we are in the regime $Re \approx 10^3$ or 10^4 (the exact number depends on how rough the boundary of the pipe is), the flow appears qualitatively different yet again. This is shown on the right of Figure 4.1. Any clearly defined wobbles have vanished. The flow just looks messy. This is the turbulent regime.

Our goal in this section is to start to understand these kinds of instabilities. Although Poiseuille and Couette are the simplest examples of laminar flows, it turns out that understanding their instabilities is not so straightforward. For this reason, we start by looking at specific instabilities in other contexts.

Poiseuille flow for low Reynolds number (on the left), where the flow is laminar; for medium Reynolds number (in the middle), where the instability starts to reveal itself; and for high Reynolds number (on the right), where the flow is turbulent.

Our method to analyse instabilities will mirror the analysis of waves in Chapter 3. That is: we start with some background flow and perturb it. The perturbations that we call waves oscillate back and forth about the original flow. In contrast, the unstable perturbations that we will meet here grow without bound. These are known as *linear instabilities*. Our linear analysis only shows the beginning of the instability, rather than the end point of the flow, but with some imaginative thinking (and the help of experiment!) we can figure out the qualitative form of the final flow.

With these successes in hand, we then turn to understand the stability of seemingly simpler flows, like Couette flow and Poiseuille flow. As we mentioned above, it turns out that understanding the fate of these flows is somewhat harder. We will succeed in giving some general results about when flows of this kind are stable against linear perturbations and when they are not. However, we will see that these results do not stand up particularly well against either experiment or numerical simulation. This is because, ultimately, these flows have more complicated *non-linear instabilities* where an arbitrarily small perturbation is innocuous, but the instability manifests itself only when the perturbation grows to a certain size.

4.1 Kelvin–Helmholtz Instability

Our first set-up is straightforward. We have two fluids, with densities $\rho_1 \leq \rho_2$. The lighter fluid sits on top and travels with constant speed U_1 in the x-direction. The heavier fluid sits at the bottom and travels with constant speed $U_2 \neq U_1$ in the same direction. Initially, there is an interface between the two at $z = 0$ as shown in Figure 4.2.

The initial velocity profile has a discontinuity at $z = 0$. In reality, this will be smoothed out by a thin layer in which viscosity is important. For our purposes, we'll neglect this and study the flow using the Euler equation for

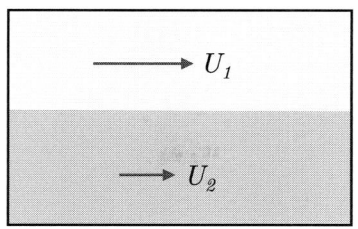 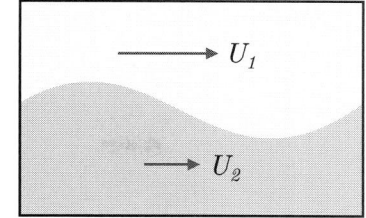

The Kelvin–Helmholtz instability between two moving fluids. The flow in the left-hand figure is unstable and turns into that on the right.

each fluid

$$\rho_i \left(\frac{\partial \mathbf{u}_i}{\partial t} + \mathbf{u}_i \cdot \nabla \mathbf{u}_i \right) = -\nabla P_i + \rho_i \mathbf{g} \quad \text{with} \ \ i = 1, 2 \ . \tag{4.1}$$

Note that we've included the effect of gravity on the right-hand side. It turns out that this won't be necessary to exhibit the physics that we're interested in, but it does have an interesting role to play.

We want to understand whether the initial flow is stable. (Spoiler: it won't be!) In particular, we'll look at perturbations in which the interface moves as

$$z = \eta(x, t) = \eta_0 e^{ikx - i\omega t} \ . \tag{4.2}$$

In other words, we're looking for an instability in which the interface develops waves.

The analysis is almost identical to that of Section 3.1 where we first met surface waves. We look for 2d flows, with $\mathbf{u} = \mathbf{u}(x, z, t)$. We will assume that the flow remains irrotational after the perturbation, so we introduce velocity potentials and write

$$\mathbf{u} = \begin{cases} \mathbf{U}_1 + \nabla \phi_1 & \text{for } z > \eta \\ \mathbf{U}_2 + \nabla \phi_2 & \text{for } z < \eta \end{cases} \ . \tag{4.3}$$

where $\mathbf{U}_i = U_i \hat{\mathbf{x}}$. The requirement that the fluid is incompressible then means that we must, once again, solve the Laplace equation

$$\nabla^2 \phi_1 = \nabla^2 \phi_2 = 0 \ . \tag{4.4}$$

As usual, all the subtleties lie in the boundary conditions. We require that the fluid returns to its initial state asymptotically, so $\phi_1 \to 0$ as $z \to \infty$ and $\phi_2 \to 0$ as $z \to -\infty$. On the interface, we impose the free boundary condition that we described previously in (3.13)

$$u_z = \frac{D\eta}{Dt} \; . \tag{4.5}$$

This now reads

$$\left.\frac{\partial \phi_i}{\partial z}\right|_{z=\eta} = \frac{\partial \eta}{\partial t} + \left(U_i + \left.\frac{\partial \phi_i}{\partial x}\right|_{z=\eta}\right)\frac{\partial \eta}{\partial x} \quad \text{for} \; i = 1, 2 \; . \tag{4.6}$$

Finally, the pressure forces on either side of the interface must balance the surface tension

$$P_2(x, \eta) - P_1(x, \eta) = -\gamma\frac{\partial^2 \eta}{\partial x^2} \tag{4.7}$$

with γ the surface tension. Rather like gravity, it will turn out that the presence of surface tension is unnecessary to explain the main physics point that we're interested in, but has an interesting implication later on.

To implement the pressure difference equation, we follow the analysis of Section 3.1.2 and use Bernoulli's principle. The analogue of our earlier result (3.17) is now

$$\rho_i\frac{\partial \phi_i}{\partial t} + \frac{1}{2}\rho_i|\mathbf{U}_i + \nabla\phi_i|^2 + P_i + \rho_i gz = f_i(t) \quad \text{for} \; i = 1, 2 \tag{4.8}$$

with $f_i(t)$ independent of space. The condition (4.7) then becomes

$$\rho_1\left(\frac{\partial \phi_1}{\partial t} + \frac{1}{2}|\mathbf{U}_1 + \nabla\phi_1|^2 + gz\right)_{z=\eta}$$

$$- \rho_2\left(\frac{\partial \phi_2}{\partial t} + \frac{1}{2}|\mathbf{U}_2 + \nabla\phi_2|^2 + gz\right)_{z=\eta} = \tilde{f}(t) - \gamma\frac{\partial^2 \eta}{\partial x^2} \tag{4.9}$$

for some spatially independent function $\tilde{f}(t)$. This, then, is our goal: solve the Laplace equations subject to (4.6) and (4.9).

The Linearised Approximation

As for surface waves, we make progress by assuming that the amplitude of the perturbation is small. For us, this means

$$k\eta_0 \ll 1 \; . \tag{4.10}$$

This allows us to linearise the boundary conditions (4.6) and (4.9). The first (4.6) becomes

$$\left.\frac{\partial \phi_i}{\partial z}\right|_{z=0} = \frac{\partial \eta}{\partial t} + \left.U_i\frac{\partial \eta}{\partial x}\right|_{z=0} \quad \text{for} \; i = 1, 2 \tag{4.11}$$

while (4.9) becomes

$$\rho_1 \left(\frac{\partial \phi_1}{\partial t} + U_1 \frac{\partial \phi_1}{\partial x} \right)_{z=0} - \rho_2 \left(\frac{\partial \phi_2}{\partial t} + U_2 \frac{\partial \phi_2}{\partial x} \right)_{z=0} + g(\rho_1 - \rho_2)\eta$$
$$= \tilde{f}(t) - \gamma \frac{\partial^2 \eta}{\partial x^2} \ . \tag{4.12}$$

Now we have something eminently more achievable on our hands: solve the Laplace equations subject to (4.11) and (4.12).

4.1.1 The Simplest Instability

To illustrate the key idea, we first ignore both gravity and surface tension. This means that we set $g = \gamma = 0$ in (4.12). We've already shown our hand for the kind of perturbation (4.2) of the interface that we're looking for. We augment this with a commensurate wavy solution for the velocity perturbations

$$\phi_i(x, z, t) = \hat{\phi}_i(z) e^{ikx - i\omega t} \quad \text{and} \quad \eta(x, t) = \eta_0 e^{ikx - i\omega t} \ . \tag{4.13}$$

The Laplace equations tell us that

$$\frac{d^2 \hat{\phi}_i}{dz^2} = k^2 \hat{\phi}_i \quad \Longrightarrow \quad \begin{cases} \hat{\phi}_1 = A_1 e^{-kz} \\ \hat{\phi}_2 = A_2 e^{+kz} \end{cases} \tag{4.14}$$

where the solutions have been chosen so that $\hat{\phi}_1 \to 0$ as $z \to +\infty$ and $\hat{\phi}_2 \to 0$ as $z \to -\infty$. The two boundary conditions in (4.11) then tell us that

$$-kA_1 = (-i\omega + ikU_1)\eta_0 \quad \text{and} \quad +kA_2 = (-i\omega + ikU_2)\eta_0 \tag{4.15}$$

while the fact that $\tilde{f}(t)$ in (4.12) is independent of x and z means that

$$\rho_1(-i\omega + ikU_1)A_1 = \rho_2 (-i\omega + ikU_2) A_2 \ . \tag{4.16}$$

We can eliminate A_1, A_2, and η_0 to get a quadratic in ω

$$(\rho_1 + \rho_2)\omega^2 - 2k\omega(\rho_1 U_1 + \rho_2 U_2) + k^2(\rho_1 U_1^2 + \rho_2 U_2^2) = 0 \ . \tag{4.17}$$

The roots of this quadratic give us the dispersion relation

$$\omega = \frac{k}{\rho_1 + \rho_2} \left[(\rho_1 U_1 + \rho_2 U_2) \pm i\sqrt{\rho_1 \rho_2} |U_1 - U_2| \right] \ . \tag{4.18}$$

The key piece of physics is sitting in that factor of i. This is the telltale sign of an instability. To see this, we substitute this frequency into the expression (4.2) for the interface, to learn that the perturbations of the interface evolve as

$$z = \eta(x, t)$$
$$= \eta_0 \exp\left(ik \left(x - \frac{\rho_1 U_1 + \rho_2 U_2}{\rho_1 + \rho_2} t \right) \pm \frac{\sqrt{\rho_1 \rho_2}}{\rho_1 + \rho_2} |U_1 - U_2| kt \right) \ . \quad (4.19)$$

The first term comes with an oscillatory factor of i and tells us that the disturbance propagates with velocity

$$v = \frac{\rho_1 U_1 + \rho_2 U_2}{\rho_1 + \rho_2} \ . \quad (4.20)$$

But our real interest lies in the second term. Here there's no factor of i: the oscillatory time-dependent behaviour $e^{-i\omega t}$ becomes exponential growth or decay when ω is complex. We learn that the perturbation grows exponentially with time. Or, more precisely, one perturbation grows and the other decays. The existence of the growing mode means that the original flow is unstable. This is the *Kelvin–Helmholtz instability*.

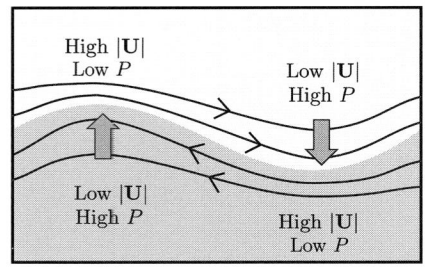

The next question that we can ask is: Why does this happen? Some intuition comes from boosting to a frame where the disturbance is stationary and then thinking about the streamlines, as shown in the figure. The streamlines of the upper fluid are more clustered together near the peaks and so, because of mass conservation, the fluid must travel faster there. Bernoulli's theorem then tells us that the pressure is smaller. But, in the trough, the streamlines are less closely packed, the fluid moves slower and so the pressure is greater.

This story is reversed for the lower fluid. The pressure is now greater in the peak and lower in the trough. Of course, the pressure is continuous across the interface itself – this was one of our initial boundary conditions (4.7) – but there is a pressure difference over the whole perturbation and this drives the crest upwards and the trough downwards. This is the reason for the instability.

In fact, this same effect also explains the existence of the decaying mode. If we set up a perturbation where the crest is travelling downwards, and the trough is travelling upwards, then the pressure differences now act to decelerate the amplitude and the disturbance grows smaller over time.

Viscosity Acts as a UV Cut-Off

For a fixed background fluid flow, the instability (4.19) grows as $\sim e^{Ukt}$. This means that the small wavelength modes, with k large, are more unstable. In fact, taken at face value the instability for very small wavelengths (i.e. very large k) grows without bound. What should we make of this? Are the equations telling us that the continuum description of fluids will ultimately break down, and the interface should be thought of in terms of individual atoms?

Thankfully, no. There are (at least) two mechanisms within the continuum description that halt the runaway behaviour for large k. One of these is surface tension, and we will describe this shortly. But even in the absence of surface tension, viscosity does the job. We should replace the discontinuity in the velocity profile with an appropriate boundary layer. Our analysis above is then valid only for wavenumbers $k \lesssim |U_1 - U_2|/\nu$. For wavelengths smaller than this, the perturbation is stabilised by the effects of viscosity. This is a common theme in fluid dynamics, and one that we will meet again in Chapter 5 when we discuss turbulence: viscosity acts as a UV cut-off.

In the present context, even in the absence of viscosity it is more realistic to model the boundary with a finite-size interface over which the velocity varies continuously. This is known as a *finite-depth shear layer*. We will not extend our analysis to consider this situation here.

4.1.2 Rolling Up the Vortex Sheet

There is another way to think about the instability, this time in terms of vorticity. If we integrate around a rectangular contour in the (x, z)-plane that crosses the interface, with sides of length L in the x-direction, then the initial flow has circulation

$$\Gamma = \oint \mathbf{u} \cdot d\mathbf{x} = (U_1 - U_2)L . \tag{4.21}$$

This means that there must be a corresponding vorticity. The flows on either side of the interface are irrotational, so this vorticity must be localised at the interface itself. In the initial flow, the interface has constant vorticity per unit length. For this reason, the interface is referred to as a *vortex sheet*.

The vorticity points in the direction $\boldsymbol{\omega} = |\boldsymbol{\omega}|\hat{\mathbf{y}}$, where $\hat{\mathbf{y}}$ points out of the page in the previous figures. We'll refer to the magnitude of vorticity as $|\boldsymbol{\omega}|$ rather than ω to avoid confusion with the frequency. Despite the name $|\boldsymbol{\omega}|$

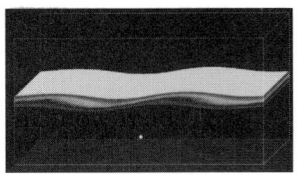 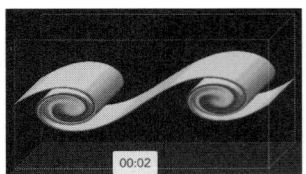 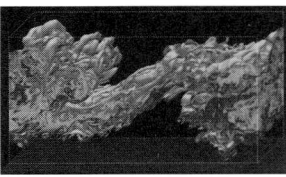

Fig. 4.3 When non-linear effects are taken into account, the Kelvin–Helmholtz instability first rolls up the vortex sheet, before ultimately becoming turbulent. Credit: W. D. Smyth.

can, like the velocity field, be complex. Indeed, it is given by

$$|\boldsymbol{\omega}| = (U_1 - U_2)\delta(z) \tag{4.22}$$

with $\delta(z)$ a delta function, localising the vorticity to the interface. We can view the instability as a deformation of this vortex sheet. Evaluated on the linearised solution, we have

$$|\boldsymbol{\omega}| = \left[(U_1 - U_2) + \frac{\partial \phi_1}{\partial x}\bigg|_{z=0^+} - \frac{\partial \phi_2}{\partial x}\bigg|_{z=0^-} \right] \delta(z - \eta)$$

$$= \left[(U_1 - U_2) + (-2\omega + k(U_1 + U_2))\eta_0 e^{ikx - i\omega t} \right] \delta(z - \eta) \ . \tag{4.23}$$

It's simplest to illustrate the physics if we restrict to the case $\rho_1 = \rho_2$, so that we have two identical fluids travelling at different speeds. Then, using the dispersion relation (4.18), this becomes

$$|\boldsymbol{\omega}| = \left[(U_1 - U_2) - i|U_1 - U_2|k\eta \right] \delta(z - \eta) \tag{4.24}$$

with $\eta = \eta_0 e^{ikx - i\omega t}$ the unstable boundary. We see that, as the perturbation grows, the vorticity is no longer constant in space. It tracks the development of the interface but, because of that factor of i, is $\pi/2$ out of phase, corresponding to

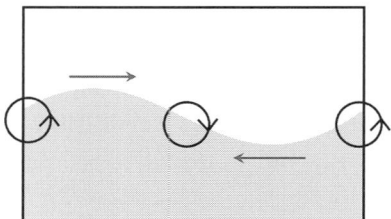

a quarter of a wavelength. This means that there is no change to the vorticity at the maxima and minima, but the vorticity alternates clockwise/anticlockwise at the midway points of the perturbation, as shown in the figure above. This acts to push the crests up, and the troughs down.

Our analysis throughout this section has been done in the linearised approximation, where the perturbation is small. However, the vortex picture gives us a good sense for what happens as the perturbation grows. If we sit in the frame where the disturbance doesn't propagate, then the upper fluid moves to the right, and the lower fluid moves to the left. The vorticity is advected in this direction, and so accumulates at the midpoints between the peaks and troughs, shown by the circles in the previous figure.

As the perturbation grows, so too does the vorticity density in this region, with the result that the perturbation starts to curl around, or "roll up". The end result is the distinctive and beautiful rolling-wave feature of the Kelvin–Helmholtz instability that can be seen in the cloud  formations shown in the picture. A simulation of the instability is shown in in Figure 4.3.

4.1.3 Gravity Helps. Surface Tension Helps Too

With our basic understanding of the instability, we can now repeat the analysis with gravity and surface tension turned on. Both manifest themselves in the boundary condition (4.12). After some algebra, the steps that previously led us to (4.17) now give

$$(\rho_1 + \rho_2)\omega^2 - 2\omega k(\rho_1 U_1 + \rho_2 U_2) + k^2(\rho_1 U_1^2 + \rho_2 U_2^2) = \gamma k^3 + gk(\rho_2 - \rho_1) \ .$$

Solving this as a quadratic in ω gives the dispersion relation

$$\omega = \frac{1}{\rho_1 + \rho_2}\left[k(\rho_1 U_1 + \rho_2 U_2)\right.$$
$$\left. \pm \sqrt{(\rho_1 + \rho_2)(\gamma k^3 + gk(\rho_2 - \rho_1)) - \rho_1\rho_2 k^2(U_1 - U_2)^2}\right] \ .$$

This reduces to (4.18) when $g = \gamma = 0$. The presence of gravity and surface tension means that the function under the square root is no longer negative for all k. This is telling us that some wavelengths are no longer unstable, while others are.

First, suppose that we have gravity $g \neq 0$ but negligible surface tension $\gamma = 0$. In this case, there is an instability only if

$$k > \frac{\rho_2^2 - \rho_1^2}{\rho_1\rho_2}\frac{g}{(U_1 - U_2)^2} \ . \tag{4.25}$$

We see that the long wavelength (small k) modes are no longer unstable. Heuristically, these modes are heavier and so the effect of gravity pulling them down wins over the runaway instability. However, gravity does not stabilise the high k, short wavelength modes.

Next suppose that we have surface tension $\gamma \neq 0$, but gravity is negligible so $g = 0$. Then there is an instability only if

$$k < \frac{\rho_1 \rho_2}{\rho_1 + \rho_2} \frac{(U_1 - U_2)^2}{\gamma} \ . \tag{4.26}$$

This time the short wavelength (large k) modes are stabilised. This too makes intuitive sense: the surface tension means that you pay a large cost in energy when the interface has large gradients, so small wavelength perturbations are rescued from the instability.

Alternatively, we can view (4.25) and (4.26) as conditions on how large the velocity difference $|U_1 - U_2|$ must be to initiate an instability for a fixed k. If we have both $g, \gamma \neq 0$ then there is no instability at all for small velocity differences. The instability only kicks in when there is some value of k such that the frequency ω has an imaginary part. This is true provided

$$(U_1 - U_2)^2 > 2 \frac{\rho_1 + \rho_2}{\rho_1 \rho_2} \sqrt{(\rho_2 - \rho_1)\gamma g} \ . \tag{4.27}$$

Moreover, there is a maximally unstable mode for which ω has the largest imaginary value. This is $k_\star = \sqrt{g(\rho_2 - \rho_1)/\gamma}$. The associated wavelength $\lambda = 2\pi/k_\star$ is the scale at which the instability will tend to develop.

We can do some order of magnitude estimates. For wind blowing above water, we have $\rho_1 = \rho_{\text{air}} \approx 1 \text{ kg m}^{-3}$ and $\rho_2 = \rho_{\text{water}} \approx 10^3 \text{ kg m}^{-3}$. The surface tension turns out to be given by $\gamma \approx 0.07 \text{ J m}^{-2}$. This gives a critical wind speed of $U_{\text{wind}} \approx 7 \text{ m s}^{-1}$. This is the minimum speed needed before the wind becomes responsible for making waves. It is, it turns out, in the ballpark of the average speed across the ocean. The corresponding unstable mode number is around $k \approx 400 \text{ m}^{-1}$, corresponding to a wavelength of a couple of centimetres. In other words, the Kelvin–Helmholtz instability due to wind can cause only tiny capillary waves (and, even here, there is a large amount of debate). The Kelvin–Helmholtz instability is not responsible for the great rolling waves on the ocean.

4.1.4 The Rayleigh–Taylor Instability

There's a particularly simple application of the machinery above. We consider two immiscible fluids, now both stationary, but with the heavier one on top. It's unlikely to come as a surprise to learn that this situation is unstable. But how does the heavy fluid succeed in moving past the lighter fluid? If they're truly immiscible, then they can't just swap places. Instead, something more complicated must be going on.

We can trivially import our previous dispersion relation, now with $U_1 = U_2 = 0$ and $\rho_1 > \rho_2$, to address this question. The frequency of the perturbation is now

$$\omega = \pm \sqrt{\frac{\gamma k^3 - gk(\rho_1 - \rho_2)}{\rho_1 + \rho_2}} \, . \tag{4.28}$$

We see that the long wavelength modes, with $k^2 < g(\rho_1 - \rho_2)/\gamma$, are unstable. We introduce the generalisation of the *capillary length* that we met earlier in (3.62)

$$l_c = \sqrt{\frac{\gamma}{g(\rho_1 - \rho_2)}} \, . \tag{4.29}$$

The most unstable mode occurs for $3k^2 = 1/l_c^2$. This determines the size of the perturbation at which the upper layer descends into the lower. This is the *Rayleigh–Taylor instability*, as illustrated in the image from a simulation shown in the figure to the right.

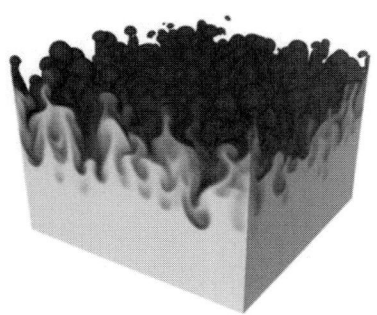

Taylor's contribution to the story above was to invoke the equivalence principle. The mathematics remains unchanged when a heavier fluid is accelerated through a lighter fluid, rather than pulled down due to gravity. This arises, among other places, in nuclear, volcanic, and supernovae explosions,

4.2 Centrifugal Instability

Our next instability arises for inviscid, rotating fluids. It is known as the *centrifugal instability*. There are a number of situations in which this occurs, but here we will focus on the simplest. We have a fluid sitting between two cylinders, both of which are aligned along the z-axis. The inner cylinder has radius R_1 and the outer cylinder radius $R_2 > R_1$. One, or both, of these cylinders is set spinning. This is the kind of situation that arises in circular Couette flow that we saw in Section 2.2. We want to know if it's stable.

We will work in cylindrical polar coordinates (r, θ, z). In these coordinates, we have $\mathbf{u} = u_r \hat{\mathbf{r}} + u_\theta \hat{\boldsymbol{\theta}} + u_z \hat{\mathbf{z}}$ where

$$\hat{\mathbf{r}} = \begin{pmatrix} \cos\theta \\ \sin\theta \\ 0 \end{pmatrix} \quad \text{and} \quad \hat{\boldsymbol{\theta}} = \begin{pmatrix} -\sin\theta \\ \cos\theta \\ 0 \end{pmatrix} . \tag{4.30}$$

We also have

$$\nabla = \hat{\mathbf{r}} \frac{\partial}{\partial r} + \frac{\hat{\boldsymbol{\theta}}}{r} \frac{\partial}{\partial \theta} + \hat{\mathbf{z}} \frac{\partial}{\partial z} . \tag{4.31}$$

When we write the Euler equation in these coordinates, the $\mathbf{u} \cdot \nabla \mathbf{u}$ term picks up contributions from $\partial \hat{\mathbf{r}} / \partial \theta = \hat{\boldsymbol{\theta}}$ and $\partial \hat{\boldsymbol{\theta}} / \partial \theta = -\hat{\mathbf{r}}$. The result is that the components of the Euler equation take the form

$$\rho \left(\frac{Du_r}{Dt} - \frac{u_\theta^2}{r} \right) = -\frac{\partial P}{\partial r} \tag{4.32}$$

$$\rho \left(\frac{Du_\theta}{Dt} + \frac{u_r u_\theta}{r} \right) = -\frac{\partial P}{\partial \theta} \tag{4.33}$$

$$\rho \frac{Du_z}{Dt} = -\frac{\partial P}{\partial z} . \tag{4.34}$$

Here the material derivative is defined by

$$\frac{D}{Dt} = \frac{\partial}{\partial t} + u_r \frac{\partial}{\partial r} + \frac{u_\theta}{r} \frac{\partial}{\partial \theta} + u_z \frac{\partial}{\partial z} . \tag{4.35}$$

In addition, we have the incompressibility condition

$$\nabla \cdot \mathbf{u} = 0 \quad \Longrightarrow \quad \frac{1}{r} \frac{\partial (r u_r)}{\partial r} + \frac{1}{r} \frac{\partial u_\theta}{\partial \theta} + \frac{\partial u_z}{\partial z} = 0 . \tag{4.36}$$

This set of equations admits a class of particularly simple axisymmetric solutions with $u_r = u_z = 0$ and

$$u_\theta = r\Omega(r) \tag{4.37}$$

for arbitrary angular velocity $\Omega(r)$. This solution is sustained by a pressure $P(r)$ such that $P' = \rho r \Omega^2$. We would like to ask the question: Is this flow stable to perturbations?

Rayleigh's Criterion

We will look at perturbations that are, themselves, axisymmetric, so that they have no angular dependence. First, we present a heuristic argument, originally due to Rayleigh, for when the flow might be unstable. This argument starts by noting that, for any axisymmetric flow with $\mathbf{u}(r, z)$ and

$P(r,z)$, the angular equation (4.33) takes the form

$$\frac{D(ru_\theta)}{Dt} = 0 \ . \tag{4.38}$$

This is just the statement of angular momentum conservation, where the angular momentum density is ρl with

$$l = ru_\theta(r) = r^2\Omega(r) \ . \tag{4.39}$$

Now consider the kinetic energy of a parcel of fluid at some radius r. This is

$$E = \frac{\rho}{2}u_\theta^2 = \frac{\rho}{2}\frac{l^2}{r^2} \ . \tag{4.40}$$

This l^2/r^2 behaviour is familiar as the angular momentum barrier that arises in orbital problems in both classical and quantum mechanics.

Here is Rayleigh's slick argument. Look at two rings of fluid, sitting within the full flow, one at radius r_1 and the other at radius $r_2 > r_1$. The kinetic energy of these two rings is

$$E_{\text{initial}} = \frac{\rho}{2}\left(\frac{l_1^2}{r_1^2} + \frac{l_2^2}{r_2^2}\right) \tag{4.41}$$

with $l_a = l(r_a)$, $a = 1,2$. Now suppose that some dynamical process happens which causes these two rings of fluid to swap position. Because angular momentum is conserved, the l_a must remain fixed, so the energy after this swap takes place is

$$E_{\text{final}} = \frac{\rho}{2}\left(\frac{l_1^2}{r_2^2} + \frac{l_2^2}{r_1^2}\right) \ . \tag{4.42}$$

The question is: Have we increased or decreased the energy in making this swap? If the energy has decreased, that's a sign that the flow will be unstable. To answer this, we just need to look at

$$\Delta E = E_{\text{final}} - E_{\text{initial}} = \frac{\rho}{2}\left(l_1^2 - l_2^2\right)\left(\frac{1}{r_2^2} - \frac{1}{r_1^2}\right) \ . \tag{4.43}$$

With $r_2 > r_1$, the second bracket is always positive. Which means that the flow is *unstable* if $l_1^2 < l_2^2$, or

$$\frac{dl^2}{dr} = \frac{d}{dr}\left(r^4\Omega^2\right) < 0 \quad \text{for instability} \ . \tag{4.44}$$

This is *Rayleigh's criterion* for the centrifugal instability.

Conversely, if $dl^2/dr > 0$ then the argument above suggests that the flow will be stable to linear, axisymmetric perturbations. It doesn't, however, say anything about perturbations that have an angular dependence.

4.2.1 Perturbation Analysis

We can put the intuitive argument above on firmer footing by looking at small perturbations around the solution (4.37). We again restrict to axisymmetric perturbations, of the form

$$\mathbf{u} = r\Omega(r)\hat{\boldsymbol{\theta}} + \tilde{\mathbf{u}}(r)e^{ikz - i\omega t} \tag{4.45}$$

with $\tilde{u}(r)$ assumed to be suitably small. As in the previous section, we will find an instability if ω is imaginary. We also shift the pressure $P(r) \rightarrow P(r) + \tilde{P}(r)$. We then linearise the equations of motion and the continuity equation. With our $e^{ikz - i\omega t}$ ansatz, every time derivative is replaced by $\partial/\partial t \rightarrow -i\omega$ while every z-derivative is replaced by $\partial/\partial z \rightarrow +ikz$.

There is a sensible order to do this calculation. First we take the u_z equation of motion (4.34) which, upon linearisation, gives

$$\rho\frac{\partial \tilde{u}_z}{\partial t} = -\frac{\partial \tilde{P}}{\partial z} \quad \Longrightarrow \quad \tilde{P} = \frac{\rho\omega}{k}\tilde{u}_z . \tag{4.46}$$

Next we take the equation of motion for u_r which, upon linearisation, gives

$$\rho\left(\frac{\partial \tilde{u}_r}{\partial t} - 2\Omega(r)\tilde{u}_\theta\right) = -\frac{\partial \tilde{P}}{\partial r}$$

$$\Longrightarrow \quad -i\omega\tilde{u}_r - 2\Omega(r)\tilde{u}_\theta = -\frac{\omega}{k}\frac{d\tilde{u}_z}{dr} \tag{4.47}$$

where, in the second expression, we've used (4.46). The we take the equation of motion for u_θ

$$\frac{\partial \tilde{u}_\theta}{\partial t} + \frac{d(r\Omega)}{dr}\tilde{u}_r + \Omega(r)\tilde{u}_r = 0$$

$$\Longrightarrow \quad \tilde{u}_\theta = -\frac{i}{\omega}\left(2\Omega + r\frac{d\Omega}{dr}\right)\tilde{u}_r$$

$$\Longrightarrow \quad \left(\omega^2 - 4\Omega^2 - 2r\Omega\frac{d\Omega}{dr}\right)\tilde{u}_r = -i\frac{\omega^2}{k}\frac{d\tilde{u}_z}{dr} \tag{4.48}$$

where, in the second line, we've substituted the expression for $\tilde{u}_\theta$ into (4.47). Now we've got an expression for $\tilde{u}_r$ and $\tilde{u}_z$, but we've yet to invoke the linearised version of the continuity equation (4.36) which reads

$$\frac{1}{r}\frac{d(r\tilde{u}_r)}{dr} + ik\tilde{u}_z = 0 \quad \Longrightarrow \quad ik\frac{d\tilde{u}_z}{dr} = -\frac{d}{dr}\left(\frac{1}{r}\frac{d(r\tilde{u}_r)}{dr}\right) . \tag{4.49}$$

We now eliminate $\tilde{u}_z$ to get a second order differential equation just for $\tilde{u}_r$

$$\frac{d}{dr}\left(\frac{1}{r}\frac{d}{dr}(r\tilde{u}_r)\right) = \frac{k^2}{\omega^2}\left(\omega^2 - 4\Omega^2 - 2r\Omega\frac{d\Omega}{dr}\right)\tilde{u}_r . \tag{4.50}$$

This looks rather daunting! Recall that we want to know if this equation has solutions for certain values of k and ω, and we'll claim instability whenever there is a solution that has $\omega^2 < 0$. But it's certainly going to be challenging to search for explicit functions $\tilde{u}_r$ that solve this equation. Instead, we will do something more indirect. This technique will make another appearance in Section 4.5 when we discuss instabilities of shear flows.

To start, we make things simpler by introducing the function

$$\Phi(r) = \frac{1}{r^3} \frac{d}{dr} \left(r^4 \Omega^2 \right) . \tag{4.51}$$

Note that the function that is differentiated, $l^2 = r^4 \Omega^2$, is the angular momentum and so, by (4.44), we expect instability if $\Phi(r) < 0$ for some r. We will soon see that this is the case. For now, we just need to note that working with $\Phi(r)$ simplifies our equation somewhat, with (4.50) becoming

$$\frac{d}{dr} \left(\frac{1}{r} \frac{d}{dr} (r\tilde{u}_r) \right) = \frac{k^2}{\omega^2} \left(\omega^2 - \Phi(r) \right) \tilde{u}_r . \tag{4.52}$$

At this stage, we need to remember our original set-up. Our fluid is sitting between two cylinders, with radii R_1 and $R_2 > R_1$. To understand the stability properties of the flow, we multiply (4.52) by $r\tilde{u}_r^\star$, where $\tilde{u}_r^\star$ is the complex conjugate of $\tilde{u}_r$, and integrate between these two cylinders. We have

$$\int_{R_1}^{R_2} dr \; r\tilde{u}_r^\star \frac{d}{dr} \left(\frac{1}{r} \frac{d}{dr} (r\tilde{u}_r) \right) = \frac{k^2}{\omega^2} \int_{R_1}^{R_2} dr \; r \left(\omega^2 - \Phi(r) \right) |\tilde{u}_r|^2 . \tag{4.53}$$

We integrate the first term by parts, throwing away the boundary term because we necessarily have $\tilde{u}_r = 0$ when $r = R_1, R_2$. Then we have

$$\int_{R_1}^{R_2} dr \; \left(\frac{1}{r} \left| \frac{d(r\tilde{u}_r)}{dr} \right|^2 + k^2 r |\tilde{u}_r|^2 \right) = \frac{k^2}{\omega^2} \int_{R_1}^{R_2} dr \; r\Phi(r) |\tilde{u}_r|^2 . \tag{4.54}$$

Now this is something that we can work with! The left-hand side is necessarily positive (because $r > 0$), so the right-hand side must also be. This means that if $\Phi(r) > 0$ everywhere in the flow, then $\omega^2 > 0$ as well and the flow is stable against these perturbations. If, in contrast, $\Phi(r) < 0$ somewhere for some region of r, then the integral on the right-hand side can be negative and, correspondingly, $\omega^2 < 0$. This is the signal of an instability. We learn that we expect the flow to be unstable if

$$\Phi(r) = \frac{1}{r^3} \frac{d}{dr} \left(r^4 \Omega^2 \right) < 0 . \tag{4.55}$$

This coincides with the previous, more hand-waving, Rayleigh criterion for instability (4.44).

4.2.2 Circular Couette Flow Revisited

We can apply our stability criterion to the circular Couette flow, also known as Taylor–Couette flow, discussed in Section 2.2. In that situation, the inner and outer cylinders sit at radii R_1 and $R_2 > R_1$ respectively, and are spinning with angular velocities Ω_1 and Ω_2. We found that the fluid flow takes the form (2.46),

$$\Omega = A + \frac{B}{r^2} \qquad (4.56)$$

where the no-slip boundary condition requires (2.47)

$$A = \frac{\Omega_2 R_2^2 - \Omega_1 R_1^2}{R_2^2 - R_1^2} \quad \text{and} \quad B = (\Omega_1 - \Omega_2)\frac{R_1^2 R_2^2}{R_2^2 - R_1^2} \; . \qquad (4.57)$$

The condition (4.55) for instability of the flow now reads

$$\Phi(r) = 4A^2 \left(1 + \frac{B}{Ar^2}\right) < 0 \; . \qquad (4.58)$$

We take $\Omega_1, \Omega_2 > 0$. Then, after a little bit of algebra, we can rewrite this criterion for instability as

$$\frac{1 - R_1^2/R_2^2}{\Omega_2/\Omega_1 - R_1^2/R_2^2} < 0 \quad \implies \quad \Omega_1 R_1^2 > \Omega_2 R_2^2. \qquad (4.59)$$

We see that the fluid is unstable if the angular momentum of the inner cylinder exceeds that of the outer cylinder. This means, for example, that if we set the outer cylinder spinning, while the inner one is stationary, we expect that the flow will be stable. Conversely, if we set the inner cylinder spinning, keeping the outer one fixed, then we will see an instability.

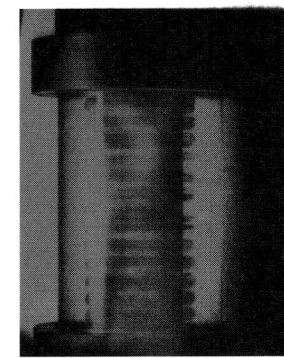

The discussion above is missing something important: viscosity. The analysis can be redone including the effects of viscosity. This is significantly harder, and was performed in a famous paper by G.H. Taylor in 1923. In this case, one finds that if the outer cylinder is fixed, then there is a critical value of Ω_1 beyond which the flow is unstable. In the same paper, Taylor also performed the experiments, demonstrating the instability. A photograph from the paper, exhibiting the instability, is shown in the figure.

4.2.3 G.I. Taylor (1886–1975)

> Every Sunday morning, Rutherford, F.W. Aston, R.H. Fowler and I used to
> play golf ... I remember one Sunday when Aston had just got results with his
> first mass spectrograph he spent the first 5 holes discussing with Rutherford
> what the new atoms he had discovered should be called. By the time we had
> got to the 6th green it was evident that the party behind us had become really
> impatient, so Rutherford and Aston made a snap decision and adopted the
> name "isotopes".
>
> *G.I. Taylor, name dropping.*

Geoffrey Ingram Taylor, or "G.I." to his friends, was the pre-eminent fluid
dynamicist of the twentieth century. So many things are named after this
guy that you begin to wonder what everyone else was doing. If you ever
want to bluff your way in fluid dynamics, just append the name Taylor to
something wet and the chances are you'll sound convincing.

The centrifugal instability described above was one of Taylor's many
achievements. This was the first time that a stability analysis, using the
full Navier–Stokes equations, had been performed and found to agree perfectly with observations. It is considered one of the landmark papers in fluid
mechanics and Taylor did both the theory and the experiment. His paper
doesn't shy away from showmanship, and he goes to great lengths to explain
how Kelvin, Rayleigh, Sommerfeld, Orr, and others had failed to understand
the instability of laminar flow, even quoting Orr as saying

> It would seem improbable that any sharp criterion for stability of fluid motion will
> ever be arrived at mathematically.

Taylor then goes on to blow them all out of the water.

4.3 Jet Instability

We've all seen it. The stream of liquid starts out unblemished and pure. But,
as it falls, ripples start to appear on the surface and these grow until the
stream ultimately disintegrates into individual droplets. This is known as the
Plateau–Rayleigh instability and our goal in this section is to understand it.
Fluid dynamicists refer to the stream of liquid as a *jet* and we'll adopt this
terminology.

In most situations the column of liquid forms in the first place because it falls under gravity. But it will turn out that the most prominent role in the development of the instability is played by surface tension. For this reason, we'll start by ignoring gravity and assuming that we are simply given a cylindrical jet, meaning a column of flowing liquid. Then, in Section 4.3.1, we'll see how gravity changes things. We ignore viscosity throughout this section.

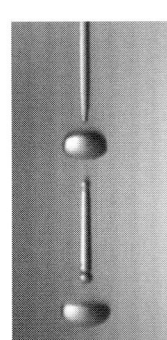

We can go to a frame where the fluid is stationary, so $\mathbf{u} = 0$, and with constant pressure P_0. The first question that we have to ask is: What keeps the jet together? Why doesn't the liquid just fly off in random directions? The answer is surface tension.

As we learned in Section 3.1.5, the surface tension allows for a pressure difference between the liquid and the outside air. For simplicity, we assume that the external pressure is vanishing. We take the jet to lie in the z-direction and have radius R_0. We use the radial coordinate $r = \sqrt{x^2 + y^2}$ so, following (3.2), the surface is given by $F(r) = r - R_0 = 0$. The curvature of a circle of radius R_0 is simply $1/R_0$ so, from (3.57), the surface tension γ balances the pressure inside the jet when

$$P_0 = \frac{\gamma}{R_0} \ . \tag{4.60}$$

Now we perturb. We will consider capillary surface waves that run along the jet, so that the radius is displaced by

$$R(z,t) = R_0 + \tilde{R}e^{ikz - i\omega t} \ . \tag{4.61}$$

This means that we're anticipating wave-like behaviour. But, as in the previous sections, this will be unstable if ω is imaginary. We look for solutions in which the velocity profile and pressure have the same behaviour,

$$\mathbf{u}(\mathbf{x}, t) = \left[u_r(r)\hat{\mathbf{r}} + u_z(r)\hat{\mathbf{z}} \right] e^{ikz - i\omega t}$$
$$P(\mathbf{x}, t) = P_0 + \tilde{P}(r)e^{ikz - i\omega t} \ . \tag{4.62}$$

Note that each of the perturbations is a function of the radial direction r, as well as exhibiting wave-like behaviour in the z-direction.

The incompressibility condition $\nabla \cdot \mathbf{u} = 0$ tells us that

$$\frac{1}{r}\frac{d(ru_r)}{dr} + iku_z = 0 \ . \tag{4.63}$$

After linearising, the two components of the Euler equation are

$$-i\omega u_r = -\frac{1}{\rho}\frac{d\tilde{P}}{dr} \quad \text{and} \quad -i\omega u_z = -\frac{ik}{\rho}\tilde{P} \ . \tag{4.64}$$

We can combine our three equations into a single, second order equation for u_r

$$r^2\frac{d^2 u_r}{dr^2} + r\frac{du_r}{dr} - (1 + k^2 r^2)u_r = 0 \ . \tag{4.65}$$

This is a standard differential equation, with solutions given by the modified Bessel functions $I_1(kr)$ and $K_1(kr)$. Of these, $K_1(x)$ diverges as $x \to 0$ and so is not appropriate for our needs. Meanwhile, $I_1(x) \to 0$ as $x \to 0$ but grows exponentially for large x, as shown in the figure. This is the solution for us. We learn that the radial velocity profile is given by

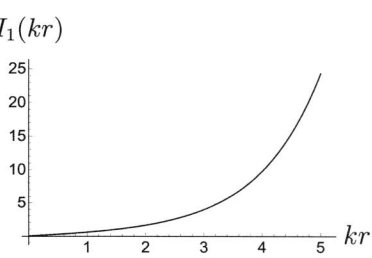

$$u_r(r) = VI_1(kr) \tag{4.66}$$

for some constant V. We still have two boundary conditions to impose on the surface of the jet, which is now defined by the constraint

$$F(r, z, t) = r - R(z, t) = 0 \ . \tag{4.67}$$

The first is the usual boundary condition (3.6) for a free surface, which tells us that the velocity of the fluid normal to the surface must track the surface itself. After linearisation, this gives

$$\frac{DF}{Dt} \approx \mathbf{u} \cdot \hat{\mathbf{r}} - \frac{\partial R}{\partial t} = 0 \quad \Longrightarrow \quad V \approx -\frac{i\omega\tilde{R}}{I_1(kR_0)} \tag{4.68}$$

and determines the unknown constant V in terms of ω and k. The second boundary condition is the requirement that the pressure is balanced by the surface tension. There are now two contributions to the curvature of the surface. The first is because the surface is curved in the (x, y)-plane and gives a term like (4.60). The second comes from the additional curvature from (3.57), arising from waves in the z-direction. Combined, these give

$$P(R(t)) \approx P_0 + \tilde{P}(R(t))e^{ikz - i\omega t} = \frac{\gamma}{R(t)} + \gamma k^2 \tilde{R}e^{ikz - i\omega t} \ . \tag{4.69}$$

We can linearise the $R(t)$ terms in the second equation. We have $\tilde{P}(R(t)) \approx \tilde{P}(R_0)$ and, by Taylor expanding, $1/R(t) \approx 1/R_0 - (\tilde{R}/R_0^2)e^{ikz-i\omega t}$. The terms involving just the background variables P_0 and R_0 cancel by virtue of (4.60), and we're left with a condition that relates the perturbation of the pressure $\tilde{P}$ to the perturbation of the radius $\tilde{R}$

$$\tilde{P}(R_0) = \frac{\tilde{R}\gamma}{R_0^2}(k^2 R_0^2 - 1) \ . \tag{4.70}$$

Already here we can see that there is a difference between short wavelength and long wavelength perturbations. For short wavelengths, with $kR_0 > 1$, the pressure increases at the surface. But for long wavelengths, with $kR_0 < 1$, the pressure decreases at the surface. This will turn out to be important.

We can get another expression for the pressure perturbation everywhere within the jet from the radial component of the Euler equation (4.64). This tells us that

$$\frac{d\tilde{P}}{dr} = i\omega\rho u_r = i\omega\rho V I_1(kr) \quad \Longrightarrow \quad \tilde{P}(r) = \frac{i\omega\rho V}{k}I_0(kr) \tag{4.71}$$

where we've made use of the Bessel function identity $I_1 = dI_0/dx$ to integrate the first expression. (The function $I_0(x)$ also grows exponentially as $x \to \infty$, but has the value $I_0(0) = 1$ at the origin.) Equating (4.70) and (4.71), and using our expression (4.68) for the constant V, we find

$$\omega^2 = \frac{k\gamma}{\rho R_0^2}\frac{I_1(kR_0)}{I_0(kR_0)}(k^2 R_0^2 - 1) \ . \tag{4.72}$$

This is the dispersion relation that we were looking for. For large k, the ratio of Bessel functions is $I_1(x)/I_0(x) \to 1$ as $x \to \infty$ and so we get waves propagating with the dispersion $\omega \sim k^{3/2}$. This is the same dispersion relation that we found for surface capillary waves. (See the large k limit of (3.63).) This is to be expected: waves with a small wavelength have no knowledge of their global surrounding. They're unaware if they're sitting on a stream of liquid or on a rolling ocean.

More interesting for our purposes is the long wavelength behaviour of the dispersion relation. The frequency is imaginary when

$$kR_0 < 1 \ . \tag{4.73}$$

This means that perturbations with wavelength bigger than the radius R_0 of the jet will grow exponentially quickly.

The plot to the right shows the function $|\omega| \sim \sqrt{x(1-x^2)I_1(x)/I_0(x)}$ in the unstable regime, corresponding to $0 \leq x \equiv kR_0 \leq 1$. We see that there is a maximally unstable mode at around $kR_0 \approx 0.7$, corresponding to a wavelength

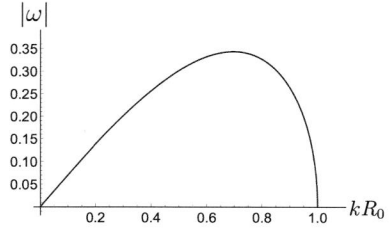

$$\lambda_{\max} \approx 9R_0 \ . \qquad (4.74)$$

The associated time that it takes the perturbation to grow is

$$T \sim \frac{1}{|\omega|} \approx \frac{1}{0.34}\sqrt{\frac{\rho R_0^3}{\gamma}} \qquad (4.75)$$

where the 0.34 is seen to be roughly the maximum of the graph. For water, $\rho \approx 10^3 \ \mathrm{kg\,m^{-3}}$ and $\gamma \approx 0.07 \ \mathrm{J\,m^{-2}}$. A stream of water from a tap has a radius of roughly 0.5 cm, and, from the analysis above, is expected to decay in around 0.1 seconds.

4.3.1 Gravity Makes the Flow Thinner

We've neglected gravity in the above analysis, despite the fact that gravity is clearly responsible for the formation of the most familiar streams of liquid. We now remedy this. In fact, the most important ramification of gravity is not in the evolution of the perturbations, but in the form of the stream itself. That's what we will focus on here.

Our result is a simple application of Bernoulli's theorem that we first met in Section 1.2.4. This says that, in a gravitational field with potential $\Phi = -\rho g z$, the quantity

$$H = \frac{1}{2}\rho|\mathbf{u}|^2 + P - \rho g z \qquad (4.76)$$

is constant. (Here we're measuring z so that it increases in a downwards direction.) Clearly we expect the component of the velocity $U = u_z$ to increase with z simply because the fluid is being accelerated downwards. But now we have to reconcile this with the fact that the pressure must balance the surface tension as in (4.60)

$$P = \frac{\gamma}{R} \ . \qquad (4.77)$$

As we will see, this requires that both the radius $R(z)$ and the velocity $U(z)$ depend on z. This follows because of mass conservation, which tells us that the flux of fluid along the stream is

$$\text{flux} = \pi R^2 U \tag{4.78}$$

and this too must be constant.

Suppose that the fluid has velocity U_0 and radius R_0 at $z = 0$. Then flux conservation gives $R_0^2 U_0 = R(z)^2 U(z)$ and Bernoulli's principle becomes

$$\frac{1}{2}\rho U_0^2 + \frac{\gamma}{R_0} = \frac{1}{2}\rho\frac{R_0^4 U_0^2}{R(z)^4} + \frac{\gamma}{R(z)} - \rho g z . \tag{4.79}$$

On rearranging,

$$R(z)^4 \left(\frac{1}{2}\rho U_0^2 + \frac{\gamma}{R_0} + \rho g z\right) - \gamma R(z)^3 - \frac{1}{2}\rho R_0^4 U_0^2 = 0 . \tag{4.80}$$

This is still kind of messy, but the physics is clear: as z increases, $R(z)$ must decrease roughly as $z^{-1/4}$, at least for large z. There is also a limit in which we can make more progress. We define the dimensionless *Weber number* as

$$We = \frac{\rho R_0 U_0^2}{\gamma} . \tag{4.81}$$

The limit of $We \to \infty$ is where surface tension effects are no longer important. In this limit we can drop the γ terms in (4.80) and we have

$$R(z) = R_0 \left(1 + \frac{2gz}{U_0^2}\right)^{-1/4} \tag{4.82}$$

where we see very clearly the advertised $z^{-1/4}$ behaviour for large z. From our analysis of the Plateau–Rayleigh instability, we know that, as the radius of the stream gets thinner, the wavelength at which the instability kicks in gets shorter.

4.4 Rayleigh–Bénard Convection

In this section we will discuss flows due to buoyancy, driven by differences in temperature. This is known as *convection*.

We will focus on a layer of fluid, sandwiched between two plates held at different temperatures, the bottom hot and the top cold. Our expectation is that this temperature difference will force the flow upwards and our goal is to understand how this happens.

We assume that the flow is divergence-free, so

$$\nabla \cdot \mathbf{u} = 0 \ . \tag{4.83}$$

As we'll see shortly, this assumption isn't entirely innocent when we have heat conduction and will induce a small amount of hand-wringing below. The equation that describes how the temperature $T(\mathbf{x}, t)$ of the fluid evolves is the heat equation

$$\frac{DT}{Dt} = \left(\frac{\partial}{\partial t} + \mathbf{u} \cdot \nabla \right) T = \alpha \nabla^2 T \tag{4.84}$$

with α the thermal diffusivity. We already met this equation in (3.202) when discussing sound waves. There was an additional $\nabla \cdot \mathbf{u}$ term in (3.202) but we've set this to zero on the grounds that we're dealing with an incompressible fluid.

Our other equation is, of course, Navier–Stokes

$$\rho \left(\frac{\partial \mathbf{u}}{\partial t} + \mathbf{u} \cdot \nabla \mathbf{u} \right) = -\nabla P - \rho(\mathbf{x}, t) g \hat{\mathbf{z}} + \mu \nabla^2 \mathbf{u} \ . \tag{4.85}$$

We've included a gravitational force and viscosity. Both will be important for what is to come. We now have five equations (two scalar and one vector) for six unknowns: $\mathbf{u}$, P, ρ, and T. We must augment these with an equation of state, relating ρ, P, and T. When discussing sound waves we relied heavily on the ideal gas equation of state. Here we are dealing with a layer of liquid and our equation of state will be different.

4.4.1 The Boussinesq Approximation

As always, we will invoke some approximation to allow us to solve the equations. This one is more subtle than most.

We start by assuming that the temperature difference between the upper and lower plates is small. Mathematically, this means that if we think of $\rho = \rho(T)$, then the difference in density is well approximated by a Taylor expansion

$$\rho(\mathbf{x}, t) \approx \rho_0 \left(1 - \beta(T(\mathbf{x}, t) - T_0) \right) \tag{4.86}$$

with β constant. We take T_0 and ρ_0 to be the temperature and density at the bottom plate. Above the plate, we expect $T \leq T_0$ and so, if $\beta > 0$, then the density will get greater as we go up. This is what we expect: increased temperature causes fluids to expand. In addition, the form (4.86) assumes that the density is independent of pressure.

At this point, it turns out that not all of our assumptions are mutually consistent! In particular, if ρ and T are linearly related, as in (4.86), then they should solve the same equation, which means that, from (4.84), the density should obey

$$\frac{D\rho}{Dt} = \alpha\nabla^2\rho \; . \tag{4.87}$$

But mass conservation means that we necessarily have

$$\frac{\partial\rho}{\partial t} + \nabla\cdot(\rho\mathbf{u}) = 0 \quad\Longrightarrow\quad \frac{D\rho}{Dt} + \rho\nabla\cdot\mathbf{u} = 0 \; . \tag{4.88}$$

We learn that the relationship (4.86) is tantamount to the requirement that

$$\rho\nabla\cdot\mathbf{u} = -\alpha\nabla^2\rho \; . \tag{4.89}$$

In other words, heat diffusion is incompatible with the assumption of a divergence-free flow $\nabla\cdot\mathbf{u} = 0$.

Our strategy is to just bluff our way through this impasse. We will have our cake and eat it by pretending that the density ρ is actually constant except when it's not. More precisely, we will assume that the changes in density due to temperature are much smaller than the initial density ρ_0. We will make the substitution $\rho = \rho_0$ in all places except one: the exception is the term $g\rho$ in the Navier–Stokes equation. This is the term that will ultimately govern the physics for the simple reason that we have orchestrated a situation in which the denser, colder fluid sits on top. This means that there's a balance of forces at play in the problem, in which the heavier fluid wants to sink, but is kept afloat by the temperature gradient. It is this buoyancy force that will be important in driving the instability.

Another, perhaps more palatable, way of saying this is that we're in a situation in which the changes in the density are small, but the gravitational acceleration is large enough to make these changes important in the buoyancy force. This collection of ideas is known as the *Boussinesq approximation*. Despite the obvious hand-waving at play here, this approximation can be made rigorous as a kind of "$g \to \infty$, $\rho \to 0$ with $g\rho$ finite" limit.

Practically, this means that we work with a divergence-free flow $\nabla\cdot\mathbf{u} = 0$ and the temperature is governed by the heat diffusion equation (4.84), while the Navier–Stokes equation becomes

$$\frac{\partial\mathbf{u}}{\partial t} + \mathbf{u}\cdot\nabla\mathbf{u} = -\frac{1}{\rho_0}\nabla P - g\left(1 - \beta(T - T_0)\right)\hat{\mathbf{z}} + \nu\nabla^2\mathbf{u} \tag{4.90}$$

where the kinematic viscosity is $\nu = \mu/\rho_0$. There are now five equations (two scalar, one vector) for five unknowns: $\mathbf{u}$, P, and T.

We now start by solving these equations for the heat flow between two plates. The lower plate sits at $z = 0$ and has temperature T_0; the upper plate sits at $z = d$ and has temperature $T_0 - \Delta T$. If the plates are separated by a distance d, then there is a simple, time-independent, solution to our equations with $\mathbf{u} = 0$ and

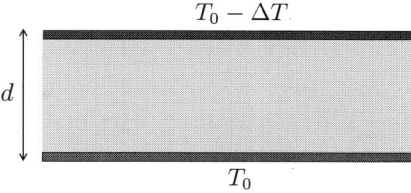

$$T = T_0 - \Delta T \frac{z}{d} \quad \text{and} \quad P = P_0 - \rho_0 g \left(z + \frac{\beta \Delta T}{2d} z^2 \right) . \qquad (4.91)$$

In this solution, the heat transfer is not due to the motion of the fluid itself. In other words, there is conduction but no convection. We would like to understand if this situation is stable against perturbations.

It turns out, unsurprisingly given the general theme of this chapter, that it won't be. The solution (4.91) is unstable to the fluid moving, so that heat is transferred by convection as well as conduction. The end point of this instability looks something like the situation on the right, with the fluid arranging itself into rolls known as *Bénard cells*, within each of which the fluid rises and falls.

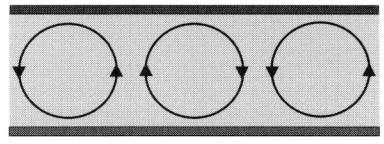

Before doing the full analysis, we can get some sense for why this happens. The key, as so often, is to look at the vorticity $\boldsymbol{\omega} = \nabla \times \mathbf{u}$. The physics is actually clearest before we make the Boussinesq approximation. We take the curl of the Navier–Stokes equation (4.85) (after dividing by the density ρ) to get

$$\frac{\partial \boldsymbol{\omega}}{\partial t} + \mathbf{u} \cdot \nabla \boldsymbol{\omega} = \frac{1}{\rho^2} \nabla \rho \times \nabla P + \boldsymbol{\omega} \cdot \nabla \mathbf{u} + \nu \nabla^2 \boldsymbol{\omega} . \qquad (4.92)$$

The novelty is the first term on the right-hand side. (A similar term survives after the Boussinesq approximation if we take the curl of (4.90).) This is telling us that, if the gradient of the density lies in a different direction to the gradient of pressure, then it will drive the creation of vorticity. As we will see below, the perturbations that have this property give rise to the convection cells. (As an aside, this term also arises in the context of internal gravity waves where it gives rise to so-called *baroclinic vorticity*.)

We can also try to figure out under what circumstances the instability will occur. Let's suppose that the width of each cell is d, the same as the height of the fluid. At this point we're just guessing this on the grounds that d is the only length scale in the game, but it will turn out that the size of the cell is indeed set by the depth. We've already noted that the heavier fluid sits on top. This means that the potential energy gained if the fluid can somehow right itself is

$$\text{P.E.} \sim gd(d^3 \Delta \rho) \sim gd^4 \rho_0 \beta \Delta T \ . \tag{4.93}$$

But to flip over, the fluid has to overcome some viscous forces. If it flips over at speed U, then the viscous force per unit volume is $\mu \nabla^2 \mathbf{u} \sim \mu U / d^2$. This means that the actual force is $\mu U d$ and the work done, which is $\int F \, dz$, scales as

$$\text{work done} \sim \mu U d^2 \ . \tag{4.94}$$

Comparing these two expressions, the gain in potential energy is sufficient to overcome the work done against viscous forces when

$$\frac{gd^2 \beta \Delta T}{\nu U} \gg 1 \ . \tag{4.95}$$

We can also get an order of magnitude estimate for the speed U at which this takes place. From the heat equation (4.84), the time scale at which heat diffuses is $\tau_{\text{diff}} \sim d^2/\alpha$. Meanwhile, if the fluid flips over at speed U then it takes a time $\tau \sim d/U$. If we equate these we get $U \sim \alpha/d$. The inequality above then becomes a condition on a dimensionless constant

$$Ra = \frac{g\beta d^3 \Delta T}{\nu \alpha} \gg 1 \ . \tag{4.96}$$

This is called the *Rayleigh number*. This back of the envelope analysis above suggests that we will find an instability when $Ra \gg 1$. We'll see that this is confirmed in our subsequent analysis.

4.4.2 Perturbation Analysis

We will perturb the background solution (4.91), turning on a velocity field $\mathbf{u}$ together with temperature and pressure perturbations that we denote as $\delta T(\mathbf{x}, t)$ and $\delta P(\mathbf{x}, t)$ respectively. These will all be considered small in the sense that we will linearise the equations of motion. The temperature equation (4.84) becomes

$$\frac{\partial \delta T}{\partial t} - \frac{\Delta T}{d} u_z = \alpha \nabla^2 \delta T \tag{4.97}$$

and the Navier–Stokes equation becomes

$$\frac{\partial \mathbf{u}}{\partial t} = -\frac{1}{\rho_0} \nabla \delta P + g \beta \, \delta T \, \hat{\mathbf{z}} + \nu \nabla^2 \mathbf{u} \, . \tag{4.98}$$

Meanwhile, the velocity perturbations remain incompressible, $\nabla \cdot \mathbf{u} = 0$.

Dimensional Analysis

We can illustrate the physics in these equations if we write them in terms of dimensionless variables. For this, we'll need to rescale both space and time as

$$\tilde{\mathbf{x}} = \frac{\mathbf{x}}{d} \quad \text{and} \quad \tilde{t} = \frac{\alpha}{d^2} t \, . \tag{4.99}$$

This means, in particular, that our lower and upper boundaries are at $\tilde{z} = 0$ and $\tilde{z} = 1$ respectively. We also rescale our dynamical variables

$$\tilde{T} = \frac{\delta T}{\Delta T} \, , \quad \tilde{P} = \frac{d^2}{\alpha^2 \rho_0} \delta P \, , \quad \tilde{\mathbf{u}} = \frac{d}{\alpha} \mathbf{u} \, . \tag{4.100}$$

Then the temperature perturbation equation (4.97) becomes

$$\frac{\partial \tilde{T}}{\partial \tilde{t}} - \tilde{u}_z = \tilde{\nabla}^2 \tilde{T} \tag{4.101}$$

while the perturbed Navier–Stokes equation (4.98) exhibits two dimensionless coefficients

$$\frac{\partial \tilde{\mathbf{u}}}{\partial \tilde{t}} = -\tilde{\nabla} \tilde{P} + Ra \, Pr \, \tilde{T} \hat{\mathbf{z}} + Pr \, \tilde{\nabla}^2 \tilde{\mathbf{u}} \, . \tag{4.102}$$

The first of these is the Rayleigh number Ra that we already met in (4.96). The second is the *Prandtl number*,

$$Pr = \frac{\nu}{\alpha} \, . \tag{4.103}$$

Unlike most of our other dimensionless numbers, the Prandtl number is a property only of the liquid, not of the flow. It has the value $Pr \approx 0.7$ for air and for most gases. It is $Pr \approx 7$ for water at room temperature.

Our goal is to solve (4.101) and (4.102) subject to suitable boundary conditions. Clearly we require $\tilde{u}_z = 0$ at $\tilde{z} = 0$ and $\tilde{z} = 1$ so that nothing flows into either plate. Also $\tilde{T} = 0$ at $\tilde{z} = 0$ and $\tilde{z} = 1$ as the plates are at a fixed temperature.

In addition, it is natural to impose the no-slip boundary condition on both plates, since our liquid is viscous. It may be natural, but we're not going to do it! The reason is simply laziness. (One is supposed to say "mathematical convenience".) The calculation below is challenging enough and by the time we get to solve the equations life is much easier if we impose the somewhat unphysical requirement that there is no stress on the plate, meaning

$$\frac{\partial \tilde{u}_x}{\partial \tilde{z}} = \frac{\partial \tilde{u}_y}{\partial \tilde{z}} = 0 \quad \text{at} \quad \tilde{z} = 0, 1 \ . \tag{4.104}$$

In fact it turns out that this is the correct boundary condition to impose on a free surface, rather than a rigid plate. As we'll see (in the discussion following (4.114)), to do the calculation we really need a boundary condition on $\tilde{u}_z$. We get this by writing the boundary condition above as

$$\frac{\partial}{\partial \tilde{z}} \left(\frac{\partial \tilde{u}_x}{\partial \tilde{x}} + \frac{\partial \tilde{u}_y}{\partial \tilde{y}} \right) = 0 \quad \Longrightarrow \quad \frac{\partial^2 \tilde{u}_z}{\partial \tilde{z}^2} = 0 \quad \text{at} \quad \tilde{z} = 0, 1 \quad (4.105)$$

where we've invoked the incompressibility condition $\tilde{\nabla} \cdot \tilde{\mathbf{u}} = 0$.

You May Wish to Roll Up Your Sleeves

We first take the curl of (4.102) to get

$$\frac{\partial \tilde{\boldsymbol{\omega}}}{\partial \tilde{t}} = Ra\,Pr\,\tilde{\nabla}\tilde{T} \times \hat{\mathbf{z}} + Pr\,\tilde{\nabla}^2 \tilde{\boldsymbol{\omega}} \tag{4.106}$$

where $\tilde{\boldsymbol{\omega}} = \tilde{\nabla} \times \tilde{\mathbf{u}}$. Curiously, it turns out that the best way to proceed is to take yet another curl. This then gives

$$\frac{\partial \tilde{\nabla}^2 \tilde{\mathbf{u}}}{\partial \tilde{t}} = Ra\,Pr \left(\hat{\mathbf{z}}\,\tilde{\nabla}^2 \tilde{T} - \tilde{\nabla} \left(\frac{\partial \tilde{T}}{\partial \tilde{z}} \right) \right) + Pr\,\tilde{\nabla}^4 \tilde{\mathbf{u}} \tag{4.107}$$

where we've used $\tilde{\nabla} \times (\tilde{\nabla} \times \tilde{\mathbf{u}}) = -\tilde{\nabla}^2 \tilde{\mathbf{u}}$ for an incompressible flow with $\tilde{\nabla} \cdot \tilde{\mathbf{u}} = 0$. This is now fourth order in spatial derivatives. We'll focus on the z-component, which is

$$\frac{\partial \tilde{\nabla}^2 \tilde{u}_z}{\partial \tilde{t}} = Ra\,Pr \left(\frac{\partial^2 \tilde{T}}{\partial \tilde{x}^2} + \frac{\partial^2 \tilde{T}}{\partial \tilde{y}^2} \right) + Pr\,\tilde{\nabla}^4 \tilde{u}_z \ . \tag{4.108}$$

This is to be solved in conjunction with the temperature equation (4.101).

We'll look for solutions using separation of variables. Furthermore, we'll anticipate the instability by looking for solutions that grow exponentially in time as $e^{\Gamma \tilde{t}}$,

$$\tilde{u}_z(\mathbf{x}, t) = V(\tilde{z})\,X(\tilde{x}, \tilde{y})e^{\Gamma \tilde{t}} \quad \text{and} \quad \tilde{T}(\mathbf{x}, t) = \theta(\tilde{z})\,X(\tilde{x}, \tilde{y})e^{\Gamma \tilde{t}} \ . \tag{4.109}$$

Note that we've assumed that both the velocity field and temperature have
the same spatial dependence $X(x, y)$ in the x- and y-directions. With this
ansatz, the temperature equation (4.101) becomes

$$\frac{1}{\theta}\left[\frac{d^2\theta}{d\tilde{z}^2} - \Gamma\theta + V\right] = -\frac{1}{X}\left[\frac{\partial^2 X}{\partial\tilde{x}^2} + \frac{\partial^2 X}{\partial\tilde{y}^2}\right] . \qquad (4.110)$$

Now, the left-hand side depends only on $\tilde{z}$ and the right-hand side depends
only on $\tilde{x}$ and $\tilde{y}$, which means that both sides must actually be constant.
For the solution to be bounded in the x- and y- directions, this constant
should be positive. We'll call it K^2. We then have

$$\tilde{\nabla}_2^2 X \equiv \frac{\partial^2 X}{\partial\tilde{x}^2} + \frac{\partial^2 X}{\partial\tilde{y}^2} = -K^2 X \quad\Longrightarrow\quad X(\tilde{x}, \tilde{y}) \sim e^{iK_x\tilde{x}+iK_y\tilde{y}} \quad (4.111)$$

where $K^2 = K_x^2 + K_y^2$. We already start to see the ripples forming in the
(x, y)-plane. Reverting back to dimensionful coordinates, we see that these
ripples will have wavelength $\lambda = 2\pi d/K$ and so the constant K is simply
the (magnitude of the) dimensionless wavenumber.

Our next task is to relate K to the instability constant Γ. This ultimately
comes from equating the left-hand side of (4.110) to the same constant K^2,

$$\left(\frac{d^2}{d\tilde{z}^2} - \Gamma - K^2\right)\theta = -V . \qquad (4.112)$$

This equation relates the temperature profile $\theta(\tilde{z})$ to the velocity profile
$V(\tilde{z})$.

We can get another equation relating these variables by substituting the
same ansatz (4.109) into the Navier–Stokes equation (4.108). We replace
$\tilde{\nabla}_2^2 X$ with the expression (4.111) to get

$$\Gamma\left(\frac{d^2}{d\tilde{z}^2} - K^2\right)V = -Ra\,Pr\,K^2\theta + Pr\left(\frac{d^2}{d\tilde{z}^2} - K^2\right)^2 V . \quad (4.113)$$

We can eliminate θ in this equation using (4.112). To do this, we have to
act with $(d^2/d\tilde{z}^2 - \Gamma - K^2)$. We have

$$\left(\frac{d^2}{d\tilde{z}^2} - K^2\right)\left(\frac{d^2}{d\tilde{z}^2} - \Gamma - K^2\right)\left(\Gamma - Pr\left(\frac{d^2}{d\tilde{z}^2} - K^2\right)\right)V$$
$$= Ra\,Pr\,K^2 V . \qquad (4.114)$$

Good? Good. After all of this, we're left with a sixth order differential equation. What can I say? Sometimes, physics just isn't pretty. (In fairness, my
friends who solve this equation for a living tell me that it is, in fact, quite

beautiful. And not just beautiful, but also self-adjoint. Eye of the beholder and all that.)

We want to solve this subject to the boundary condition $V = 0$ and, from (4.105), $d^2V/d\tilde{z}^2 = 0$ at the two boundaries $\tilde{z} = 0, 1$. We also require $\theta = 0$ at $\tilde{z} = 0, 1$ and, from (4.113), we can see that this is only consistent if, in addition, we require $d^4V/d\tilde{z}^4 = 0$ at the boundaries. This means that we must solve (4.114) subject to the six boundary conditions

$$V = \frac{d^2V}{d\tilde{z}^2} = \frac{d^4V}{d\tilde{z}^4} = 0 \quad \text{at} \quad \tilde{z} = 0, 1 \; . \tag{4.115}$$

This, as we shall now see, isn't so bad.

Before we do the thing that isn't so bad, it's worth pausing to reconsider these boundary conditions. Recall that the condition (4.105) was born more out of sloth than physical necessity. With a rigid boundary, it would have been more appropriate to impose the no-slip condition, which leads to the requirement $dV/d\tilde{z} = 0$ at the boundaries rather than $d^2V/d\tilde{z}^2 = 0$. It turns out that we have to work harder to find such solutions.

Returning to the boundary conditions (4.115), the solutions are simply sine functions

$$V(\tilde{z}) = \sin(n\pi\tilde{z}) \tag{4.116}$$

with $n \in \mathbb{Z}$. The somewhat daunting looking equation (4.114) then becomes the algebraic condition

$$\left(n^2\pi^2 + K^2\right)\left(n^2\pi^2 + \Gamma + K^2\right)\left(n^2\pi^2 + K^2 + \frac{\Gamma}{Pr}\right) = Ra\,K^2 \; . \tag{4.117}$$

This is the analogue of our previous dispersion relations, now telling us how the characteristic (inverse) time of the instability Γ relates to the wavelength. For some fixed $n \in \mathbb{Z}$, there is clearly a solution to this equation with $\Gamma > 0$ when the Rayleigh number is sufficiently large,

$$Ra > \frac{(n^2\pi^2 + K^2)^3}{K^2} \; . \tag{4.118}$$

This confirms the intuition that we saw previously that the instability should only kick in when Ra is big enough.

The smallest value of Ra for which there is a solution occurs when $n = 1$. In this case, we minimise the function above

$$\frac{d}{dK^2}\left[\frac{(\pi^2 + K^2)^3}{K^2}\right] = 0 \quad \Longrightarrow \quad K^2 = \frac{\pi^2}{2} \; . \tag{4.119}$$

From our previous discussion, this corresponds to the formation of a cell with size $\sim 2\sqrt{2}d$. Plugging this value back into (4.118), we have an instability provided that

$$Ra > \frac{27\pi^4}{4} \approx 660 \ . \tag{4.120}$$

For values of Ra just above this number, only the $n = 1$ mode is unstable. As Ra increases, more and more K modes become unstable as well as modes of higher n. As usual, the most unstable mode is that with largest Γ and this determines the preferred choice of K.

Recall that the calculation above was done using the somewhat unrealistic free boundary conditions (4.104) at each plate. The physics is not qualitatively changed if we instead impose a no-slip boundary condition on each plate. (Nor, indeed, if we have a no-slip condition on one plate and a free boundary condition on the other.) In the case of no-slip boundary conditions on both plates, it turns out that the flow is unstable only for $Ra \gtrsim 1700$, with cells of size $\sim 2d$ forming.

Rayleigh–Bénard convection cells provide a beautiful example of spontaneous pattern formation. They can be seen in many places, from the granular structure on the surface of the Sun, to the swirls in a bowl of miso soup.

4.5 Instabilities of Inviscid Shear Flows

In this section, we turn our attention to instabilities of shear flows. That is, flows that take the form

$$\mathbf{u}(\mathbf{x}, t) = U(z)\hat{\mathbf{x}} \tag{4.121}$$

together with some pressure field $P(\mathbf{x})$. We'll consider this velocity field over some finite width

$$-h \leq z \leq +h \tag{4.122}$$

with suitable boundary conditions imposed on the edges $z = \pm h$.

The Kelvin–Helmholtz instability of Section 4.1 is an instability of a shear flow, albeit with the shear hiding in the velocity discontinuity. But there are many more examples of interest. In particular, both Couette flow and Poiseuille flow described in Section 2.2 are of this form. Our goal is to understand when instabilities of such flows may arise. We'll make progress, albeit limited. In particular, it turns out that we will not be able to demonstrate the instability of either of these basic flows, even though they are both unstable at suitably high Reynolds number. We will, however, be able to understand why this is a difficult problem!

We give the shear flow a nudge with perturbations of the form

$$\mathbf{u}(\mathbf{x}, t) = U(z)\hat{\mathbf{x}} + \tilde{\mathbf{u}}(z)e^{ik_x x + ik_y y - i\omega t} \qquad (4.123)$$

together with some perturbation of the pressure

$$\delta P(\mathbf{x}) = \rho \tilde{P}(z)e^{ik_x x + ik_y y - i\omega t} \ . \qquad (4.124)$$

Note that we've included a factor of ρ on the right-hand side to ensure that later equations are a little simpler. It does mean, however, that $\tilde{P}$ does not have the dimensions of pressure.

As in our previous examples, we take the wavenumbers k_x and k_y to be some fixed real numbers and we choose them to be positive, $k_x, k_y > 0$. We then use the equations of motion to determine ω. In general, ω can be complex (as, indeed, can $\tilde{u}_z$ and $\tilde{P}_z$) and an instability occurs when we find $\text{Im}(\omega) > 0$. We will now derive some general statements about when the original flow is unstable.

We first met shear flows when we introduced viscosity in Chapter 2, with the shear induced by the no-slip boundary condition. Nonetheless, we will first study instabilities in the context of the Euler equation

$$\frac{\partial \mathbf{u}}{\partial t} + \mathbf{u} \cdot \nabla \mathbf{u} = -\frac{1}{\rho}\nabla P \ . \qquad (4.125)$$

We will subsequently add viscosity to the discussion in Section 4.6.

The linearised perturbation equations read

$$i(k_x U - \omega)\tilde{u}_x + \tilde{u}_z \frac{dU}{dz} = -ik_x \tilde{P} \qquad (4.126)$$

$$i(k_x U - \omega)\tilde{u}_y = -ik_y \tilde{P} \qquad (4.127)$$

$$i(k_x U - \omega)\tilde{u}_z = -\frac{d\tilde{P}}{dz} \ . \qquad (4.128)$$

These should be augmented with the incompressibility condition

$$ik_x \tilde{u}_x + ik_y \tilde{u}_y + \frac{d\tilde{u}_z}{dz} = 0 \ . \qquad (4.129)$$

We've allowed for wave-like perturbations in both the x- and y-directions and a general perturbation in the z-direction. In fact, if we're looking for the onset of instabilities, then we can ignore motion in the y-direction and focus just on the two-dimensional setting. This follows from the following.

Claim: Perturbations in the y-direction do not induce further instabilities of the flow.

Proof: Define a velocity in the diagonal direction in the (x, y)-plane by

$$v = \frac{1}{K} \left(k_x \tilde{u}_x + k_y \tilde{u}_y \right) \quad \text{with} \quad K^2 = k_x^2 + k_y^2 . \tag{4.130}$$

We can combine (4.126) and (4.127) into the equation

$$i(KU - \omega')v + \tilde{u}_z \frac{dU}{dz} = -iK\tilde{P}' \tag{4.131}$$

where the primes do not denote derivatives but, instead, are a rescaled frequency and a rescaled pressure perturbation

$$\omega' = \frac{K}{k_x} \omega \quad \text{and} \quad \tilde{P}' = \frac{K}{k_x} \tilde{P} . \tag{4.132}$$

In terms of these rescaled variables, the perturbation equation (4.128) in the z-direction and the incompressibility condition (4.129) become

$$i(KU - \omega')\tilde{u}_z = -\frac{d\tilde{P}'}{dz} \quad \text{and} \quad iKv + \frac{d\tilde{u}_z}{dz} = 0 . \tag{4.133}$$

Equations (4.131) and (4.133) describe an effective 2d system, with various quantities rescaled as above. In particular, the frequency is rescaled so that $|\omega'| > |\omega|$. But, as in previous examples, the question of whether the flow is unstable will boil down to whether the imaginary part of the frequency is non-zero. Clearly, we have $\text{Im}\,\omega' \neq 0$ whenever $\text{Im}\,\omega \neq 0$. The fact that $|\omega'| > |\omega|$ tells us that the growth of instability is faster in the v-direction, but doesn't change when the instability kicks in. □

This result is a baby version of Squire's theorem which holds for viscous flows and says that a 3d mode, with a $k_y \neq 0$ component, experiences a lower Reynolds number than the 2d mode with $k_y = 0$.

4.5.1 Rayleigh's Criterion

Because our interest will lie in identifying the conditions under which flows are unstable, the claim above affords us the option of simplifying the situation by setting $\tilde{u}_y = k_y = 0$. We'll also take this opportunity to write $k_x = k$. This leaves us with the perturbation equations (4.126), (4.127), and

(4.129). We can eliminate $\tilde{u}_x$ using the incompressibility condition (4.129), leaving us with two equations

$$-\left(U - \frac{\omega}{k}\right)\frac{d\tilde{u}_z}{dz} + \tilde{u}_z\frac{dU}{dz} = -ik\tilde{P} \qquad (4.134)$$

$$i(kU - \omega)\tilde{u}_z = -\frac{d\tilde{P}}{dz} . \qquad (4.135)$$

Next, we can eliminate the pressure perturbation at the expense of differentiating the first of the equations above. In doing so, there are two terms of the form $(dU/dz)(d\tilde{u}_z/dz)$ that cancel and we are left with a second order differential equation for $\tilde{u}_z$,

$$\left[\left(U - \frac{\omega}{k}\right)\left(\frac{d^2}{dz^2} - k^2\right) - \frac{d^2U}{dz^2}\right]\tilde{u}_z = 0 . \qquad (4.136)$$

This is the *Rayleigh equation*. Our goal is to solve this equation for a given background flow $U(z)$ and wavenumber k, subject to the boundary condition $\tilde{u}_z = 0$ at $z = \pm h$, which ensures that the perturbations don't flow into the edges of the channel. The solution contains both the profile of the boundary perturbation $\tilde{u}_z$ as well as the corresponding frequency ω.

That factor of $(U - \omega/k)$ in front of the differential operator makes the Rayleigh equation somewhat more daunting than, say, the Schrödinger equation in quantum mechanics. It means that the differential equation has a singular point at any z such that $U(z) = c$ where $c = \omega/k$ is real. These are known as *critical layers*.

We'd like to know when solutions to the Rayleigh equation (4.136) have $\text{Im}(\omega) \neq 0$, signalling an instability. One possibility is to simply pick a choice of initial flow $U(z)$ and solve the eigenvalue equation (4.136). That is often hard. Here, instead, we derive some simple, general conditions that any flow must obey if it is to have a linear instability.

We write the Rayleigh equation as

$$\frac{d^2\tilde{u}_z}{dz^2} - k^2\tilde{u}_z - \frac{U''}{U - \omega/k}\tilde{u}_z = 0 . \qquad (4.137)$$

We now multiply both sides by the complex conjugate of the velocity perturbation, $\tilde{u}_z^\star$, and integrate across the width of flow in the z-direction. After integrating by parts, and using the boundary condition $\tilde{u}_z = 0$ at $z = \pm h$, we have

$$\int_{-h}^{+h} dz \left(\left|\frac{d\tilde{u}_z}{dz}\right|^2 + k^2|\tilde{u}_z|^2\right) = -\int_{-h}^{+h} dz \frac{U''}{U - \omega/k}|\tilde{u}_z|^2 . \qquad (4.138)$$

The left-hand side is real and positive. This means that the imaginary part of the right-hand side must vanish. The only complex quantity on the right-hand side is the frequency ω, so this tells us that

$$\text{Im}\left[\int_{-h}^{+h} dz\, \frac{U''}{U - \omega/k}|\tilde{u}_z|^2\right] = \frac{\text{Im}(\omega)}{k}\int_{-h}^{+h} dz\, \frac{U''}{|U - \omega/k|^2}|\tilde{u}_z|^2$$
$$= 0 . \tag{4.139}$$

How can this equation be satisfied? One way is if $\text{Im}(\omega) = 0$, in which case the flow is marginally stable. But they're not the perturbations that we care about. Instead, we're interested in the situation in which $\text{Im}(\omega) \neq 0$ and the flow is unstable. This can only happen if

$$\int_{-h}^{+h} dz\, \frac{U''}{|U - \omega/k|^2}|\tilde{u}_z|^2 = 0 . \tag{4.140}$$

But both $|\tilde{u}_z|^2$ and the denominator are manifestly positive. The only possible way that the equality can hold is if U'' changes sign at some value of $-h < z < h$, so that it is positive in some region of the flow and negative in another. This means that there must be an inflection point

$$\frac{d^2 U}{dz^2} = 0 \quad \text{for some } -h < z < h . \tag{4.141}$$

This is the *Rayleigh criterion* for linear instability of an inviscid flow. It is not a sufficient criterion. But it is necessary.

4.5.2 Fjortoft's Criterion

There is more that we can say about the condition for instability, at least under certain circumstances. Our equation (4.138) tells us that

$$\text{Re}\left[\int_{-h}^{+h} dz\, \frac{U''}{U - \omega/k}|\tilde{u}_z|^2\right] < 0$$
$$\implies \int_{-h}^{+h} dz\, \frac{U''(U - \text{Re}(\omega)/k)}{|U - \omega/k|^2}|\tilde{u}_z|^2 < 0 . \tag{4.142}$$

If there is an instability, then (4.140) must hold and this means that the $\text{Re}(\omega)$ term in the above expression actually vanishes when integrated over the width of the channel. So we have the requirement

$$\int_{-h}^{+h} dz\, \frac{U''U}{|U - \omega/k|^2}|\tilde{u}_z|^2 < 0 . \tag{4.143}$$

It's not immediately obvious that this buys us anything beyond the Rayleigh criterion. This is because we have little information about most of the variables in equation (4.143). We assume that we're given some starting flow

$U(z)$ and we're searching for a perturbation with some fixed wavenumber k. But both ω and $\tilde{u}_z$ are fixed by the Rayleigh equation (4.136) and we've yet to solve that. In particular, for a given k it may well be that $\tilde{u}_z(z)$ is peaked at some value of z, but we have no idea where. So it's difficult to see what new information can be buried in the inequality (4.143)

Nonetheless, there's a clever trick that does allow us to extract more information from (4.143). The equation (4.138) that previously allowed us to drop the $\text{Re}(\omega)$ term from the integral also allows us to add any other constant to the numerator, so we can equally well write (4.143) as

$$\int_{-h}^{+h} dz \ \frac{U''(U - U_\star)}{|U - \omega/k|^2} |\tilde{u}_z|^2 < 0 \qquad (4.144)$$

for any constant $U_\star$ of our choice. At this point, we need two further assumptions to make progress. We assume that:

- There is just a single Rayleigh inflection point at $z = z_\star$.

- The background flow $U(z)$ is monotonically increasing, so $U'(z) \geq 0$ for all z.

With these two assumptions, we take $U_\star = U(z_\star)$. Together with the assumption of monotonicity of U, this ensures that the factor $(U - U_\star)$ flips sign at $z = z_\star$. This, of course, is the same place that $U''(z)$ flips sign so, for this particular choice, the product $U''(U - U_\star)$ has the same sign for all values of z. The inequality (4.143) tells us that this sign must be negative:

$$U''(U - U_\star) \leq 0 \quad \forall \ z \qquad (4.145)$$

with equality only when $z = z_\star$. This is the *Fjortoft criterion*.

We can also phrase this criterion in terms of the vorticity of the background flow, $\boldsymbol{\omega} = \nabla \times \mathbf{u}$, whose magnitude is

$$|\boldsymbol{\omega}| = \Omega = \frac{dU}{dz} \ . \qquad (4.146)$$

The Rayleigh criterion tells us that there is a point in the flow where the vorticity is stationary, with $d\Omega/dz = 0$. The Fjortoft criterion tells us that, away from this inflection point,

$$\frac{d\Omega}{dz}(U - U_\star) < 0 \ . \qquad (4.147)$$

Suppose that, away from this point, $U'(z) > 0$ so that $\Omega > 0$. Then for $z < z_\star$ we have $U - U_\star < 0$ and (4.147) tells us that Ω is increasing. Similarly, for $z > z_\star$, (4.147) says that the vorticity Ω must be decreasing. In other

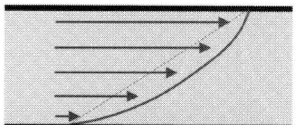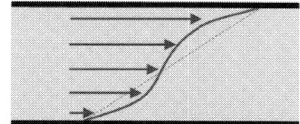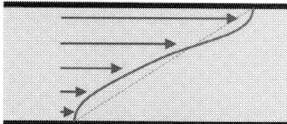

Fig. 4.4   The first flow is stable by Rayleigh, the second by Fjortoft. Only the third may be unstable.

words, the Fjortoft criterion is telling us that, for a flow to be unstable, the vorticity Ω must have a *maximum* at $z = z_\star$. The same conclusion is reached if $U'(z) < 0$: the magnitude of the vorticity $|\Omega|$ must have a maximum for the flow to be unstable. The Kelvin–Helmholtz instability of Section 4.1 is an (admittedly singular) example of this, with the vorticity a maximum at the velocity discontinuity.

Three examples of shear flows are shown in Figure 4.4. The first has no point where $U''(z) = 0$ and so is stable by the Rayleigh criterion. The second has an inflection point but the vorticity is a minimum there: it is stable by the Fjortoft criterion. The third has a point where the vorticity is maximum and obeys both criteria. Only this one may have a linear instability. And even then, it is not guaranteed. Our conditions are necessary, not sufficient.

You might have noticed that our original Poiseuille flow of Section 2.2 is stable according to the Rayleigh criterion, with $U''(z) > 0$ everywhere. That's something of a surprise because we showed photographs in Figure 4.1 of Poiseuille flow disintegrating as the Reynolds number is cranked up. But the Rayleigh and Fjortoft criteria apply only to inviscid flows and it turns out that to fully understand instabilities we need to include the effects of viscosity. We'll turn to this in Section 4.6. (The Couette flow is something of a special case since it has $U''(z) = 0$ everywhere and we'll consider it separately in Section 4.5.4.)

4.5.3 Howard's Semi-circle Theorem

There's yet more information to be extracted from the Rayleigh equation. To do this, we make the change of variables that combines the perturbation $\tilde{u}_z$ with the background flow U,

$$V = \frac{\tilde{u}_z}{U - \omega/k} \, . \qquad (4.148)$$

We should be a little careful because it may be that the new variable V is ill-defined if, for some value of z, we have $U(z) = \omega/k$. This is a general peril that lurks in the Rayleigh equation. Clearly, however, this will not be a problem if we have an unstable perturbation with ω complex because $U(z)$ is real.

The Rayleigh equation (4.136) is

$$\left[(U - \omega/k) \left(\frac{d^2}{dz^2} - k^2 \right) - \frac{d^2 U}{dz^2} \right] (U - \omega/k) V = 0 \ . \qquad (4.149)$$

After expanding out, this becomes

$$(U - \omega/k)^2 \frac{d^2 V}{dz^2} + 2(U - \omega/k) \frac{dU}{dz} \frac{dV}{dz} - k^2 (U - \omega/k)^2 V = 0 \quad (4.150)$$

which we can then write as

$$\frac{d}{dz} \left[\left(U - \frac{\omega}{k} \right)^2 \frac{dV}{dz} \right] - k^2 \left(U - \frac{\omega}{k} \right)^2 V = 0 \ . \qquad (4.151)$$

Now we play the same trick as before: we multiply by the complex conjugate $V^\star$ and integrate over the width of the flow. After integrating by parts, and using the boundary condition $\tilde{u}_z = 0$ at $z = \pm h$, we have

$$\int_{-h}^{+h} dz \ \left(U - \frac{\omega}{k} \right)^2 Q = 0 \quad \text{with} \quad Q = \left| \frac{dV}{dz} \right|^2 + k^2 |V|^2 \ . \qquad (4.152)$$

Clearly the quantity Q is non-negative for all z. If the flow is unstable then $\text{Im}(\omega) \neq 0$, and the imaginary part of the equation is

$$\frac{2 \, \text{Im}(\omega)}{k} \int_{-h}^{+h} dz \ \left(U - \frac{\text{Re}(\omega)}{k} \right) Q = 0 \ . \qquad (4.153)$$

So the quantity in brackets must change sign. This is only possible if

$$k U_{\text{min}} < \text{Re}(\omega) \leq k U_{\text{max}} \ . \qquad (4.154)$$

This can also be viewed as a bound on the phase velocity of the disturbance, $c = \text{Re}(\omega)/k$. The phase velocity is bounded by the velocities of the background flow: $U_{\text{min}} < c < U_{\text{max}}$.

The real part of (4.152) reads

$$\int_{-h}^{+h} dz \ \left((Uk - \text{Re}(\omega))^2 - \text{Im}(\omega)^2 \right) Q = 0 \ . \qquad (4.155)$$

There's a way to massage this into something more useful by removing the explicit factors of U in the integrand. To remove the term quadratic in U, we make use of the trivial fact that $(U - U_{\text{min}})(U_{\text{max}} - U) \geq 0$, so we have

$$\int_{-h}^{+h} dz \ (U - U_{\text{min}})(U_{\text{max}} - U) k^2 Q \geq 0 \ . \qquad (4.156)$$

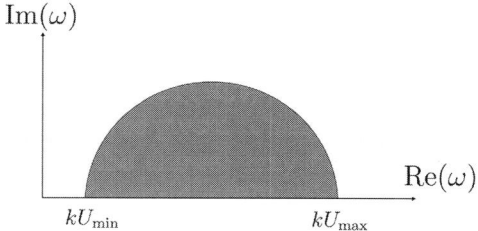

Fig. 4.5 The range of allowed unstable behaviour in the complex ω plane.

Adding (4.155) and (4.156), and using (4.158), we have

$$\int_{-h}^{+h} dz \left(Uk\Big(-2\,\mathrm{Re}(\omega) + k(U_{\max} - U_{\min})\Big) \right.$$
$$\left. + \mathrm{Re}(\omega)^2 - \mathrm{Im}(\omega)^2 - U_{\max}U_{\min}k^2\right)Q \geq 0\ . \qquad (4.157)$$

But we can remove terms linear in U since, by (4.153), we have

$$\int_{-h}^{+h} dz\, Uk\,Q = \int_{-h}^{+h} dz\, \mathrm{Re}(\omega)Q\ . \qquad (4.158)$$

The upshot is that we get the inequality

$$\int_{-h}^{+h} dz \left(\mathrm{Re}(\omega)^2 + \mathrm{Im}(\omega)^2 - \mathrm{Re}(\omega)(U_{\max} - U_{\min})k + U_{\max}U_{\min}k^2\right)Q \leq 0\ .$$

Clearly the object in the large brackets must be non-positive, meaning that the real and imaginary parts of the frequency are bounded by

$$\left(\mathrm{Re}(\omega) - \frac{k}{2}(U_{\max} - U_{\min})\right)^2 + \mathrm{Im}(\omega)^2 \leq \frac{k^2}{4}(U_{\max} + U_{\min})^2\ . \qquad (4.159)$$

This is *Howard's semi-circle theorem*. It puts a bound on the instability in the complex ω-plane in the form of a semi-circle as shown in Figure 4.5. In particular, this condition gives a bound on the growth rate $\mathrm{Im}(\omega)$ of the instability.

4.5.4 Couette Flow Revisited

Couette flow is special because, as we saw in Section 2.2, the velocity field is linear: $U(z) \sim z$. This means that $U''(z) = 0$ for all z and the Rayleigh criterion (4.140) appears to be toothless. Which leads us to ask: Is the flow actually unstable?

In this case, the background flow is simple enough that we can just go ahead and solve the equations. The Rayleigh equation (4.136) reads

$$\left(\alpha z - \frac{\omega}{k}\right)\left(\frac{d^2}{dz^2} - k^2\right)\tilde{u}_z = 0 \qquad (4.160)$$

where we've taken the background flow to be $U(z) = \alpha z$ for constant α. If we look for smooth solutions then the background perturbation must be given by

$$\left(\frac{d^2}{dz^2} - k^2\right)\tilde{u}_z = 0 \quad \Longrightarrow \quad \tilde{u}_z = A \sinh k(z - z_0) \ . \qquad (4.161)$$

But we still need to impose the boundary condition $\tilde{u}_z = 0$ at $z = \pm h$. And that can only hold if $A = 0$.

This is a little confusing! It seems to be telling us that there can be no perturbations of Couette flow, stable or otherwise. But that doesn't sound right. What if you splash it?! Surely something must happen. What?

In fact, to find perturbations we must relax the condition that $\tilde{u}_z$ is smooth. Suppose that we admit the solution

$$\tilde{u}_z = \begin{cases} A_+ \sinh k(z - h) & z > z_c \\ A_- \sinh k(z + h) & z < z_c \end{cases} \ . \qquad (4.162)$$

This immediately satisfies the boundary condition requirement $\tilde{u}_z = 0$. We still need the velocity perturbation to be continuous, even if it's not differentiable, and this imposes the condition

$$A_+ \sinh k(z_c - h) = A_- \sinh k(z_c + h) \ . \qquad (4.163)$$

This should be thought of as fixing, say A_+, in terms of A_- given the position z_c of the jump.

How worried should we be about the discontinuity in du/dz? The derivative in the Rayleigh equation (4.160) will certainly hit it, giving infinity. But we can rescue the situation if the prefactor vanishes. This holds if the frequency ω is given by

$$\omega = \alpha z_c k \ . \qquad (4.164)$$

This is rather nice. We find that we have a continuous spectrum of stable, wave-like modes, with the discontinuity propagating at the same phase velocity as the background flow at that point, $\omega/k = \alpha z_c = U(z_c)$.

The modes that we have found are marginal, in the sense that they neither

grow nor decay with time. But we've only looked at first order in perturbation theory and it may well be that higher order corrections change the
story. Although we will not show it here, it turns out that these marginal
modes are not rendered unstable by higher order terms. Nor are they unstable when we include viscosity. Couette flow is stable to arbitrarily small
perturbations. In other words, Couettte flow is linearly stable.

However, linear stability is not the same thing as actual stability! Experiment and numerics show that Couette flow becomes unstable and develops
turbulence for $Re \gtrsim 400$. This is a *non-linear instability*, meaning that it
only kicks in when the amplitude of perturbations is large enough. We will
briefly discuss this in Section 4.7.

4.6 Instabilities of Viscous Shear Flows

In this section, we'll repeat the analysis of the stability of shear flows, this
time with viscosity added.

The set-up is the same as in Section 4.5, with a flow over a finite width
$|z| \leq h$, with suitable boundary conditions (such as no-slip) imposed on the
edges. The initial flow $\mathbf{U}$, together with the perturbations $\tilde{\mathbf{u}}$, take the form

$$\mathbf{u}(\mathbf{x}, t) = U(z)\hat{\mathbf{x}} + \tilde{\mathbf{u}}(z)e^{ik_x x + ik_y y - i\omega t} \ . \tag{4.165}$$

In addition, there is a perturbation of the pressure field $P(\mathbf{x})$

$$\delta P(\mathbf{x}) = \rho \tilde{P}(z)e^{ik_x x + ik_y y - i\omega t} \tag{4.166}$$

where we've again included a factor of ρ on the right-hand side to simplify
subsequent equations.

This time we will perturb the full Navier–Stokes equation

$$\frac{\partial \mathbf{u}}{\partial t} + \mathbf{u} \cdot \nabla \mathbf{u} = -\frac{1}{\rho}\nabla P + \nu \nabla^2 \mathbf{u} \ . \tag{4.167}$$

The linearised perturbation equations are

$$i(k_x U - \omega)\tilde{u}_x + \tilde{u}_z \frac{dU}{dz} = -ik_x \tilde{P} + \nu \left(\frac{d^2}{dz^2} - k_x^2 - k_y^2 \right) \tilde{u}_x \tag{4.168}$$

$$i(k_x U - \omega)\tilde{u}_y = -ik_y \tilde{P} + \nu \left(\frac{d^2}{dz^2} - k_x^2 - k_y^2 \right) \tilde{u}_y \tag{4.169}$$

$$i(k_x U - \omega)\tilde{u}_z = -\frac{d\tilde{P}}{dz} + \nu \left(\frac{d^2}{dz^2} - k_x^2 - k_y^2 \right) \tilde{u}_z \tag{4.170}$$

and the incompressibility condition is, as before

$$ik_x\tilde{u}_x + ik_y\tilde{u}_y + \frac{d\tilde{u}_z}{dz} = 0 \ . \tag{4.171}$$

We would like to derive a generalisation of the Rayleigh equation (4.136), now including the effects of viscosity. This is an equation just for $\tilde{u}_z$.

To achieve this, we first take the divergence of the Navier–Stokes equation which, after using the incompressibility condition $\nabla \cdot \mathbf{u} = 0$, gives

$$\frac{1}{\rho}\nabla^2 P = -2\frac{dU}{dz}\frac{\partial u_z}{\partial x} \quad\Longrightarrow\quad \left(\frac{d^2}{dz^2} - k_x^2 - k_y^2\right)\tilde{P} = -2ik_x\frac{dU}{dz}\tilde{u}_z \ . \tag{4.172}$$

If we now take ∇^2 of the third equation (4.170) (which really means multiplying by $(d^2/dz^2 - k_x^2 - k_y^2)$, then we get a fourth order differential equation for $\tilde{u}_z$

$$\left[i(k_xU - \omega)\left(\frac{d^2}{dz^2} - k_x^2 - k_y^2\right) - ik_x\frac{d^2U}{dz^2} - \nu\left(\frac{d^2}{dz^2} - k_x^2 - k_y^2\right)^2\right]\tilde{u}_z = 0 \ . \tag{4.173}$$

This is the *Orr-Sommerfeld equation*. If we set $k_y = 0$ then, in the limit $\nu \to 0$ (which, as always, should really be thought of as the limit $Re \to \infty$), it reduces to the Rayleigh equation (4.136).

We want to impose the no-penetration boundary condition $\tilde{u}_z = 0$ and also the no-slip boundary condition $\tilde{u}_x = \tilde{u}_y = 0$ at the edges of the flow. From (4.171), no-slip implies that $\tilde{u}'_z = 0$. But that still leaves the other linear combination of $\tilde{u}_x$ and $\tilde{u}_y$, which we recognise as the vorticity normal to the edge,

$$\zeta = \boldsymbol{\omega} \cdot \hat{\mathbf{z}} = ik_x\tilde{u}_y - ik_y\tilde{u}_x \ . \tag{4.174}$$

We'll impose $\zeta = 0$ at the boundary. But we can also look at the dynamics of this component of vorticity in the interior of the flow. By taking suitable linear combinations of (4.168) and (4.169) (k_y of the first minus k_x of the second) we get

$$\left[i(k_xU - \omega) - \nu\left(\frac{d^2}{dz^2} - k_x^2 - k_y^2\right)\right]\zeta = ik_y\frac{dU}{dz}\tilde{u}_z \ . \tag{4.175}$$

This is the *Squire equation*. We should solve it in conjunction with the Orr-Sommerfeld equation (4.173), with the velocity perturbations $\tilde{u}_z$ acting as a source for the vorticity perturbations ζ.

In the limit $\nu = 0$ we have no possibility of interesting vorticity fluctuations since ζ is fully determined by the velocity perturbation $\tilde{u}_z$. However, the presence of viscosity turns the algebraic relation into a differential equation and this has more interesting solutions. In particular, we can have vorticity perturbations $\zeta \neq 0$ even when $\tilde{u}_z = 0$. These are known as *Squire modes*.

Claim: Squire modes are always damped. This means that the vorticity doesn't lead to an instability unless driven by the velocity perturbation $\tilde{u}_z$.

Proof: The proof uses the same kind of trick that we saw when deriving the Rayleigh criterion. We multiply the homogeneous Squire equation by the complex conjugate $\zeta^\star$ and integrate over the width of the flow. After integrating by parts and imposing the boundary condition $\zeta = 0$ at $z = \pm h$, we have

$$\nu \int_{-h}^{+h} dz \left(\left|\frac{d\zeta}{dz}\right|^2 + (k_x^2 + k_y^2)|\zeta|^2 \right) = -\int_{-h}^{+h} dz \; i(k_x U - \omega)|\zeta|^2 \; . \quad (4.176)$$

Taking the real part gives

$$\nu \int_{-h}^{+h} dz \left(\left|\frac{d\zeta}{dz}\right|^2 + (k_x^2 + k_y^2)|\zeta|^2 \right) = -\mathrm{Im}(\omega) \int_{-h}^{+h} dz \; |\zeta|^2 \; . \quad (4.177)$$

Clearly the left-hand side is positive if $\zeta \neq 0$. So the right-hand side must also be positive, which means that $\mathrm{Im}(\omega) < 0$. But the fluctuations evolve as $e^{-i\omega t} \sim e^{\mathrm{Im}(\omega)t}$ so for $\mathrm{Im}(\omega) < 0$ these are decaying modes. $\square$

4.6.1 Poiseuille Flow Revisited

To finish this section, we return to Poiseuille flow, restricted to a plane. Recall from Section 2.2 that Poiseuille flow is driven by a constant pressure gradient, with the background velocity field taking the parabolic form

$$U(z) = \alpha(h^2 - z^2) \quad (4.178)$$

for some constant α. We define the Reynolds number for this flow to be

$$Re = \frac{U_{\mathrm{max}}h}{\nu} = \frac{\alpha h^3}{\nu} \; . \quad (4.179)$$

To determine the stability of Poiseuille flow, our task is clear. We look for perturbations with $k_y = 0$ and $k_x = k$ and try to solve the Orr-Sommerfeld equation

$$\left[i(kU - \omega)\left(\frac{d^2}{dz^2} - k^2\right) - ik\frac{d^2U}{dz^2} - \nu\left(\frac{d^2}{dz^2} - k^2\right)^2\right]\tilde{u}_z = 0 \ . \quad (4.180)$$

We view this as an eigenvalue equation, with both the eigenfunction $\tilde{u}_z$ and the eigenvalue ω to be determined. The flow is unstable if there exists solutions with $\text{Im}(\omega) > 0$.

This is not a computation that can be done analytically. (It was the topic of Heisenberg's PhD thesis, where he introduced an approximation method before he gave up and found an easier topic to work on for his postdoc, namely discovering quantum mechanics.) The first accurate result was achieved by Orzag in 1971 in a tour-de-force numerical computation. One finds that there is no instability for low Reynolds number, but Poiseuille flow becomes linearly unstable for any Reynolds number beyond the critical value

$$Re > Re_{\text{crit}} \approx 5772 \ . \quad (4.181)$$

The range of unstable k-values for a given Reynolds number takes the form of the dark shaded region shown in the figure.

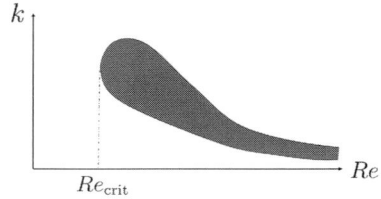

There's something rather surprising about this result. By now we're used to viscosity acting as a dampening force, causing perturbations to die away. But here it plays the opposite role! Without viscosity, Poiseuille flow is linearly stable. Viscosity causes it to become unstable at high Reynolds number. The idea that viscosity, or more generally diffusion, can drive instabilities was first suggested by Turing in a context that we will briefly describe in the next section.

This is far from the last word on instabilities. The linear analysis sketched above suggests that Couette flow should be stable for all Re while Poiseuille flow should break down only for $Re \approx 6000$. Neither of these agrees well with experiment. Instead, Couette flow is turbulent by $Re \approx 400$ while Poiseuille flow is typically turbulent by $Re \approx 2000$. This shows up the failure of our simple linear analysis. To understand what's really going on, we can take a lesson from the theory of dynamical systems.

4.7 Lessons from Dynamical Systems

There are two surprises in the search for instabilities in the simplest shear flows. The first is that Poiseuille flow appears to be unstable because of the viscosity term.

The second is that Couette flow is stable against infinitesimal perturbations, at least for suitably small Reynolds number. But real world fluids are not. This is because these flows are non-linearly unstable, meaning that the amplitude of the perturbation needs to grow suitably large before the instability kicks in. For example, for Couette flow it is thought that the amplitudes should scale as $\mathcal{O}(1/Re)$ as Re increases in order for the instability to occur.

Both of these surprises have analogues in the much simpler setting of dynamical systems. Understanding this behaviour can give us intuition for the fluid case. So, in this section, we change direction slightly and study some simple features of dynamical systems. First, we describe a phenomenon known as the *Turing instability* in which diffusion renders a stable fixed point unstable. (Recall that, in the Navier–Stokes equation, the viscosity term is the diffusion of momentum.)

Next, we describe a reasonably simple mechanism that drives perturbations to have large amplitudes. These are not full instabilities, and don't show up in the kind of linearised $e^{ikx-i\omega t}$ analysis that we've been doing so far in this chapter. Instead, the amplitudes of certain modes grow for some time, before settling back down again. This is known as *transient growth*. It is this growth of modes that ultimately pushes the system over the edge to instability. In this section, we explain how the basic idea of transient growth arises. We won't do this in fluid mechanics, but instead in a much simpler setting of dynamical systems.

4.7.1 Turing Instability

Diffusion makes things spread out. (We will devote Chapter 7 to explaining this statement in detail.) It is therefore a surprise to learn that, in certain circumstances, diffusion can cause homogeneous systems to become unstable. Like many of the best results in physics, we can see this by diagonalising a 2×2 matrix.

We start with a dynamical system with two degrees of freedom, u and v.

For now, suppose that these depend only on time, $u(t)$ and $v(t)$, although we will shortly promote them to be variables of space as well. A dynamical system is a pair of coupled, first order differential equations that tell us how u and v evolve,

$$\frac{\partial u}{\partial t} = f(u, v) \quad \text{and} \quad \frac{\partial v}{\partial t} = g(u, v) \tag{4.182}$$

for some functions $f(u, v)$ and $g(u, v)$.

A *fixed point* of this system $(u^\star, v^\star)$ obeys $f(u^\star, v^\star) = g(u^\star, v^\star) = 0$. To determine whether this fixed point is stable or unstable, we look at perturbations

$$u(t) = u^\star + \hat{u}(t) \quad \text{and} \quad v(t) = v^\star + \hat{v}(t) \tag{4.183}$$

and look at the linearised equation of motion for $\hat{\mathbf{u}} = (\hat{u}, \hat{v})$,

$$\frac{\partial \hat{\mathbf{u}}}{\partial t} = A\hat{\mathbf{u}} \quad \text{with} \quad A = \begin{pmatrix} f_u & f_v \\ g_u & g_v \end{pmatrix}_{(u^\star, v^\star)}. \tag{4.184}$$

Here we've written the components of the Jacobian matrix as $f_u = \partial f / \partial u$ and so on.

Stability is then determined by the eigenvalues of the Jacobian matrix A. If the real parts of both eigenvalues are negative, then the fixed point is stable. If the real part of one of the eigenvalues is positive, then the fixed point is unstable.

For a 2×2 matrix, it's often simplest to work with the trace (which is the sum of the two eigenvalues) and the determinant (which is the product) rather than with the eigenvalues directly. The fixed point is stable if

$$\text{Tr}\, A = f_u + g_v < 0$$
$$\det A = f_u g_v - f_v g_u > 0 . \tag{4.185}$$

In what follows, we will assume that we are dealing with a stable fixed point.

So far, our variables depend only on time. Now we let them vary in both time and space, so we have $u(\mathbf{x}, t)$ and $v(\mathbf{x}, t)$. In addition to the original dynamical system (4.182), we will add some diffusion into the story, so that the equations of motion are now

$$\frac{\partial u}{\partial t} = f(u, v) + D_1 \nabla^2 u \quad \text{and} \quad \frac{\partial v}{\partial t} = g(u, v) + D_2 \nabla^2 v . \tag{4.186}$$

We have two diffusion constants, D_1 and D_2. Equations of this kind, that mix a dynamical system with diffusion, are known as *reaction-diffusion equations*. We will see that, under the right circumstances, these additional diffusion terms can render the stable fixed point $(u^\star, v^\star)$ unstable.

It's certainly true that the fixed point is stable to homogeneous perturbations because that just takes us back to the analysis we did before. So we should look at inhomogeneous perturbations of the form

$$u(\mathbf{x}, t) = u^\star + \hat{u}(t)e^{i\mathbf{k}\cdot\mathbf{x}} \quad \text{and} \quad v(\mathbf{x}, t) = v^\star + \hat{v}(t)e^{i\mathbf{k}\cdot\mathbf{x}} \;. \qquad (4.187)$$

Now the linearised equations of motion become

$$\frac{\partial \hat{\mathbf{u}}}{\partial t} = A_{\text{new}} \hat{\mathbf{u}} \quad \text{with} \quad A_{\text{new}} = A - \begin{pmatrix} D_1 k^2 & 0 \\ 0 & D_2 k^2 \end{pmatrix} \;. \qquad (4.188)$$

To understand stability, we must look at the eigenvalues of the modified Jacobian A_{new}. Your first inclination is to think that the diffusion terms only make the system more stable. After all, they both appear with negative signs in the matrix. But that's not the way eigenvalues work! The modified Jacobian has trace

$$\text{Tr}\, A_{\text{new}} = \text{Tr}\, A - (D_1 + D_2)k^2 < 0 \;. \qquad (4.189)$$

So the trace is indeed more negative. This is what our intuition was telling us. However, the determinant is not so straightforward. It is

$$\begin{aligned} \det A_{\text{new}} &= (f_u - D_1 k^2)(g_v - D_2 k^2) - f_v g_u \\ &= \det A - k^2(D_1 g_v + D_2 f_u) + D_1 D_2 k^4 \;. \end{aligned} \qquad (4.190)$$

The system is stable only if $\det A_{\text{new}} > 0$. But we see that this is not guaranteed. There's that middle term which could possibly turn the determinant negative.

To see if this happens, we write

$$\det A_{\text{new}} = ak^4 - bk^2 + c \;. \qquad (4.191)$$

This dips below zero for some region where $k^2 > 0$ only if $b > 0$ and the quadratic has real roots. This, in turn, requires that the discriminant $b^2 - 4ac$ is positive. Translated back to the expression (4.190), we learn that the fixed point becomes unstable for some window of k provided that

$$D_1 g_v + D_2 f_u > \sqrt{4 D_1 D_2 (f_u g_v - f_v g_u)} \;. \qquad (4.192)$$

This is the condition for *Turing instability*.

Note that the system is stable to both long wavelength modes (with $k \ll 1$)

and short wavelength modes (with $k \gg 1$). The long wavelength modes are close to the homogeneous system which is known to be stable. The short wavelength modes are eliminated quickly by diffusion. The instability occurs only in some intermediate regime,

$$k_\star^2 - \Delta < k^2 < k_\star^2 + \Delta \tag{4.193}$$

with

$$k_\star = \sqrt{\frac{b}{2a}} \quad \text{and} \quad \Delta = \frac{\sqrt{b^2 - 4ac}}{2a} . \tag{4.194}$$

You could imagine starting with a system that does not exhibit the Turing instability, and then slowly varying parameters until you reach the phase transition at $b^2 = 4ac$. At this point, the modes with wavenumber

$$k_\star = \left(\frac{c}{a}\right)^{1/4} = \left(\frac{\det A}{D_1 D_2}\right)^{1/4} \tag{4.195}$$

are the first to become unstable.

Instability Requires Different Diffusivities

The Turing instability feels counterintuitive. At heart, the idea is that the matrix A has eigenvalues with negative real parts, but the matrix A_{new} does not. It feels like subtracting things off the diagonal should only decrease both eigenvalues. But, as we have seen, it's possible to decrease one and increase the other. That's the algebraic crutch on which the instability relies

For this to happen, however, it's important that $D_1 \neq D_2$. Indeed, if $D_1 = D_2 = D$ then we have $A_{\text{new}} = A - k^2 D \mathbb{1}$. If the eigenvalues of A are λ_i then the eigenvalues of A_{new} are $\lambda_i - k^2 D$ and both decrease.

Another way of seeing this is to define the ratio of diffusivities

$$d = \frac{D_1}{D_2} . \tag{4.196}$$

Then the condition for instability (4.192) becomes

$$f_u + d g_v > 2\sqrt{d \det A} . \tag{4.197}$$

The right-hand side is clearly greater than zero so we must have

$$f_u + d g_v > 0 . \tag{4.198}$$

Meanwhile, we know that

$$\text{Tr} \, A = f_u + g_v < 0 . \tag{4.199}$$

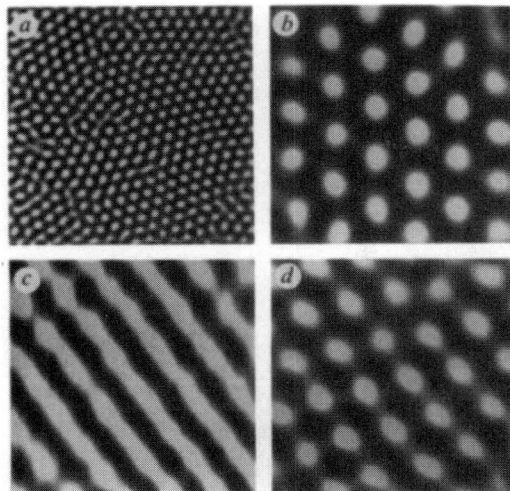

Fig. 4.6   Turing instabilities seen in chemical reactions; taken from Ouyang and Swinney [66].

So clearly $d = 1$ does not do the job. For an instability to kick in, the diffusivities must be suitably different. Moreover, this simple algebra highlights what's actually going on. For the instability to occur, one of f_u and g_v must be positive and the other negative. Suppose that we have $f_u > 0$ and $g_v < 0$. Then the variable u is called an *activator* because it wants to increase things while the variable v is called the *inhibitor* because it wants to reduce things. The result (4.199) is telling us that, in the homogeneous situation, the inhibitor wins. But if we take $d \ll 1$, so the inhibitor diffuses faster than the activator, then the result (4.197) is telling us that, ultimately, the activator can triumph.

Pattern Formation

Our analysis above only finds the instability. As always, the next question is: What becomes of it? We know that the system will necessarily become inhomogeneous and we might expect that this happens at a characteristic wavelength

$$\lambda_\star \approx \frac{2\pi}{k_\star} \, . \tag{4.200}$$

To understand the resulting pattern, we need to study the full non-linear behaviour of a system. And this typically means doing numerics. Some of the resulting patterns, due to the Turing instability in chemical reactions, are shown in Figure 4.6.

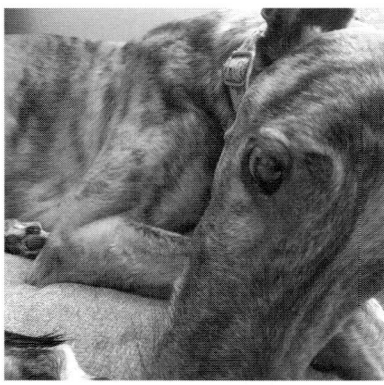

Fig. 4.7 This is Eddie, possibly Turing unstable, definitely not Turing complete.

It has been suggested that many other patterns seen in nature, including animal coats like the one shown in Figure 4.7, can be traced to a Turing diffusion instability of the kind described above.

4.7.2 Transient Growth

We can also gain some intuition for the phenomenon of non-linear instability using a simple dynamical system. Again, we take a two-component vector, $\mathbf{u}(t) = (u(t), v(t))$. This time, we can focus only on the time dynamics and we need only a linear equation of motion of the form

$$\dot{\mathbf{u}} = A\mathbf{u} \tag{4.201}$$

with A a 2×2 matrix that may be complex. We can solve this by looking at the eigenvectors and eigenvalues of the matrix

$$A\mathbf{e}_a = \lambda_a e_a \quad \text{with} \quad a = 1, 2 . \tag{4.202}$$

The dynamics depends on the properties of the matrix A.

The simplest situation arises when A is Hermitian, so $A^\dagger = A$. This is a situation that is familiar from quantum mechanics. We know that $\lambda_a \in \mathbb{R}$ and the eigenvectors are necessarily orthogonal if the eigenvalues are distinct. Indeed, we can always choose them to be orthonormal

$$\mathbf{e}_a^\dagger \cdot \mathbf{e}_b = \delta_{ab} . \tag{4.203}$$

In fact, something similar happens if the matrix A is *normal*. A normal matrix obeys $A^\dagger A = AA^\dagger$. It is a weaker condition than Hermiticity. In this case, the eigenvalues are generically complex, $\lambda_a \in \mathbb{C}$, but we again have orthonormal eigenvectors (4.203).

For a normal matrix, we construct the solution to (4.202) by decomposing $\mathbf{u}(t)$ in terms of eigenvectors,

$$\mathbf{u}(t) = \alpha(t)\mathbf{e}_1 + \beta(t)\mathbf{e}_2 \quad \Longrightarrow \quad \mathbf{u}(t) = \alpha_0 e^{\lambda_1 t}\,\mathbf{e}_1 + \beta_0 e^{\lambda_2 t}\,\mathbf{e}_2 \ . \ (4.204)$$

Now we can start to address questions of stability. For example, if A is Hermitian, so that $\lambda_a \in \mathbb{R}$, we could say that the origin is stable provided that $\lambda_a < 0$, so that $\mathbf{u}(t) \to 0$ as $t \to \infty$.

Things are more interesting when the matrix A is *non-normal*, so $A^\dagger A \neq AA^\dagger$. We will be interested in situations where the eigenvectors are linearly independent, so still span $\mathbb{C}^2$, but they are not, in general, orthogonal. In fact, as we will see, the most interesting situation arises when the eigenvectors are almost parallel.

As an example, consider the 2×2 matrix with

$$A = \begin{pmatrix} -\epsilon_1 & 1 \\ 0 & -\epsilon_2 \end{pmatrix} \ . \tag{4.205}$$

As the notation suggests, we will think about ϵ_1 and ϵ_2 as small numbers. This matrix has eigenvalues $-\epsilon_1$ and $-\epsilon_2$ and eigenvectors

$$\mathbf{e}_1 = \begin{pmatrix} 1 \\ 0 \end{pmatrix} \quad \text{with} \quad A\mathbf{e}_1 = -\epsilon_1 \mathbf{e}_1 \ ,$$

$$\mathbf{e}_2 = \begin{pmatrix} 1 \\ \epsilon_1 - \epsilon_2 \end{pmatrix} \quad \text{with} \quad A\mathbf{e}_2 = -\epsilon_2 \mathbf{e}_2 \ . \tag{4.206}$$

Note that we've made no attempt to normalise $\mathbf{e}_2$. The eigenvectors are linearly independent provided that $\epsilon_1 \neq \epsilon_2$. We will make this assumption but, with ϵ_1 and ϵ_2 small, it's clear that we're flirting with danger here. The matrix A is almost degenerate. This is reflected in the fact that the two eigenvectors are almost parallel.

So far, it's not clear what this buys us. After all, we can still write down the general solution (4.204),

$$\mathbf{u}(t) = \alpha_0 e^{-\epsilon_1 t}\mathbf{e}_1 + \beta_0 e^{-\epsilon_2 t}\mathbf{e}_2 \ . \tag{4.207}$$

If we take $\epsilon_1, \epsilon_2 > 0$ then it's clear that this is a stable system by our previous criterion, with $\mathbf{u}(t) \to 0$ as $t \to \infty$.

However, this hides an interesting feature! Just because the magnitude decays at late times, doesn't necessarily mean that it always decays. To check

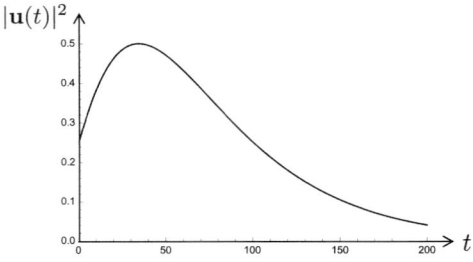

Fig. 4.8 An example of transient growth, with $|\mathbf{u}(t)|^2$ plotted for $\epsilon_1 = 0.03$, $\epsilon_2 = 0.01$, and $\gamma = -1.5$.

this, we need to look at early times. We scale out the overall magnitude of the vector by writing the relative coefficients as

$$\beta_0 = \gamma \alpha_0 \ . \tag{4.208}$$

The magnitude of the vector $\mathbf{u}(t)$ is then given by

$$|\mathbf{u}(t)|^2 = \alpha_0^2 \left(e^{-2\epsilon_1 t} + 2\gamma e^{-(\epsilon_1 + \epsilon_2)t} + \gamma^2 \left(1 + (\epsilon_1 - \epsilon_2)^2 \right) e^{-2\epsilon_2 t} \right) \tag{4.209}$$

$$= \text{constant} - 2\alpha_0^2 \left(\epsilon_1 + \gamma(\epsilon_1 + \epsilon_2) + \gamma^2 (1 + (\epsilon_1 - \epsilon_2)^2)\epsilon_2 \right) t + \dots$$

where we focussed on the linear growth. There's an overall minus sign, which suggests that magnitude does decay. However, we see that it might be possible to pick $\gamma < 0$ such that this decay is actually a growth. This happens if the term in large brackets is negative,

$$\gamma^2 (1 - (\epsilon_1 - \epsilon_2)^2)\epsilon_2 + \gamma(\epsilon_1 + \epsilon_2) + \epsilon_1 < 0 \ . \tag{4.210}$$

Dropping terms of order ϵ^2, it's simple to check that the magnitude initially grows provided that

$$-\frac{\epsilon_1}{\epsilon_2} < \gamma < -1 \quad \text{for} \quad \epsilon_1 > \epsilon_2 \ . \tag{4.211}$$

If $\epsilon_1 < \epsilon_2$, the range for growth is $-1 < \gamma < -\epsilon_1/\epsilon_2$. Note that the upper end of the window (4.211) is the "hard to reach" vectors, which are orthogonal to $\mathbf{e}_1$ and almost orthogonal to $\mathbf{e}_2$. Such vectors have the property that, to construct $\mathbf{u}$ of magnitude $\mathcal{O}(1)$, we need coefficients of $\mathcal{O}(1/\epsilon)$.

The magnitude $|\mathbf{u}(t)|^2$ is plotted in Figure 4.8 for values that lie inside the window (4.211). We see clearly that the magnitude grows, before finally decaying. This is *transient growth*.

How much growth can we expect to get this way? Returning to (4.209), we can differentiate with respect to time to see when the growth will end. We will work to leading order, setting $(\epsilon_1 - \epsilon_2)^2 \approx 0$. Then the turning point sits at

$$
\frac{d|\mathbf{u}|^2}{dt} \approx -2\alpha_0^2 e^{-2\epsilon_1 t}\left(\epsilon_1 + (\epsilon_1 + \epsilon_2)\gamma e^{(\epsilon_1 - \epsilon_2)t} + \epsilon_2\gamma^2 e^{2(\epsilon_1 - \epsilon_2)t}\right) = 0 \ . \ (4.212)
$$

This is the same quadratic that we already solved, just with γ replaced by $\gamma e^{(\epsilon_1 - \epsilon_2)t}$. If we have $\epsilon_1 > \epsilon_2$, then the growth occurs until

$$
e^{(\epsilon_1 - \epsilon_2)t_{\max}} = -\frac{\epsilon_1}{\epsilon_2\gamma} \ . \tag{4.213}
$$

We see that the growth continues until a time of order $t_{\max} \sim 1/(\epsilon_1 - \epsilon_2)$. At this point, the magnitude of $\mathbf{u}$ has grown to

$$
\frac{|\mathbf{u}(t_{\max})|^2}{|\mathbf{u}(0)|^2} = e^{-2\epsilon_1 t_{\max}} \frac{1 - (\epsilon_1/\epsilon_2)^2}{(1+\gamma)^2} \ . \tag{4.214}
$$

It's clear that this is maximised if we take $\gamma \to -1$, which is the upper limit of our transient growth window. Naively, it looks as if we get an infinite amount of growth if we do this. But we were slightly careless when computing (4.211), dropping the $\mathcal{O}(\epsilon^2)$ terms. Taking a little more care, we find that the upper end of the window (for $\epsilon_1 > \epsilon_2$) is

$$
\gamma < -1 - \epsilon_2^2 + \epsilon_1\epsilon_2 + \dots \ . \tag{4.215}
$$

We see that the growth scales parametrically as $|\mathbf{u}(t_{\max})|^2/|\mathbf{u}(0)|^2 \sim 1/\epsilon^2$. The closer the non-normal matrix is to being degenerate, the larger the growth.

4.7.3 Very Briefly, Transient Growth in Fluids

So far we've seen that some simple dynamical systems, governed by non-normal matrices, have the property that they exhibit transient growth. Moreover, for certain initial conditions, this growth can be large, scaling as $\sim 1/\epsilon^2$ where ϵ is a small parameter that tells us how close our non-normal matrix is to degenerating completely.

What does this have to do with fluids? We won't give any mathematical details here. Just some words. The linearised fluctuations of a viscous fluid are governed by the Orr-Sommerfeld (4.173) and Squire (4.175) equations. These can be written in terms of a differential operator. The key observation is that this operator is non-normal, meaning that its eigenfunctions are complete, but not orthogonal. In this way, it exhibits transient growth very similar to the simple matrix system that we described above. It is thought that transient growth pushes the system over the edge, driving the non-linear instability to turbulence.

In contrast, the linearised fluctuation operators that arose for the Rayleigh–Bénard instability of Section 4.4 are normal (in fact, self-adjoint). In that case, there is no transient growth.

5 Turbulence

When the speed of a fluid increases beyond some critical value, things have a tendency to go a bit squirly. The calm, serene laminar flows that we've seen in earlier chapters become unstable and are replaced by something messy and dirty, with the fluid moving in seemingly random directions, tripping over itself, with eddies forming and stretching before disintegrating into smaller eddies. This is turbulent flow.

There is every reason to believe that turbulent flow is correctly described by the Navier–Stokes equation, not least because numerical simulations of the equation exhibit turbulence in all its glory. (In this context, the simulations go by the name *DNS*, standing for *direct numerical simulation*.) But understanding the full details of turbulent flow remains, to put it mildly, a formidable problem. Turbulence kicks in when the Reynolds number is greater than some critical value $Re > Re_{\text{crit}}$. The exact number depends on the kind of flow we're looking at, but a ballpark figure is

$$Re_{\text{crit}} \approx 10^3 \ . \tag{5.1}$$

At these speeds, the advective term $(\mathbf{u} \cdot \nabla)\mathbf{u}$ in the Navier–Stokes equation is important. This is the only non-linear term in the equation and it drives the system to a chaotic state, with the motion wildly dependent on the initial conditions, and streamlines that start nearby diverging. The challenge is to understand this motion.

This challenge is, it turns out, hard. Despite many decades of study, turbulence remains poorly understood. In contrast to previous chapters, we will not attempt to find explicit solutions to the Navier–Stokes equations. Instead, we will retreat and look at averaged properties of flows. This might seem like a strange thing to do. After all, the Navier–Stokes equation is, at the end of the day, just a differential equation and, as such, its behaviour is entirely deterministic. Nonetheless, turbulent motion appears to be random. This can be traced to the sensitive dependence on initial conditions that characterises chaotic systems. To proceed, we will embrace this randomness and work in a statistical sense. Rather than trying to analyse any specific solution, we will instead try to extract properties of appropriately averaged solutions.

Our goals in this chapter will be limited. We won't look at any specific turbulent flows, such as boundary layers or wakes. Instead, we will just try to understand some very general properties that are shared by all turbulent flows, at least in some regime. Nor will we study the interesting behaviour that happens for flows around the critical Reynolds number Re_{crit}, where instabilities develop. Instead we focus on what happens with $Re \gg Re_{\mathrm{crit}}$, a regime known as *fully developed turbulence*. Even here, there is much that is not understood. Indeed, it's not even clear what the right variables are to characterise turbulent flows. Here we offer only the basics of the subject.

5.1 Mean Flow

As we mentioned above, to understand turbulence it's necessary to think on a more probabilistic level about the Navier–Stokes equation. But given that the Navier–Stokes equation is purely deterministic, it's not obvious what this means. If we're going to think about averaged properties, the first question we should ask is: What are we actually averaging over?

There are different answers that we could give to this. One way to proceed is to average over different initial conditions to the Navier–Stokes equation. We could pick some collection of initial conditions, all of which look similar. Because of the chaotic nature of the equation, each will give rise to very different solutions. We could then try to figure out average properties of these solutions. This is known as the *ensemble average* and is similar to the philosophy underlying statistical mechanics.

Alternatively, we could do something that feels more physical. A turbulent velocity field $\mathbf{u}(\mathbf{x}, t)$ varies rapidly in both space and time and we could choose to average over either of these. There is a general expectation (although no proof) that, for a typical flow, it doesn't matter which average we choose: the temporal, spatial, and ensemble averages should all give the same answer. This goes by the name of the *ergodic hypothesis*.

Here we will average over time (because it turns out to be the simplest). We decompose the complicated turbulent flow $\mathbf{u}(\mathbf{x}, t)$ into an averaged, mean flow $\mathbf{U}(\mathbf{x}, t)$ together with some fluctuations $\delta\mathbf{u}(\mathbf{x}, t)$

$$\mathbf{u} = \mathbf{U} + \delta\mathbf{u} \tag{5.2}$$

where to define the mean flow we average over some time scale T

$$\mathbf{U}(\mathbf{x}, t) = \langle\mathbf{u}(\mathbf{x}, t)\rangle := \frac{1}{T} \int_t^{t+T} dt' \ \mathbf{u}(\mathbf{x}, t') \ . \tag{5.3}$$

This is called *Reynolds averaging*. There are two options for how to think about the time scale T:

- We could simply take $T \to \infty$. In this case, we have a steady mean flow $\mathbf{U}(\mathbf{x})$ that does not itself depend on time.

- Alternatively, we may have a situation in which there are two different time scales in the flow. The turbulent fluctuations occur over some short time scale τ_{short}, which is superimposed on some averaged flow which takes place over some much longer time scale τ_{long}. In this case we could take $\tau_{\text{short}} \ll T \ll \tau_{\text{long}}$ to get a mean velocity field $\mathbf{U}(\mathbf{x}, t)$ which varies only over the long time scale.

In what follows, we'll adopt the second of these. This isn't for any particularly well-motivated physical reason, but simply because it's not much more effort to do this and it obviously includes the $T \to \infty$ situation as a special case in which $\mathbf{U}(\mathbf{x})$ is stationary.

Since our mean flow is $\mathbf{U} = \langle\mathbf{u}\rangle$, the complicated velocity fluctuations are $\delta\mathbf{u} = \mathbf{u} - \langle\mathbf{u}\rangle$. By construction, this means that the average of the fluctuations vanishes

$$\langle\delta\mathbf{u}\rangle = 0 \ . \tag{5.4}$$

Importantly the Reynolds averaging commutes with spatial differentiation, so if our fluid is incompressible then both the mean flow and the fluctuations must be separately incompressible

$$\nabla \cdot \mathbf{u} = 0 \quad \Longrightarrow \quad \langle\nabla \cdot \mathbf{u}\rangle = 0$$
$$\Longrightarrow \quad \nabla \cdot \mathbf{U} = 0 \quad \Longrightarrow \quad \nabla \cdot \delta\mathbf{u} = 0 \ . \tag{5.5}$$

We do a similar averaging for other fields, including the pressure, which we write as

$$P = \langle P \rangle + \delta P \tag{5.6}$$

with $\langle P \rangle$ defined by a time average in the same way as (5.2). Again, we have designed things so that the average fluctuation necessarily vanishes: $\langle\delta P\rangle = 0$.

We will actually explore the averaged properties of the Navier–Stokes equation twice in this chapter. Our focus in this section will be on deriving an equation for the mean flow $\mathbf{U}$ after integrating out the fluctuations. This won't take us particularly far, not least because it feels like we're throwing out the baby and keeping the bath water since, often, the fluctuations are much more interesting than the mean flow! Nonetheless, we include this approach because it gives some intuition for the difficulties involved. Moreover, this approach is useful when modelling situations where there is clearly some overarching mean flow, like transport down a pipe, with turbulence bubbling away underneath (it turns out that this often takes place in a regime of Reynolds numbers $10^3 \lesssim Re \lesssim 10^5$) and it will allow us to define some commonly used concepts such as "Reynolds stress" and "eddy viscosity". Then, in Section 5.3, we will retrace the same steps, this time focussing on the fluctuations themselves. It's only in this second approach that we'll start to make some real progress.

5.1.1 The Reynolds Averaged Navier–Stokes Equation

If we substitute the decomposition (5.2) and (5.6) into the Navier–Stokes equation, we have

$$\rho \left(\frac{\partial (\mathbf{U} + \delta \mathbf{u})}{\partial t} + (\mathbf{U} + \delta \mathbf{u}) \cdot \nabla (\mathbf{U} + \delta \mathbf{u}) \right)$$
$$= -\nabla (\langle P \rangle + \delta P) + \mu \nabla^2 (\mathbf{U} + \delta \mathbf{u}) . \qquad (5.7)$$

We've neglected any further forces acting on the fluid, such as the forcing term needed to drive turbulence, but they can be added as needed. Now we average this equation and use the fact that $\langle \delta \mathbf{u} \rangle = \langle \delta P \rangle = 0$. We need to be a little careful with the time derivative term: we have

$$\left\langle \frac{\partial (\delta \mathbf{u})}{\partial t} \right\rangle = \frac{1}{T} \int_t^{t+T} dt' \, \frac{\partial (\delta \mathbf{u})}{\partial t'} = \frac{1}{T} \left[\delta \mathbf{u}(\mathbf{x}, t+T) - \delta \mathbf{u}(\mathbf{x}, t) \right]$$
$$= \frac{1}{T} \frac{\partial}{\partial t} \int_t^{t+T} dt' \, \delta \mathbf{u}(\mathbf{x}, t')$$
$$= \frac{\partial \langle \delta \mathbf{u} \rangle}{\partial t} = 0 \qquad (5.8)$$

where, again, we've used the fact that $\langle \delta \mathbf{u} \rangle = 0$. We're then left with

$$\rho \left(\frac{\partial \mathbf{U}}{\partial t} + \mathbf{U} \cdot \nabla \mathbf{U} + \langle \delta \mathbf{u} \cdot \nabla \delta \mathbf{u} \rangle \right) = -\nabla \langle P \rangle + \mu \nabla^2 \mathbf{U} . \qquad (5.9)$$

This is *almost* the Navier–Stokes equation for the averaged velocity $\mathbf{U}$. The only difference is the term $\delta \mathbf{u} \cdot \nabla \delta \mathbf{u}$, quadratic in the fluctuations, that wasn't killed by averaging. We take this over to the right-hand side and treat it as part of the stress tensor, writing the Navier–Stokes equation for the averaged flow in the form (2.21)

$$\rho\left(\frac{\partial U_i}{\partial t} + U_j\frac{\partial U_i}{\partial x_j}\right) = \frac{\partial \sigma_{ij}}{\partial x_j} \; . \tag{5.10}$$

This is the *Reynolds averaged Navier–Stokes* equation. The stress tensor on the right-hand side is

$$\sigma_{ij} = -\langle P\rangle\delta_{ij} + \mu\left(\frac{\partial U_i}{\partial x_j} + \frac{\partial U_j}{\partial x_i}\right) - \rho\langle \delta u_i \delta u_j\rangle \tag{5.11}$$

where we've used the fact that $\nabla \cdot \delta \mathbf{u} = 0$ in writing it in this form. We see that, in this approximation, the role of the fluctuations is to guide the mean flow through the additional term

$$R_{ij} = \rho\langle \delta u_i \delta u_j\rangle \tag{5.12}$$

in the stress tensor. This is known as the *Reynolds stress* or, sometimes, the *turbulent stress*. (Actually, more often the Reynolds stress is defined without the factor of ρ, even though that isn't, strictly, a stress.) So if we want to understand how the mean flow flows, we need to understand something about the variance of the fluctuations $\langle \delta u_i \delta u_j\rangle$.

Finding Closure

Our next task is to get an expression for this extra contribution to the stress tensor R_{ij}. To this end, if we subtract the averaged Navier–Stokes equation (5.10) from our starting point (5.7), we have

$$\rho\Big(\frac{\partial(\delta \mathbf{u})}{\partial t} + (\mathbf{U}\cdot\nabla)\delta\mathbf{u} + (\delta\mathbf{u}\cdot\nabla)\mathbf{U}$$
$$+ (\delta\mathbf{u}\cdot\nabla)\delta\mathbf{u} - \langle\delta\mathbf{u}\cdot\nabla\delta\mathbf{u}\rangle\Big) = -\nabla\delta P + \mu\nabla^2\delta\mathbf{u} \; . \tag{5.13}$$

If we multiply this by $\delta\mathbf{u}$, we get the following expression for the tensor $\delta u_i \delta u_j$

$$\rho\Big(\frac{\partial(\delta u_i \delta u_j)}{\partial t} + U_l\frac{\partial(\delta u_i \delta u_j)}{\partial x_l} + \delta u_i \delta u_l\frac{\partial U_j}{\partial x_l} + \delta u_j \delta u_l\frac{\partial U_i}{\partial x_l}$$
$$+ \frac{\partial}{\partial x_l}(\delta u_l \delta u_i \delta u_j) + \delta u_i\frac{\partial R_{lj}}{\partial x_l} + \delta u_j\frac{\partial R_{li}}{\partial x_l}\Big)$$
$$= -\delta u_i\frac{\partial(\delta P)}{\partial x_j} - \delta u_j\frac{\partial(\delta P)}{\partial x_i} + \mu^2\left(\delta u_i\nabla^2\delta u_j + \delta u_j\nabla^2\delta u_i\right) \; . \tag{5.14}$$

We now take the average to get an equation for the Reynolds stress tensor that, schematically, takes the form

$$\frac{\partial R_{ij}}{\partial t} + (\mathbf{U} \cdot \nabla) R_{ij} = -\rho \frac{\partial}{\partial x_l} \langle \delta u_l \delta u_i \delta u_j \rangle + \text{other stuff} \qquad (5.15)$$

where the other stuff includes other averages such as $\langle \delta P \, \delta \mathbf{u} \rangle$. The key point is that we can get ourselves an equation for R_{ij}, but it involves a three-point average $\langle \delta \mathbf{u}^3 \rangle$. And if we try to get an equation for $\langle \delta \mathbf{u}^3 \rangle$ then you probably won't be surprised to hear that it involves $\langle \delta \mathbf{u}^4 \rangle$, and so on. We find that we have an infinite hierarchy of equations. This is not unusual in physics when doing this kind of analysis. (An analogous situation will arise in Chapter 6 when deriving the Boltzmann equation where it is called the BBGKY hierarchy.) Within the context of turbulence, this is known as the *closure problem*: the set of equations don't close and keep forcing you to look at the next order in fluctuations.

What to do about it? Well, there is no mathematically well-defined way to truncate this infinite series of equations. Nor is there a physical reason to expect some simplification to occur. Turbulence is a strongly coupled problem and, to do things properly, you really need to worry about this infinite series of equations. Of course, that's not particularly practical. So to proceed, the usual strategy is present some hand-waving physical argument and then just to make something up. These made-up approaches have a name: they are collectively called "closure models". There are many.

Here's the simplest example of a made-up thing, due to Boussinesq. Suppose that, for some reason, the three-point averages $\langle \delta \mathbf{u}^3 \rangle$ and higher are actually unimportant. Then we can look for an expression for the Reynolds stress R_{ij} that depends only on the mean flow $\mathbf{U}$. One, particularly simple option is to postulate that it takes the form

$$R_{ij} = -\mu_T \left(\frac{\partial U_i}{\partial x_j} + \frac{\partial U_j}{\partial x_i} \right) + \frac{2}{3} \rho K \delta_{ij} \qquad (5.16)$$

which depends on two, unknown constants μ_T and K. The latter has a nice physical interpretation: it is the kinetic energy (per unit mass) in the fluctuations

$$K = \frac{1}{2} \langle \delta \mathbf{u} \cdot \delta \mathbf{u} \rangle \ . \qquad (5.17)$$

The former has a nice name: it is called the *turbulent viscosity*, or sometimes the *eddy viscosity*. This guess for the Reynolds stress has the nice effect of simply renormalising the stress tensor σ_{ij} on the right-hand side of the averaged Navier–Stokes equation which, from (5.11), becomes

$$\sigma_{ij} = - \left(\langle P \rangle + \frac{2}{3} \rho K \right) \delta_{ij} + (\mu + \mu_T) \left(\frac{\partial U_i}{\partial x_j} + \frac{\partial U_j}{\partial x_i} \right) . \qquad (5.18)$$

This takes the same form as the usual stress tensor, but with an effective pressure $P_{\text{eff}} = \langle P \rangle + \frac{2}{3} K$ and an effective viscosity $\mu_{\text{eff}} = \mu + \mu_T$. The end result is that this guess has led us back to the original Navier–Stokes equation, but with an extra contribution to the pressure and a shifted value of the viscosity.

There are many other, more sophisticated closure models, in which one tries to incorporate $\langle \delta \mathbf{u}^3 \rangle$ corrections and so on, and then gives up at some higher order. They may be more sophisticated, but it's not obvious that they are more right and we won't discuss them here. Instead, we will reset and go in a different direction.

5.2 The Energy Cascade

Big whorls have little whorls
Which feed on their velocity,
And little whorls have lesser whorls
And so on to viscosity.

Lewis Fry Richardson

Turbulence is one of the great problems in physics. To make progress it's clear that we're going to have to break out some pretty powerful machinery. And things don't get more powerful than dimensional analysis. In this section, we will use dimensional analysis to get a handle on one very specific property of turbulence: What happens to the energy?

The set of ideas described in this section are due to Richardson, Taylor, and Kolmogorov. These ideas culminated in a series of papers by Kolmogorov in 1941 and fluid dynamicists often refer to this argument, rather elliptically, as K41. It is, I think, one of the greatest applications of dimensional analysis in all of physics.

5.2.1 The Zeroth Law of Turbulence

To get started, we need a few facts about turbulent flows. First is the observation that turbulence is very much a dissipative phenomenon: if you leave a turbulent fluid alone, it will quickly relax back to equilibrium with

the turbulent properties dying away. This means that something must be
feeding the turbulence to keep it alive. In other words, there has to be some
injection of energy into the system. This energy could be due to some external pressure difference, some shear effect arising from gravity, or some
teaspoon stirring the tea. The details won't concern us and we'll model this
energy injection by some external force density $\mathbf{f}(\mathbf{x}, t)$, as in our original
Navier–Stokes equation (2.3). The work done (per unit time) by this force
is

$$\text{work done} = \int d^3x \, \mathbf{f} \cdot \mathbf{u} \; . \tag{5.19}$$

While energy is injected into the system by the work done, energy is also lost
to heat through dissipation. We've already seen in Chapter 2 that dissipation
in fluid mechanics is due to viscosity. The kinetic energy (per unit mass) of
the fluid is $K = \frac{1}{2} \int d^3x \, \mathbf{u}^2$. In (2.31), we found that the energy lost is due
to the integrated vorticity,

$$\frac{\partial K}{\partial t} = -\nu \int d^3x \, \left| \frac{\partial u_i}{\partial x_j} \right|^2 = -\nu \int d^3x \, |\boldsymbol{\omega}|^2 \tag{5.20}$$

where $\boldsymbol{\omega} = \nabla \times \mathbf{u}$ is the usual vorticity and, as we mentioned previously, the
integral $\int d^3x \, |\boldsymbol{\omega}|^2$ is also known as the *enstrophy*. One of the key quantities
in what follows is the dissipation rate per unit volume per unit mass, which
we write as

$$\varepsilon = \frac{\nu}{V} \int d^3x \, |\boldsymbol{\omega}|^2 \; . \tag{5.21}$$

with V the spatial volume.

Underlying our arguments is a simple, but very counterintuitive, experimental observation. In the limit of large Reynolds number, where we have
turbulent flows, the dissipation rate is constant

$$\epsilon \to \text{constant} \quad \text{as} \quad Re \to \infty \; . \tag{5.22}$$

In a nod to thermodynamics, the result (5.22) is sometimes referred to as the
zeroth law of turbulence, presumably in the hope that someone might later
come along and add further laws. It also goes by the name of the *dissipation
anomaly*.

At first glance, (5.22) may seem reasonable. But, if we think of the $Re \to$
∞ limit as $\nu \to 0$, keeping U and L fixed, then there is clearly something
strange going on because, from (5.21), we have

$$\lim_{\nu \to 0} \frac{\nu}{V} \int d^3x \, |\boldsymbol{\omega}|^2 = \text{constant} \; . \tag{5.23}$$

This means that the vorticity must diverge in this limit to compensate.

This seems to be in some mild tension with the widely held belief that the Navier–Stokes equation does not form singularities in finite time. It's not a direct contradiction because there's no statement about the evolution of a single flow in the zeroth law. Nonetheless, it serves to highlight that things can be subtle, and that the $Re \to \infty$ limit of the Navier–Stokes equation is *not* the same thing as the Euler equation where we set ν to be strictly zero, in which case there is certainly no dissipation! This is a road that we've been down before when discussing d'Alembert's paradox.

Peculiar as this situation is, we already met a hint of it when we discussed the Burgers vortex in Section 2.2.4. Recall that this is a vortex-line solution to the full Navier–Stokes equation, with vorticity (2.62) that takes the form

$$\omega(r) \sim \frac{1}{\nu} e^{-\alpha r^2 / 2\nu} \tag{5.24}$$

where α is a constant. We noted at the time that this solution has the property that the dissipation per unit length (2.65) is independent of the viscosity ν. This means that in the limit $\nu \to 0$, the dissipation remains constant while the vorticity (5.24) becomes singular.

Drag Force

The zeroth law of thermodynamics cannot (yet!) be proven from first principles, starting with the Navier–Stokes equation. The evidence for its validity is experimental and can be seen, for example, in the drag force F experienced by an object.

If the object has length L and mass M, and moves with speed U, then on dimensional grounds the drag force must take the form

$$F = \frac{MU^2}{L} f(Re) \tag{5.25}$$

for some function $f(Re)$ of the dimensionless Reynolds number $Re = UL/\nu$. This is not a function that we can reliably calculate, but a sketch of experimental data for $f(Re)$ for the drag force on a cylinder is shown in Figure 5.1. For small Re, we have

$$f(Re) \sim \frac{1}{Re} \quad \text{as } Re \to 0. \tag{5.26}$$

This ensures that the drag force scales linearly with U at low Reynolds number, in agreement with the results of Section 2.4. (See, for example,

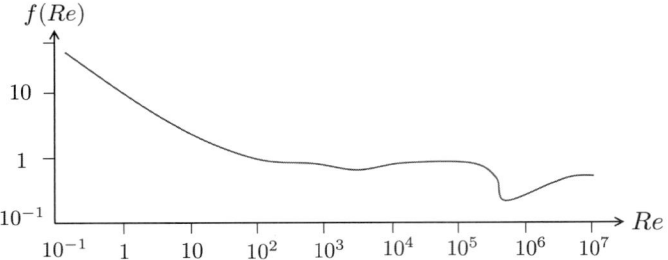

Fig. 5.1 The drag force on a cylinder, following the book by Frisch [33].

Stokes' law (2.105) for the drag force on a sphere.) More interesting for our current purpose is the behaviour at large Re. There is a strange dip at $Re \approx 10^6$. This is a manifestation of the drag crisis that we explained in Section 2.5. But, apart from this hiccup, the drag force remains roughly constant over several decades and, crucially, shows no sign of tending towards zero as $Re \to \infty$.

The work done (per unit time) in pushing against this drag force is just FU. In steady state, we can equate the work done with the energy dissipated,

$$\text{dissipation} = FU = \frac{MU^2}{L} f(Re) \ . \tag{5.27}$$

As $Re \to \infty$, the rate of dissipation tends towards a constant, independent of the viscosity ν. This is a surprise, given that dissipation only arises because of viscosity! But it seems, nonetheless, to be true. This is the essence of the zeroth law.

5.2.2 Some Dimensional Analysis

With the zeroth law in hand, we now give a qualitative description of turbulent flows. We inject energy into the system through the work done (5.19), and energy drains away through dissipation given in (5.20). In steady-state situations, where the flow does not speed up over time, these two must be equal, giving

$$\text{work done} = \rho\nu \int d^3x \ |\boldsymbol{\omega}|^2 \ . \tag{5.28}$$

This is possible only because of the zeroth law. The left-hand side does not depend on the viscosity, while the right-hand side seemingly does because there's a factor of ν sitting there. The zeroth law (5.23) tells us that this dependence is fake for turbulent flows at large Reynolds number.

The key to understanding the physics is to appreciate that the processes on either side of the equation (5.28) take place at very different scales. On the left-hand side, the driving force is a macroscopic phenomenon, typically comparable to the size of the entire system. Meanwhile, on the right-hand side, dissipation is a phenomenon that occurs at a much smaller scale. This manifests itself in (5.28) because dissipation is greatest when there are large gradients in velocity and hence a large vorticity. So energy goes in at the largest scale, and out at the smallest. We would like to put some equations to these words to make them more concrete.

We start by quantifying the work done. Suppose that the driving force takes place over some large length scale L. This is sometimes called the *outer scale*. Over this scale, the mean velocity field will vary with some magnitude U. The Reynolds number (2.71) for the flow is roughly

$$Re \sim \frac{UL}{\nu} \tag{5.29}$$

and, by assumption, we have $Re \gg 1$.

The turbulent flow is not laminar, but swirling in many directions. It's useful to think of this in terms of vorticity, with the swirling referred to as eddies. In this somewhat cartoon picture, the flow on the large, outer scale consists of eddies of size L.

Now our first stab at some dimensional analysis. The steady-state condition (5.28) equates the work done to the dissipation rate (per unit volume, per unit mass) ε given in (5.21)

$$\frac{\text{work done}}{\rho V} = \varepsilon . \tag{5.30}$$

Each side of this equation has dimension $[\varepsilon] = L^2 T^{-3}$. We must be able to express the work done in terms of the large scale, macroscopic quantities U and L. Dimensional analysis means that there is a unique possibility, namely

$$\frac{\text{work done}}{\rho V} \sim \frac{U^3}{L} . \tag{5.31}$$

Equating this to the rate of dissipation ε given in (5.21), we have

$$\frac{U^3}{L} \sim \frac{\nu}{V} \int d^3x \, |\boldsymbol{\omega}|^2 . \tag{5.32}$$

Suppose that the dissipation also comes from these same, macroscopically large velocity gradients. Then we would have $|\boldsymbol{\omega}| \sim U/L$, which would give a dissipation rate

$$|\boldsymbol{\omega}| \sim \frac{U}{L} \quad \implies \quad \varepsilon = \frac{\nu}{V} \int d^3x \, |\boldsymbol{\omega}|^2 \sim \frac{\nu U^2}{L^2} \sim \frac{1}{Re} \frac{U^3}{L} \, . \qquad (5.33)$$

But that's nowhere near enough dissipation for our steady-state condition (5.32) to hold! It's less dissipation than we need by a factor of Re and, as we have stressed, turbulent flow takes place at values of the Reynolds number $Re \gg 1$. It must be the case that dissipation arises with a vorticity $|\boldsymbol{\omega}|$ that is larger than that caused by the driving force. Which means that there must be larger gradients of the velocity field $\mathbf{u}$ and so physics taking place on some smaller scale. That, of course, agrees with experimental observations of turbulence, where there are features on many different scales. We would like to construct a simple model of this.

5.2.3 Scale Invariance

Energy is injected at some length scale L where it causes eddies of size L. But, as we have seen, these eddies don't dissipate enough energy, and structures on smaller scales must form. It's useful to think of these new length scales emerging as the original, large eddies break up into smaller ones.

There's nothing clean and simple going on here where, for example, the initial eddy neatly splits in two. Instead, as we have stressed, turbulence is a messy and complicated phenomenon and it consists of eddies of all possible sizes, at least within a range that is bounded above by the outer scale L and, as we will see shortly, bounded below by a much smaller length scale denoted η.

As the larger eddies break up, they lose energy, which is fed into the smaller eddies below them. The eddies of size r have some velocity difference $u(r)$. For now, it will suffice to think of $u(r)$ as the average velocity difference between two points,

$$u(r)^2 = \langle (\mathbf{u}(\mathbf{x} + \mathbf{r}) - \mathbf{u}(\mathbf{x}))^2 \rangle \, . \qquad (5.34)$$

By translational invariance, this result does not depend on the point $\mathbf{x}$. By rotational invariance it does not depend on the direction of $\mathbf{r}$. A more precise definition will be given in Section 5.3 where we introduce correlation functions and we will see that we should really focus on the *longitudinal* velocity difference.

The eddies of size r are being fed some energy by the bigger boys above them but, at the same time, they're losing energy as they themselves decay into the smaller eddies below.

There are (at least) two assumptions here. The first is that the transfer of energy is *local* in scale. That means that eddies of a certain size don't arise from eddies that are vastly bigger, and nor do they decay into eddies that are vastly smaller. Instead, eddies of a given size are affected only by those of neighbouring size.

This assumption is reasonable. Focus on your favourite eddy of some fixed size. Those eddies that are much bigger contribute to some background flow, sweeping your favourite eddy along, but not distorting it in any appreciable way. Meanwhile, those eddies that are much smaller could do damage, except they're not moving coherently: they are just a random background noise. It's only the eddies of comparable size that can shear the eddy you care about, causing it to stretch and twist and ultimately disintegrate into smaller (but not much smaller!) eddies. This idea of locality in scale size is made precise in a framework called the *renormalisation group* that we will meet in the book on Statistical Physics.

The second assumption is that, at least in some regime of scales, the transfer of energy from bigger to smaller eddies takes place with no dissipation at all. Indeed, we've seen that the dissipation due to the very largest eddies (5.33) is suppressed by $1/Re$, and we make the approximation that this can be ignored completely. This is essentially the statement that viscosity is irrelevant for this aspect of turbulence. This assumption means that the eddies at size r receive some energy and promptly pass it down to smaller scales. This process is known as the *energy cascade* and was first proposed by Richardson.

The eddies of size r will evolve in a significant manner over a *turnover time* of order

$$\tau(r) \sim \frac{r}{u(r)} \; . \tag{5.35}$$

This is roughly the time it takes the eddies to decay. The kinetic energy of eddies at scale r scales as $\sim u(r)^2$. These eddies hold on to this energy for a time of order $\tau(r)$ before passing it down the chain, while receiving the same from above. This energy passing through is equal to the work done (per unit time) at the top end, and to the dissipation rate, at the bottom end, meaning

$$\frac{U^3}{L} \sim \frac{u(r)^2}{\tau(r)} \sim \varepsilon \ . \tag{5.36}$$

Note that it's important that we didn't allow the viscosity ν to sneak into this formula since that carries dimensions $[\nu] = L^2 T^{-1}$ and messes up the dimensional analysis. This is a reiteration of a point that we made above: the viscosity is irrelevant for the energy cascade since no dissipation is taking place.

If we now substitute (5.35) into (5.36), we find that there is a scale invariance in the energy cascade, with the velocities of eddies of size r obeying

$$\frac{u(r)^3}{r} \sim \varepsilon \ . \tag{5.37}$$

Rearranging this, we find that the velocities of eddies of size r scale as

$$u(r) \sim (\varepsilon r)^{1/3} \ . \tag{5.38}$$

This is known as the *Kolmogorov–Obhukov* law. It is often expressed in terms of the correlations between velocities at different points,

$$\langle (\mathbf{u}(\mathbf{x} + \mathbf{r}) - \mathbf{u}(\mathbf{x}))^2 \rangle \sim (\varepsilon r)^{2/3} \ . \tag{5.39}$$

This is the *2/3 law*, named after the exponent. Later, in Section 5.3, we will give a more precise formulation of this law and see that it is better viewed as a statement about a so-called "three-point correlation function".

The idea that the same energy ε is passed down the chain, as captured in (5.36), means that the cascade is *scale-invariant*. This is seen in the simple power-law dependence (5.38) of the velocity $u(r)$, which means that the velocity difference obeys

$$u(\lambda r) = \lambda^{1/3} u(r) \ . \tag{5.40}$$

This contrasts with the scale invariance $u(\mathbf{x}) \to \lambda^{-1} \mathbf{u}(\lambda \mathbf{x})$ that we met in (2.79) when we first introduced the Navier–Stokes equation. However, this scaling was imposed by the viscosity term and, so far, we have neglected viscosity in our discussion of the energy cascade.

Richardson Diffusion

One simple implication of the result (5.38) arises if we put two particles next to each other in a turbulent flow and watch how they evolve. In many, non-turbulent, situations, the particles will separate due to a process known as *diffusion* that we will describe in detail in Chapter 7. Usually, diffusion is due to the random bombardment of underlying molecules, and the separation r between two particles has characteristic behaviour

$$r(t) \sim t^{1/2} \ . \tag{5.41}$$

But turbulent flows are different. If we think of $u(r) \sim r/t$, then (5.38) tells us that the separation between particles in a turbulent flow should grow as

$$r(t) \sim t^{3/2} \ . \tag{5.42}$$

We see that, at long times, the turbulent flow is much more efficient at separating the particles than standard diffusion. This is sometimes called *Richardson diffusion*. In some sense, it's wrong to call it diffusion at all because the mechanism is not the same. Instead, the separation of the particles is due to the chaotic nature of the flow.

5.2.4 Viscosity Brings the Cascade to a Halt

The energy cascade does not involve dissipation, merely a transfer of energy from large scales to small. But at some point this energy cascade should come to a halt and the energy ε should be dissipated into heat. To understand when this happens, we return to the statement (5.32) which says that the energy in is equal to the energy out. We saw that this certainly wasn't satisfied by the dissipation from large eddies. But now we can ask: For what scale η does this energy balance hold?

The eddies of size η have velocity differentials $u(\eta)$ and (5.32) holds if

$$\frac{\nu u(\eta)^2}{\eta^2} \sim \varepsilon \ . \tag{5.43}$$

But we also know from (5.38) how $u(\eta)$ and η are related. This gives us the *Kolmogorov scale*, also known as the *inner scale*

$$\eta \sim \left(\frac{\nu^3}{\varepsilon}\right)^{1/4} \sim \left(\frac{\nu^3 L}{U^3}\right)^{1/4} \sim \frac{L}{(Re)^{3/4}} \ . \tag{5.44}$$

Clearly $\eta \ll L$ since $Re \gg 1$. This is where energy is finally dissipated to heat.

Note that the Kolmogorov scale can also be determined by dimensional analysis: it is the unique length scale that can be formed from the energy dissipation rate ε and the viscosity ν. Alternatively, you could think of a scale-dependent Reynolds number, defined as $Re(r) = u(r)r/\nu$. The Kolmogorov scale η is where the effects of viscosity become important, in the sense that $Re(\eta) \sim 1$.

This finishes our crude, dimensional analysis approach to turbulence. The energy is injected at some scale L and dissipated at the much smaller scale $\eta = L/(Re)^{3/4}$. The scales in between, $\eta \ll r \ll L$, are called the *inertial range* and exhibit a scale-invariant energy cascade. The higher the Reynolds number, the larger the inertial range. Note that we've not really used the Navier–Stokes equation at any point in the analysis. Everything follows from the hypothesis that, in the inertial range, big eddies cascade down into smaller eddies in a way that does not involve any dissipation.

You might worry that the Kolmogorov scale η is so small that it invalidates the use of fluid mechanics! That, happily, does not happen. For this, we need to invoke a couple of results from Chapter 6 where we look at the microscopic underpinnings of fluids. The continuum fluid description is valid on scales $\gg \lambda$, where λ is the *mean free path* of the underlying molecules, meaning the distance that they travel between collisions. These molecules travel with an average speed which is comparable to the sound speed c_s. The viscosity ν is then in the ballpark of

$$\nu \sim c_s \lambda \ . \tag{5.45}$$

(See, for example, (6.35) for a naive, but simple derivation of the dynamic viscosity $\mu = \nu\rho$ in terms of the mean free path.) This means that the ratio of the Kolmogorov scale to the mean free path λ is

$$\frac{\eta}{\lambda} \sim \frac{c_s \eta}{\nu} \sim \frac{c_s}{\varepsilon^{1/4}\nu^{1/4}} \sim \frac{c_s L^{1/4}}{U^{3/4}\nu^{1/4}} \sim \frac{c_s}{U}(Re)^{1/4}. \tag{5.46}$$

where, in the middle there, we used the relation $\varepsilon \sim U^3/L$ relating the dissipation to the work done. The quantity $M = U/c_s$ is the Mach number that we met in Section 3.5. It is only valid to use the incompressible Navier–Stokes equation when $M \ll 1$. We see, perhaps surprisingly, that the ratio η/λ actually increases as the Reynolds number increases. We are not in danger of needing to resort to molecular physics.

The Energy Spectrum

We can also phrase the energy cascade in terms of wavenumbers $k \sim 1/r$. Let $E(k)\,dk$ be the kinetic energy per unit mass stored in eddies with wavenumber between k and $k + dk$. Then $E(k)$ has dimension $[E] = L^3 T^{-2}$. On

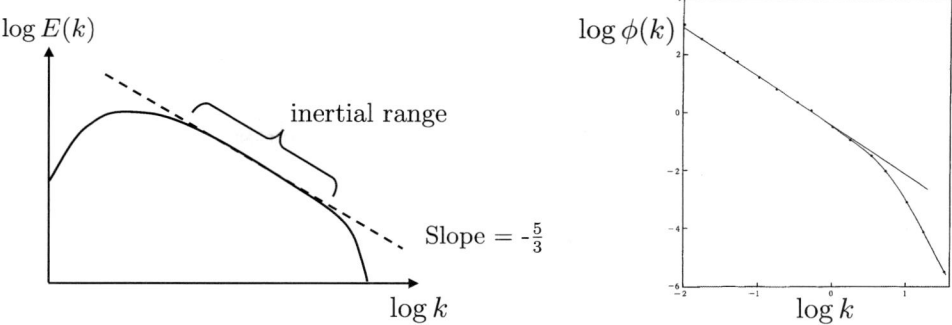

Fig. 5.2 On the left, a sketch of the expected behaviour of $E(k)$ based on dimensional analysis. The energy is injected at small k and dissipated at large k, with the characteristic $E(k) \sim k^{-5/3}$ in the inertial range. On the right, data, taken from Grant et. al. [36], The function $\phi(k)$ shown on the vertical axis is closely related to $E(k)$ described above: it is $2E = k^2 \partial^2 \phi / \partial k^2 - k \partial \phi / \partial k$.

dimensional grounds, we must have

$$E(k) \sim \varepsilon^{2/3} k^{-5/3} \ . \tag{5.47}$$

The expected behaviour is sketched on the left of Figure 5.2. This $k^{-5/3}$ behaviour matches well with experiment. The first test was done with a probe attached to a ship which sailed back and forth in a tidal channel just off Vancouver Island. The data is shown in the right-hand side of Figure 5.2, with the straight line having slope $-5/3$. This $k^{-5/3}$ exponent is directly related to the $r^{2/3}$ exponent of the 2/3 law by a Fourier transform.

5.3 Velocity Correlations

In this section, we put a little more meat on the dimensional analysis argument above. In particular, we will derive a more rigorous version of the Kolmogorov–Obhukov law (5.38) that says $u(r) \sim (\epsilon r)^{1/3}$. We will find, perhaps surprisingly, that this law really holds for more complicated correlations between velocities.

We will do this by returning to the averaging procedure that we introduced in Section 5.1, but now with a focus on the fluctuations $\delta \mathbf{u}$ rather than the mean flow $\mathbf{U}$. To do this, it's simplest if we assume that there are only fluctuations with a vanishing mean flow $\mathbf{U} = 0$. This means, of course, that $\delta \mathbf{u} = \mathbf{u}$ and we can drop the δ's. With no background flow governing the fluctuations, all points in space and time are, statistically at least, the same. This is known as homogeneous and isotropic turbulence and it offers the simplest setting where we may hope to understand a little of what's going on.

(An aside: if setting $\mathbf{U} = 0$ seems too restrictive then there is another way to think about things. We could, alternatively, zoom into some patch where $\mathbf{U}$ is approximately constant and then boost to a frame in which it vanishes. In this way, much of our analysis should hold locally even for general mean flows $\mathbf{U}(\mathbf{x})$.)

The averages that we encountered in Section 5.1 involved fluctuations at the same point in space. For example, the Reynolds stress tensor is

$$R_{ij} = \rho \langle \delta u_i(\mathbf{x}) \delta u_j(\mathbf{x}) \rangle \qquad (5.48)$$

with the two velocities evaluated at the same point. Because we no longer have a background flow $\mathbf{U}(\mathbf{x})$, the system has translational invariance. This means, among other things, that R_{ij} doesn't depend on the point $\mathbf{x}$ at which it's evaluated.

There is now a generalisation in which the correlation between velocity fields is computed at different points

$$C_{ij}(\mathbf{x}_1, \mathbf{x}_2) = \langle u_i(\mathbf{x}_1) u_j(\mathbf{x}_2) \rangle . \qquad (5.49)$$

A number of constraints on this correlation function follow simply from the symmetries of our problem which are enhanced because we're assuming $\mathbf{U} = 0$. First, translational invariance means that it is only a function of the difference $\mathbf{r} = \mathbf{x}_1 - \mathbf{x}_2$ and we write $C_{ij}(\mathbf{x}_1, \mathbf{x}_2) = C_{ij}(\mathbf{r})$. Second, isotropy, together with parity invariance, means that the six components of a general symmetric tensor are reduced to just two

$$C_{ij}(\mathbf{r}) = C_{TT}(r) \left(\delta_{ij} - \hat{r}_i \hat{r}_j \right) + C_{LL}(r) \hat{r}_i \hat{r}_j \qquad (5.50)$$

where $\hat{r}_i = r_i/r$. Here $C_{TT}(r)$ is the transverse correlation function and $C_{LL}(r)$ the longitudinal correlation function. The latter measures correlations of the velocity in the direction of the line joining $\mathbf{x}_1$ and $\mathbf{x}_2$. The fact that a tensor of this kind can be reduced to just two functions was first pointed out by Robertson, better known as the R in FRW cosmology.

Note that we could have also added an additional term proportional to $\epsilon_{ijk} \hat{r}_k$ on the right-hand side of (5.50). But this is forbidden if our system has parity invariance. Sticking with just the two terms in (5.50) means that we have parity invariance in the guise of $C_{ij}(\mathbf{r}) = C_{ij}(-\mathbf{r})$.

When $r = 0$, so the two points coincide, we have a handle on the correlation function, which must take the form

$$C_{ij}(0) = \frac{2}{3}K\delta_{ij} \tag{5.51}$$

where $K = \frac{1}{2}\langle \mathbf{u} \cdot \mathbf{u} \rangle$ is the kinetic energy (divided by the density ρ). This coincides with the expression (5.16) when $\mathbf{U} = 0$. But while (5.16) was pulled out of thin air, here (5.51) follows because, in the absence of a background mean flow, the only symmetric two-tensor that we have at our disposal is δ_{ij}.

There's one last constraint that comes from the fact that the fluid is incompressible, $\nabla \cdot \mathbf{u} = 0$, which means that

$$\frac{\partial C_{ij}}{\partial r_i} = 0 \ . \tag{5.52}$$

We can use this to relate $C_{TT}(r)$ and $C_{LL}(r)$. To do this, we recall that $\hat{r}_k = r_k/r$ and make use of the identities

$$\frac{\partial \hat{r}_k}{\partial r_i} = \frac{\delta_{ik}}{r} - \frac{r_i r_k}{r^3} \quad \Longrightarrow \quad \frac{\partial(\hat{r}_i \hat{r}_j)}{\partial r_i} = 2\frac{\hat{r}_j}{r} \ . \tag{5.53}$$

Then

$$\frac{\partial C_{ij}}{\partial r_i} = \frac{dC_{TT}}{dr}\hat{r}_i\left(\delta_{ij} - \hat{r}_i\hat{r}_j\right) + \frac{dC_{LL}}{dr}\hat{r}_i\hat{r}_i\hat{r}_j + 2(C_{LL} - C_{TT})\frac{\hat{r}_j}{r} \ . \tag{5.54}$$

The first term vanishes and we're left with the simple expression

$$C_{TT} = C_{LL} + \frac{r}{2}\frac{dC_{LL}}{dr} \ . \tag{5.55}$$

This is known as the *von Kármán relation*.

In what follows, we'll also have need for the closely related *structure function*. (Strictly, this is the *two-point structure function*, to distinguish it from higher point correlation functions that we will see below.) This looks at the correlation between the difference in the velocity fluctuations, measured at two points

$$S_{ij}(\mathbf{r}) = \langle (u_i(\mathbf{x}_1) - u_i(\mathbf{x}_2))(u_j(\mathbf{x}_1) - u_j(\mathbf{x}_2)) \rangle \ . \tag{5.56}$$

Expanding out the four terms, the structure function can be trivially expressed in terms of the correlation function as

$$S_{ij}(\mathbf{r}) = 2C_{ij}(0) - 2C_{ij}(\mathbf{r}) = \frac{4}{3}K\delta_{ij} - 2C_{ij}(\mathbf{r}) \ . \tag{5.57}$$

Note, in particular, that $S_{ij}(0) = 0$. As with the correlation function, we can decompose the structure function into transverse and longitudinal pieces

$$S_{ij}(\mathbf{r}) = S_{TT}(r)\left(\delta_{ij} - \hat{r}_i\hat{r}_j\right) + S_{LL}(r)\hat{r}_i\hat{r}_j \ . \tag{5.58}$$

Comparing to the components of the correlation function, we have

$$S_{LL}(r) = \frac{4}{3}K - 2C_{LL}(r) \tag{5.59}$$

with a similar expression for the transverse component

$$S_{TT}(r) = \frac{4}{3}K - 2C_{TT}(r) . \tag{5.60}$$

One advantage of working with the structure function is that we can make contact with the simple dimensional analysis arguments of Section 5.2. In particular, if we take the Kolmogorov–Obhukov law (5.38) at face value and apply it to the structure function, it tells us to expect

$$S_{ij}(r) \sim r^{2/3} . \tag{5.61}$$

This should hold only in the inertial range, $\eta \ll r \ll L$. Since the von Kármán relation (5.55) also holds for the structure function, it tells us that, in the inertial range, $S_{TT} = \frac{4}{3}S_{LL}$. (Actually, as part of our analysis we'll get a better understanding of the Kolmogorov–Obhukov law and see that the result $S_{ij} \sim r^{2/3}$ is not exact: nonetheless, we may hope that it's not a wildly inaccurate expectation.)

The fact that the two-point function $S_{ij}(r)$ grows with r is striking. It is telling us that fluctuations grow on larger scales, as bigger and bigger eddies come into play. This growth of the structure function is in sharp contrast to what happens in statistical field theory and critical phenomena, where correlation functions decay in space.

5.3.1 Navier–Stokes for Correlation Functions

We'll attempt to compute the correlation functions using the Navier–Stokes equation

$$\frac{\partial \mathbf{u}}{\partial t} = -(\mathbf{u} \cdot \nabla)\mathbf{u} - \frac{1}{\rho}(\nabla P - \mathbf{f}) + \nu\nabla^2\mathbf{u} . \tag{5.62}$$

We've included the driving force $\mathbf{f}$ which, as in Section 5.2, will be responsible for the injection of energy. Because averaging commutes with differentiation (both with respect to time and space), we have

$$\frac{\partial C_{ij}(\mathbf{r}; t)}{\partial t} = \langle \partial_t u_i(\mathbf{x}_1)\, u_j(\mathbf{x}_2)\rangle + \langle u_i(\mathbf{x}_1)\, \partial_t u_j(\mathbf{x}_2)\rangle$$

$$= -\frac{\partial}{\partial x_{1k}}\langle u_k(\mathbf{x}_1)u_i(\mathbf{x}_1)\, u_j(\mathbf{x}_2)\rangle - \frac{\partial}{\partial x_{2k}}\langle u_i(\mathbf{x}_1)u_k(\mathbf{x}_2)u_j(\mathbf{x}_2)\rangle$$

$$\quad - \frac{1}{\rho}\langle \partial_i P(\mathbf{x}_1)\, u_j(\mathbf{x}_2)\rangle - \frac{1}{\rho}\langle u_i(\mathbf{x}_1)\partial_j P(\mathbf{x}_2)\rangle$$

$$\quad + \frac{1}{\rho}\langle f_i(\mathbf{x}_1)u_j(\mathbf{x}_2)\rangle + \frac{1}{\rho}\langle u_i(\mathbf{x}_1)f_j(\mathbf{x}_2)\rangle$$

$$\quad + \nu\langle \nabla^2 u_i(\mathbf{x}_1)\, u_j(\mathbf{x}_2)\rangle + \nu\langle u_i(\mathbf{x}_1)\nabla^2 u_j(\mathbf{x}_2)\rangle\,. \qquad (5.63)$$

In the first line we have used the incompressibility of the fluid, $\nabla \cdot \mathbf{u} = 0$, to take the derivative outside the average. We can see immediately from this expression that, as in Section 5.1, to get an equation for the two-point correlation function $C_{ij}(\mathbf{r})$, we need to know something about the three-point function $\langle \delta u^3\rangle$. If we tried to get an equation for $\langle \delta u^3\rangle$ then we would, as before, find that it pushes us towards the four-point function $\langle \delta u^4\rangle$, and so on. This is the same closure problem that we met previously.

This time, however, there is something that we can say without going down the closure rabbit hole. First we'll sort out some of the terms in (5.63), and then return to the three-point function.

Claim: The pressure terms vanish: $\langle u_i(\mathbf{x}_1)P(\mathbf{x}_2)\rangle = 0$.

Proof: Using homogeneity and isotropy, we must have $\langle P(\mathbf{x}_1)u_i(\mathbf{x}_2)\rangle = f(r)\hat{r}_i$ for some function $f(r)$ where, as before, $\mathbf{r} = \mathbf{x}_1 - \mathbf{x}_2$. But incompressibility tells us that

$$0 = \frac{\partial}{\partial x_{1i}}\langle u_i(\mathbf{x}_1)P(\mathbf{x}_2)\rangle = f'(r) + \frac{2}{r}f(r) \quad \implies \quad f(r) = \frac{\alpha}{r^2} \qquad (5.64)$$

for some constant α. But the correlation $\langle u_i(\mathbf{x}_1)P(\mathbf{x}_2)\rangle$ should be finite as $r \to 0$. Which means that we must have $\alpha = 0$ and $\langle u_i(\mathbf{x}_1)P(\mathbf{x}_2)\rangle = 0$. □

Next, we turn our attention to the energy injection terms involving the correlation $\langle u_i(\mathbf{x}_1)f_j(\mathbf{x}_2)\rangle$. To make sense of these, we need to specify the form of the forcing term although, as explained in Section 5.2, the expectation is that the energy will cascade down to smaller scales in a way that is ultimately independent of the forcing term we choose. It turns out that things are particularly simple if we pick a random forcing term that takes the form of Gaussian white noise. We will look at white noise in some detail in Chapter 7 but, for now, it means that the forcing term is taken to be uncorrelated at different times

$$\langle f_i(\mathbf{x}_1, t_1) f_j(\mathbf{x}_2, t_2) \rangle = \delta(t_1 - t_2) \rho^2 \varepsilon_{ij}(\mathbf{r}) \tag{5.65}$$

for some choice of function $\varepsilon_{ij}(\mathbf{r})$ which we get to decide. We'll take it to be symmetric, so $\varepsilon_{ij} = \varepsilon_{ji}$, and isotropic so $\varepsilon_{ij}(\mathbf{r}) = \varepsilon_{ij}(-\mathbf{r})$. We'll shortly see how this tensor ε_{ij} is related to the work done, and, through that, to the dissipation rate ε that played such a key role in our dimensional analysis argument of Section 5.2. One important property of white noise is that the value of the force at any time t is completely uncorrelated with its value at any earlier time t'. This will be important shortly.

Claim: With the force given by the Gaussian white noise (5.65), the correlation between the force and velocity is given by

$$\langle f_i(\mathbf{x}_1) u_j(\mathbf{x}_2) \rangle = \frac{\rho}{2} \varepsilon_{ij}(\mathbf{r}) \ . \tag{5.66}$$

It's not obvious that the left-hand side is symmetric in i and j. Part of the claim is that it is.

Proof: We integrate up the Navier–Stokes equation (5.62) to get the expression for the velocity field $\mathbf{u}(\mathbf{x}, t)$,

$$\mathbf{u}(\mathbf{x}, t) = \int_0^t dt' \left[-(\mathbf{u} \cdot \nabla)\mathbf{u} - \frac{1}{\rho}(\nabla P - \mathbf{f}) + \nu \nabla^2 \mathbf{u} \right] \ . \tag{5.67}$$

Clearly the integral runs over t' with $0 \leq t' \leq t$. Substituting this into the correlation function then gives

$$\langle f_i(\mathbf{x}_1, t) u_j(\mathbf{x}_2, t) \rangle$$
$$= \int_0^t dt' \left\langle f_i(\mathbf{x}_1, t) \left[-u_k \frac{\partial u_j}{\partial x^k} - \frac{1}{\rho}\left(\frac{\partial P}{\partial x_j} - f_j \right) + \nu \nabla^2 u_j \right](\mathbf{x}_2, t') \right\rangle \ . \tag{5.68}$$

All the fields $\mathbf{u}(\mathbf{x}, t')$ and $P(\mathbf{x}, t')$ can't be affected by a force $\mathbf{f}(\mathbf{x}, t)$ that acts at a later time $t > t'$. The only contribution in this correlation function therefore comes from

$$\langle f_i(\mathbf{x}_1, t) u_j(\mathbf{x}_2, t) \rangle = \frac{1}{\rho} \int_0^t dt' \, \langle f_i(\mathbf{x}_1, t) f_j(\mathbf{x}_2, t') \rangle$$
$$= \rho \int_0^t dt' \, \delta(t - t') \varepsilon_{ij}(\mathbf{r}) \ . \tag{5.69}$$

Now, if we integrate $\int dt' \, \delta(t - t')$ over a range that includes the point $t' = t$ then the integral clearly gives 1. Here, however, the point $t' = t$ sits right at the end of the integral range. This is a bit ambiguous but there's a sensible way to think about it. In general, the integral of the delta function gives the

Heaviside step function (so called because it's heavier on one side than the other)

$$\int_0^t \delta(t') = \theta(t) := \begin{cases} 0 & \text{for } t < 0 \\ 1 & \text{for } t > 0 \end{cases} . \qquad (5.70)$$

In (5.69) we find ourselves with the step function evaluated at $\theta(0)$. The only reasonable value at the origin is $\theta(0) = 1/2$. Indeed, this follows from any standard regularisation of the step function, for example

$$\theta(x) = \lim_{\epsilon \to 0} \left(\frac{1}{2} + \frac{1}{\pi} \tan^{-1} \left(\frac{x}{\epsilon} \right) \right) \quad \Longrightarrow \quad \theta(0) = \frac{1}{2} . \qquad (5.71)$$

Substituting this into (5.69), we have

$$\int_0^t dt' \, \delta(t - t') = \frac{1}{2} \quad \Longrightarrow \quad \langle f_i(\mathbf{x}_1, t) u_j(\mathbf{x}_2, t) \rangle = \frac{\rho}{2} \varepsilon_{ij}(\mathbf{r}) . \qquad (5.72)$$

Adding the second contribution then gives the claimed result (5.66). □

From (5.66), we see that if we take the trace of the tensor $\varepsilon_{ij}(\mathbf{r})$ and evaluate it at $\mathbf{r} = 0$ we have the work done (divided by the density)

$$\varepsilon_{ii}(0) = \frac{2}{\rho} \langle \mathbf{f} \cdot \mathbf{u} \rangle = 2\varepsilon . \qquad (5.73)$$

On the right-hand side, we've identified this with the dissipation rate ε that we met in Kolmogorov's dimensional analysis argument in Section 5.2. We'll make contact with these ideas again later. For now, note that we can make the same tensor decomposition as in (5.50) and write

$$\varepsilon_{ij}(\mathbf{r}) = \varepsilon_{TT}(r) \left(\delta_{ij} - \hat{r}_i \hat{r}_j \right) + \varepsilon_{LL}(r) \hat{r}_i \hat{r}_j . \qquad (5.74)$$

This gives $\varepsilon_{LL}(r) = \hat{r}_i \hat{r}_j \varepsilon_{ij}(\mathbf{r})$. If we evaluate this tensor at $\mathbf{r} = 0$ then we must have $\varepsilon_{ij}(0) = \varepsilon_{LL}(0)\delta_{ij}$, as there is no other tensor in the game. (If you're worried that $\hat{r}_i$ is ill-defined at $\mathbf{r} = 0$, then simply evaluate everything with a fixed direction $\hat{r}_i$ and then take the limit $\mathbf{r} \to 0$.) This then gives

$$\varepsilon_{LL}(0) = \frac{2}{3}\varepsilon . \qquad (5.75)$$

This will be important in what follows. Evaluated at $\mathbf{r} = 0$, the component ε_{LL} is proportional to the work done (5.73). Under the further assumption that we have a steady state, so the work done is equal to the dissipation, we identify ε with the dissipation rate.

Let's pause to take stock. Our equation (5.63) for the correlation function has now become

$$\frac{\partial C_{ij}(\mathbf{r};t)}{\partial t} = -\frac{\partial}{\partial x_{1k}} \langle u_k(\mathbf{x}_1)u_i(\mathbf{x}_1)\,u_j(\mathbf{x}_2)\rangle - \frac{\partial}{\partial x_{2k}} \langle u_i(\mathbf{x}_1)u_k(\mathbf{x}_2)u_j(\mathbf{x}_2)\rangle$$
$$+ \; \varepsilon_{ij}(\mathbf{r}) + 2\nu\nabla^2 C_{ij}(\mathbf{r}) \tag{5.76}$$

which is starting to look a little simpler. Our next task is to better understand the structure of the three-point function.

5.3.2 The Structure of the Three-Point Function

We write the three-point function as

$$C_{ij,k}(\mathbf{x}_1, \mathbf{x}_2) = \langle u_i(\mathbf{x}_1)u_j(\mathbf{x}_1)u_k(\mathbf{x}_2)\rangle \;. \tag{5.77}$$

The comma is there in the subscripts on the left-hand side to remind us that two of the velocities are evaluated at $\mathbf{x}_1$ and the third at $\mathbf{x}_2$. (The comma doesn't mean differentiation. This isn't general relativity!) By isotropy, we must have

$$C_{ij,k}(\mathbf{x}_1, \mathbf{x}_2) = C_{ij,k}(\mathbf{r}) = C_{ji,k}(\mathbf{r}) \;. \tag{5.78}$$

And by parity invariance,

$$C_{ij,k}(\mathbf{r}) = -C_{ij,k}(-\mathbf{r}) = -\langle u_i(-\mathbf{x}_1)u_j(-\mathbf{x}_1)u_k(-\mathbf{x}_2)\rangle$$
$$= -\langle u_i(\mathbf{x}_2)u_j(\mathbf{x}_2)u_k(\mathbf{x}_1)\rangle \tag{5.79}$$

with the overall minus sign arising because the correlation function involves an odd number of velocities. In the final equality, we've invoked translational invariance and shifted all arguments by $\mathbf{x}_1 + \mathbf{x}_2$.

The tensor structure means that we can reduce the correlation function to just three functions of $r = |\mathbf{r}|$

$$C_{ij,k}(\mathbf{r}) = A(r)\delta_{ij}\hat{r}_k + B(r)\left(\delta_{ik}\hat{r}_j + \delta_{jk}\hat{r}_i\right) + D(r)\hat{r}_i\hat{r}_j\hat{r}_k \;. \tag{5.80}$$

These different functions are further related by the incompressibility of the flow. For the two-point function, this gave us the von Kármán relation (5.55). For the three-point function, we have the following.

Claim: Incompressibility gives the relations

$$B = -\frac{1}{2r}\frac{d(r^2 A)}{dr} \tag{5.81}$$

and

$$3A + 2B + D = 0 \;. \tag{5.82}$$

Proof: Incompressibility $\nabla \cdot \mathbf{u} = 0$ means that

$$\frac{\partial C_{ij,k}}{\partial (x_2)_k} = -\frac{\partial C_{ij,k}}{\partial r_k} = 0 \ . \tag{5.83}$$

We use $\partial \hat{r}_k / \partial r_i = \delta_{ik}/r - \hat{r}_i \hat{r}_k / r$ and, after a line or two of algebra, we get

$$\frac{\partial C_{ij,k}}{\partial r_k} = \delta_{ij} \left(A' + \frac{2A}{r} + \frac{2B}{r} \right) + \hat{r}_i \hat{r}_j \left(2B' - \frac{2B}{r} + D' + \frac{2D}{r} \right) \ . \tag{5.84}$$

Each of these tensor structures must individually vanish. The first gives the relation (5.81). The vanishing of the second term can be written as

$$\frac{1}{r^2} \frac{d}{dr} \left[r^2 (D + 2B) \right] = \frac{6B}{r} \ . \tag{5.85}$$

If we substitute in our expression for B in the recently proved (5.81), this becomes

$$\frac{d}{dr} \left[r^2 (D + 2B + 3A) \right] = 0 \quad \Longrightarrow \quad 3A + 2B + D = \frac{\text{constant}}{r^2} \ . \tag{5.86}$$

We can fix the constant by looking at $C_{ij,k}(0)$ which must take the value $C_{ij,k}(0) = 0$ for the simple reason that there's no invariant three-tensor with the right symmetry properties that it can equal. This means that the constant in (5.86) is actually zero and so we get (5.82). $\square$

We can combine the two relations (5.81) and (5.82) to give $D = rA' - A$. The upshot is that the three-point correlation function actually depends on just a single function $A(r)$,

$$C_{ij,k}(\mathbf{r}) = A\delta_{ij}\hat{r}_k - \frac{1}{2} \left(rA' + 2A \right) (\delta_{ik}\hat{r}_j + \delta_{jk}\hat{r}_i) + (rA' - A)\hat{r}_i\hat{r}_j\hat{r}_k \ . \tag{5.87}$$

Next, it will also be useful to introduce the three-point structure function, which is the generalisation of (5.56),

$$S_{ijk}(\mathbf{r}) = \langle (u_i(\mathbf{x}_1) - u_i(\mathbf{x}_2))(u_j(\mathbf{x}_1) - u_j(\mathbf{x}_2))(u_k(\mathbf{x}_1) - u_k(\mathbf{x}_2)) \rangle \ . \tag{5.88}$$

This is completely symmetric in all three indices. If we expand and cancel terms (remembering that we have translational invariance so $\langle \mathbf{u}(\mathbf{x}_1)^3 \rangle = \langle \mathbf{u}(\mathbf{x}_2)^3 \rangle$), then we can relate the structure function to the correlation function

$$S_{ijk}(\mathbf{r}) = -2(C_{ij,k} + C_{ik,j} + C_{jk,i}) \ . \tag{5.89}$$

If we substitute in the expression (5.87) we get

$$S_{ijk}(\mathbf{r}) = 2(A + rA')(\delta_{ij}\hat{r}_k + \delta_{ik}\hat{r}_j + \delta_{jk}\hat{r}_i) - 6(rA' - A)\hat{r}_i\hat{r}_j\hat{r}_k \ . \tag{5.90}$$

The fully longitudinal part of the structure function is defined to be

$$S_{LLL}(r) = S_{ijk}(\mathbf{r})\hat{r}_i\hat{r}_j\hat{r}_k \ . \tag{5.91}$$

From (5.90), we see that this is the same thing as the function $A(r)$, up to an overall constant

$$S_{LLL}(r) = 12A(r) \ . \tag{5.92}$$

In what follows, we'll work with $S_{LLL}(r)$ as the object that describes the three-point function.

5.3.3 The von Kármán–Howarth Equation

Now we can return to our expression (5.76) for the dynamics of the correlation function. This reads

$$\frac{\partial C_{ij}}{\partial t} = -\frac{\partial C_{ik,j}}{\partial r_k} - \frac{\partial C_{jk,i}}{\partial r_k} + \varepsilon_{ij}(\mathbf{r}) + 2\nu\nabla^2 C_{ij}(\mathbf{r}) \tag{5.93}$$

where the indices in the second $\partial C/\partial r$ term have rearranged themselves courtesy of (5.79). We focus on the longitudinal component of the two-point correlator, $C_{LL} = C_{ij}\hat{r}_i\hat{r}_j$, which obeys

$$\frac{\partial C_{LL}}{\partial t} - \varepsilon_{LL} - 2\nu\hat{r}_i\hat{r}_j\nabla^2 C_{ij} = -2\hat{r}_i\hat{r}_j\frac{\partial C_{ik,j}}{\partial r_k} \ . \tag{5.94}$$

We have a little bit of work to do to move those $\hat{r}_i\hat{r}_j$ terms inside the derivatives. On the right-hand side, we have

$$\hat{r}_i\hat{r}_j\frac{\partial C_{ik,j}}{\partial r_k} = \frac{\partial}{\partial r_k}(C_{ik,j}\hat{r}_i\hat{r}_j) - C_{ik,j}\frac{\partial(\hat{r}_i\hat{r}_j)}{\partial r_k}$$

$$= \frac{\partial}{\partial r_k}(C_{ik,j}\hat{r}_i\hat{r}_j) - \frac{1}{r}\left(C_{ii,j}\hat{r}_j + C_{ik,k}\hat{r}_i - 2C_{ij,k}\hat{r}_i\hat{r}_j\hat{r}_k\right) \tag{5.95}$$

where we've again made use of the identity $\partial\hat{r}_i/\partial r_k = \delta_{ik}/r - \hat{r}_i\hat{r}_k/r$. Now, from (5.87), we can compute the various contractions of $C_{ij,k}$ with $\hat{r}$

$$C_{ik,j}\hat{r}_i\hat{r}_j = A\hat{r}_k - \left(rA' + 2A\right)\hat{r}_k + (rA' - A)\hat{r}_k = -2A\hat{r}_k$$
$$C_{ii,j}\hat{r}_j = 3A - \left(rA' + 2A\right) + (rA' - A) = 0$$
$$C_{ik,k}\hat{r}_i = A - 2\left(rA' + 2A\right) + (rA' - A) = -rA' - 4A \tag{5.96}$$
$$C_{ij,k}\hat{r}_i\hat{r}_j\hat{r}_k = A - \left(rA' + 2A\right) + (rA' - A) = -2A \ .$$

So (5.95) becomes

$$\hat{r}_i\hat{r}_j\frac{\partial C_{ik,j}}{\partial r_k} = -2\frac{\partial}{\partial r_k}(A\hat{r}_k) + A' = -\frac{4A}{r} - A' \ . \tag{5.97}$$

We have a similar task for the $\nu\nabla^2 C_{ij}$ term in (5.94). For this, it's best to return to the expression (5.50). A slightly tedious exercise in algebra gives

$$\nabla^2 C_{ij} = \nabla^2 C_{TT} \left(\delta_{ij} - \hat{r}_i\hat{r}_j\right) + \nabla^2 C_{LL}\hat{r}_i\hat{r}_j + \frac{C_{LL} - C_{TT}}{r^2} \left(2\delta_{ij} - 6\hat{r}_i\hat{r}_j\right)$$
$$+ \frac{1}{r}\left(\frac{\partial C_{LL}}{\partial r_k} - \frac{\partial C_{TT}}{\partial r_k}\right)\left(\delta_{ik}\hat{r}_j + \delta_{jk}\hat{r}_i - 2\hat{r}_i\hat{r}_j\hat{r}_k\right) . \tag{5.98}$$

If we now contract with $\hat{r}_i\hat{r}_j$, the second line disappears and we're left with

$$\hat{r}_i\hat{r}_j\nabla^2 C_{ij} = \nabla^2 C_{LL} - \frac{4}{r^2}(C_{LL} - C_{TT}) . \tag{5.99}$$

But now we can use the von Kármán relation (5.55) to write this purely in terms of C_{LL},

$$\hat{r}_i\hat{r}_j\nabla^2 C_{ij} = \nabla^2 C_{LL} + \frac{2}{r}\frac{\partial C_{LL}}{\partial r}$$
$$= \frac{1}{r^2}\frac{\partial}{\partial r}\left(r^2\frac{\partial C_{LL}}{\partial r}\right) + \frac{2}{r}\frac{\partial C_{LL}}{\partial r}$$
$$= \frac{1}{r^4}\frac{\partial}{\partial r}\left(r^4\frac{\partial C_{LL}}{\partial r}\right) . \tag{5.100}$$

We can put these pieces back into the expression (5.94), which becomes an equation that relates the longitudinal two-point function C_{LL} to the three-point function $S_{LLL} = 12A$,

$$\frac{\partial C_{LL}}{\partial t} - \varepsilon_{LL} - \frac{2\nu}{r^4}\frac{\partial}{\partial r}\left(r^4\frac{\partial C_{LL}}{\partial r}\right) = \frac{1}{6}\left(\frac{\partial S_{LLL}}{\partial r} + \frac{4S_{LLL}}{r}\right) . \tag{5.101}$$

We can express everything in terms of the structure function using the relation (5.59), which relates $S_{LL} = \frac{4}{3}K - 2C_{LL}$ with $K = \frac{1}{2}\langle \mathbf{u} \cdot \mathbf{u} \rangle$ the average kinetic energy (divided by the density). The end result is

$$\frac{\partial S_{LL}}{\partial t} = \frac{4}{3}\frac{\partial K}{\partial t} - 2\varepsilon_{LL} + \frac{2\nu}{r^4}\frac{\partial}{\partial r}\left(r^4\frac{\partial S_{LL}}{\partial r}\right) - \frac{1}{3r^4}\frac{\partial}{\partial r}\left(r^4 S_{LLL}\right) . \tag{5.102}$$

This is the *von Kármán–Howarth equation*. It is not a closed equation, which means that we can't (yet) just sit down and solve it. Instead, it tells us how the two-point correlations of the velocity S_{LL} evolve with time and they are, as expected, determined by the three-point correlations S_{LLL}.

5.3.4 Kolmogorov's 4/5 Law

It's a short step from the von Kármán–Howarth equation to what we want. We'll focus on the static case where we've reached a kind of equilibrium, which means that we can drop the time derivatives in (5.102),

$$\frac{1}{r^4}\frac{\partial}{\partial r}\left(r^4\left(2\nu\frac{\partial S_{LL}}{\partial r}-\frac{S_{LLL}}{3}\right)\right)=2\varepsilon_{LL}(r)\ . \tag{5.103}$$

We will further assume that, as in Section 5.2, the energy is injected on a large scale L. Following (5.75), we interpret this as the statement

$$\varepsilon_{LL}(\mathbf{r})=\frac{1}{3}\varepsilon_{ii}(\mathbf{r})=\frac{2}{3}\varepsilon+\mathcal{O}\left(\frac{r}{L}\right) \tag{5.104}$$

where ε is the work done per unit mass that we met in the dimensional analysis argument of Section 5.2. We then integrate our differential equation to get

$$2\nu\frac{\partial S_{LL}}{\partial r}-\frac{1}{3}S_{LLL}=\frac{4}{15}\varepsilon r \tag{5.105}$$

where the constants of integration have been put to zero using the fact that $S_{LL}(0)=S_{LLL}(0)=0$. Rearranging, we have an expression for the three-point correlations,

$$S_{LLL}(r)=-\frac{4}{5}\varepsilon r+6\nu\frac{\partial S_{LL}}{\partial r}\ . \tag{5.106}$$

This expression is valid at all scales $r\ll L$. The second term on the right-hand side is proportional to the viscosity ν which, as we saw previously, can be associated to the Kolmogorov viscosity scale η,

$$\nu\sim\left(\varepsilon\eta^4\right)^{1/3}\ . \tag{5.107}$$

This means that, at distances $r\gg\eta$, the second term in (5.106) can be neglected. The result is that, in the inertial range, we have

$$S_{LLL}(r)=-\frac{4}{5}\varepsilon r\quad\text{for}\ \ \eta\ll r\ll L. \tag{5.108}$$

This is *Kolmogorov's 4/5 law*. (Fluid dynamicists love to name their laws after fractions. In this case, the fraction is the prefactor, rather than the exponent.)

The 4/5 law is important because the number of exact results about turbulence can be counted on one finger. This is the one. It is a more rigorous version of the Kolmogorov–Obhukov law (5.38) that says we should have $u_r\sim(\varepsilon r)^{1/3}$, but here applied to the three-point function S_{LLL} with a fixed, universal prefactor of $-4/5$. In deriving this result, we didn't have to make any of the more heuristic assumptions about the energy cascade and scale invariance that fed into the dimensional analysis argument of Section 5.2.

Numerical data giving support for the 4/5 law is shown in Figure 5.3.

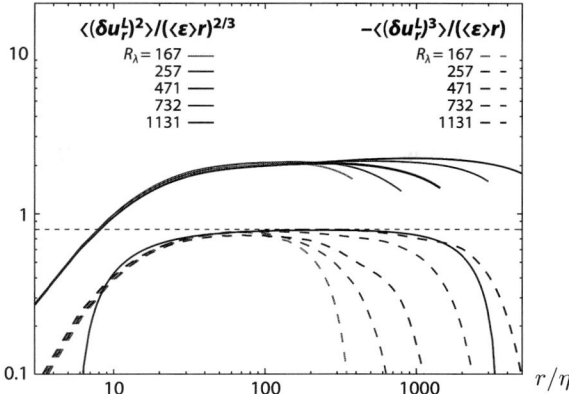

Fig. 5.3 Numerical evidence for scaling in turbulent flows, taken from Ishihara et. al. [42]. The bottom curves show $-S_{LLL}/\varepsilon r$ for various values of the Reynolds number, with the dotted horizontal line the value $4/5$. The upper curves show the corresponding result for the two-point function $S_{LL}/(\varepsilon r)^{2/3}$.

The lower curves show $S_{LLL}/\varepsilon r$ for various values of the Reynolds number. (The variable shown in the figure is $R_\lambda \approx \sqrt{Re}$.) Clearly there is an inertial range in which the correlation function plateaus on the expected value of $4/5$, shown by the horizontal dotted line. It is worth noting, however, that it does not reach this plateau until $r > 10\eta$.

Given that the $4/5$ law agrees with a naive application of the Kolmogorov–Obhukov law to the three-point function, you might wonder if we can extrapolate to other correlation functions with, for example, the two-point function taking the form $S_{LL}(r) \sim r^{2/3}$. But that's not the way correlation functions work: just because you know that $\langle u^3 \rangle \sim r$, doesn't mean that $\langle u^p \rangle \sim r^{p/3}$. We might hope that it's a decent starting point, but hope falls rather short of a proof.

Numerical results seem to back this up. The upper curves in Figure 5.3 show $S_{LL}/(\varepsilon r)^{2/3}$, again plotted for various values of the Reynolds number. We see that these curves again plateau, although along a line that is perhaps not quite as flat as the three-point function.

This leaves us in a rather strange situation in the theory of turbulence, one that I haven't seen repeated in other areas of physics. We have good evidence for scale invariance, or near scale invariance, which is very common in physics and will be discussed in some detail in the book on Statistical Physics when we talk about critical phenomena and the renormalisation group. The strange thing about turbulence is that we have a handle on the three-point correlation function which, through the 4/5 law, holds some special protected status. But we don't have a correspondingly reliable handle on the two-point function, nor indeed on any higher point functions.

5.3.5 Briefly, Intermittency

As we've seen, the three-point function has an exact scaling in the inertial range, $\langle u^3 \rangle \sim r$. And the two-point function seem to be reasonably well modelled by the related scaling $\langle u^2 \rangle \sim r^{2/3}$, but it is not precise. Numerical evidence seem to suggest that higher point functions have exponents that differ more from the naive expectation $\langle u^p \rangle \sim r^{p/3}$.

There are various ways that we could attempt to account for this in our simplistic dimensional analysis of Section 5.2. For example, we assumed that the energy cascade retains no memory of the outer scale L on which we initially inject energy. If this scale could sneak into the energy cascade, then it would infect our dimensional analysis. Written in terms of wavenumbers, the result for the energy spectrum (5.47) could be corrected to

$$E(k) \sim \varepsilon^{2/3} k^{-5/3} (kL)^\zeta \tag{5.109}$$

for some ζ known as the *intermittency exponent*, the name arising because the experimental manifestation is that turbulent flows have periods in which the velocity fluctuations are weak, interspersed with intermittent bursts in which the fluctuations are much larger. In many situations, this exponent seems to be small. But it is not known how to calculate it.

There are close similarities between this story and what's seen in so-called critical points in phase transitions. There too one sees scale invariance, and a naive dimensional analysis argument (known in that context as "mean field theory") suggests a particular value for certain exponents. But that's not the value that is seen experimentally. The flaw in that context is that the short distance UV cut-off (i.e. the atomic scale) unexpectedly sneaks in to the dimensional analysis and contributes what's known as an "anomalous critical exponent". It is entirely analogous to the intermittency exponent in turbulence, except that now this involves the long distance IR cut-off. You can read more about critical points and how to compute anomalous exponents in the book on Statistical Physics. We will not, sadly, say anything further about intermittency in turbulence.

5.3.6 Andrey Kolmogorov (1903–1987)

> Every mathematician believes that he is ahead of the others. The reason none state this belief in public is because they are intelligent people.
>
> *Kolmogorov*

Pure mathematicians rarely make fundamental and lasting contributions to physics. Kolmogorov is one of the few exceptions. In addition to his pioneering work on turbulence, he also made major contributions to the foundations of probability theory, computational complexity, topology, and, in the form of the KAM theorem, classical mechanics.

Kolmogorov spent his entire career in Moscow. He lived for almost 50 years with another mathematician, Pavel Aleksandrov, in what seems to have been a romantic relationship. Needless to say, Stalinist USSR was not the easiest place for such a domestic arrangement.

It took some time for Kolmogorov's ideas on turbulence to percolate through to the West and many scientists continued to wrestle with the problem, unaware of the breakthrough that was made in 1941. After the war, Heisenberg and others involved in the Nazi bomb effort were famously interned in Farm Hall near Cambridge when G.I. Taylor decided to pay them a visit to discuss the energy cascade in turbulence (because who among us hasn't taken advantage of a potential seminar speaker passing through town?). The papers of Kolmogorov were discovered soon after by the fluid dynamicist George Batchelor, a student of Taylor, and, in 1947, advertised to the wider world.

6 Kinetic Theory

This chapter is something of a departure from the main theme of the book. In fact, it is almost a betrayal of the philosophy of fluid mechanics. Our focus, until now, has been on a macroscopic description of what fluids can do. Now it is time to look deeper. We will revisit some of the key ideas of fluid mechanics, but viewed through the lens of the constituent atoms. This subject is known as *kinetic theory*.

It's unusual for fluid mechanics textbooks to contain a section on kinetic theory. But there is a good reason for describing the basics of kinetic theory here. This is because the punchline of this chapter is a re-derivation of the Navier–Stokes equations, now from the perspective of the underlying 10^{23} atoms. Each of these atoms is moving independently, occasionally colliding with its neighbours, and following a path that is governed by Newton's laws of motion. Yet from this extraordinarily complicated system, the simple equations of fluid mechanics emerge. In other words, this chapter is the origin story for fluids.

There is something surprising in this. The Newtonian equations governing 10^{23} atoms form a Hamiltonian system. But the Navier–Stokes equation most certainly is not a Hamiltonian system because viscosity is a type of friction. Said slightly differently, we start with equations that are invariant under time-reversal invariance, and end up with equations that are not! And that's interesting. We are learning something about the origin of irreversibility in physics.

In addition, taking a microscopic viewpoint allows us to start thinking more carefully about the parameters that sit in the Navier–Stokes equation, in particular the viscosity. From the viewpoint of a fluid dynamicist, the viscosity is something that you look up in a book. (Or, if you're so inclined, measure experimentally.) But it's interesting to ask whether we can calculate the value of a fluid's viscosity from first principles, starting from knowledge of the atomic masses and interactions. This, it turns out, is possible for gases but not for liquids. In fact, in general, liquids are difficult to understand: the constituent particles in a liquid are packed close together, constantly interacting with their neighbours, and this makes it challenging to model them a straightforward way. In contrast, the atoms in a gas interact only rarely and this, as we will see, gives us a much better handle on their properties.

The upshot is that the techniques of this chapter will allow us to derive the Navier–Stokes equation, together with a first principles accounting of the viscosity, but only for dilute gases, not for liquids.

This comes with a further consequence: gases, in contrast to liquids, are rather easy to squeeze. This means that, as we study the dynamics of the density and velocity, we will be forced to think more carefully about the temperature as well. This provides an opportunity to put on firmer footing some of the equations that we met in Section 3.4 when discussing sound waves.

6.1 Things Bumping Into Other Things

We start in this section by giving a rather elementary account of kinetic theory. It will be a sort of baby version of the subject that will help build some intuition for what's going on, but with some heuristic and slightly flaky derivations in places. With this under our belt, we will then turn to a more sophisticated account, starting in Section 6.2.

We will be dealing with dilute gases. We start by modelling these as non-interacting particles, and then think about what happens when these particles bump into each other.

Consider a gas consisting of N non-interacting particles, each of mass m, travelling with an average speed v, sitting in a cubic box of sides L and volume $V = L^3$. We would like to derive an expression for the pressure of this gas. The pressure of the gas is the force per unit area acting on the wall of the box. So we start by thinking about what happens when a particle bounces off the wall.

We can decompose the average speed of a particle v in terms of the individual components

$$v^2 = v_x^2 + v_y^2 + v_z^2 = 3v_x^2 \tag{6.1}$$

where, in the final equality, we've assumed that there's no preferred direction so $v_x^2 = v_y^2 = v_z^2$. An atom travelling with velocity component v_x in the x-

direction bounces elastically off the wall, so that it returns with velocity $-v_x$. The particle experiences a change of momentum $\Delta p_x = 2mv_x$. But, because the particle is trapped in a box, it will next hit this same wall at a time $\Delta t = 2L/v_x$ later. This means that the force on the wall due to this atom is

$$F = \frac{\Delta p_x}{\Delta t} = \frac{mv_x^2}{L} = \frac{mv^2}{3L} \ . \tag{6.2}$$

Summing over all the atoms that hit the wall, the force is

$$F = \frac{Nmv^2}{3L} \ . \tag{6.3}$$

The pressure is the force per unit area, or

$$P = \frac{Nmv^2}{3L^3} = \frac{Nmv^2}{3V} \ . \tag{6.4}$$

This relates the pressure of the gas to the average kinetic energy $\frac{1}{2}mv^2$ of the particle.

For the next step, we need to invoke a simple result from statistical mechanics. This is the following statement, relating the average kinetic energy to the temperature T of the gas

$$\frac{1}{2}mv^2 = \frac{3}{2}k_B T \ . \tag{6.5}$$

Here k_B is Boltzmann's constant, $k_B \approx 1.4 \times 10^{-23}$ J K^{-1}, which, as the units suggest, acts as a conversion factor between energy and temperature.

We will not derive (6.5) here; it needs a slightly deeper understanding of the meaning of temperature than we will give here. There is, however, a useful mnemonic that underlies (6.5): each particle has three degrees of freedom (because it can move in three directions), and at temperature T, each degree of freedom picks up $\frac{1}{2}k_B T$ worth of energy. This is known as *equipartition of energy* and we met it previously, also without justification, in Section 3.4.2.

One immediate consequence of (6.5) is that, when combined with (6.4), we get something very familiar

$$PV = Nk_B T \ . \tag{6.6}$$

This, of course, is the ideal gas law.

For the purposes of this chapter, where we work only with monatomic gases, we will take (6.5) as the definition of temperature T. A more fundamental definition of temperature, and how it reproduces (6.5), will be given in the book on Statistical Physics.

6.1.1 The Maxwell–Boltzmann Distribution

In the derivation above, we needed only the average speed of a particle in the gas. In reality, the particles in a gas have a distribution of speeds and this will be important in what follows. We call the distribution function $f(v)$.

The function $f(v)$ depends only on $v = |\mathbf{v}|$. It is a probability distribution, which means that it is normalised to

$$\int d^3v \; f(v) = 4\pi \int_0^\infty dv \; v^2 f(v) = 1 \; . \tag{6.7}$$

The middle equation above provides the best interpretation of $f(v)$: we should think of $4\pi v^2 f(v) \, dv$ as the probability that the particle has speed between v and $v + dv$.

There is a particularly slick argument, due to Maxwell, that allows us to determine the distribution function $f(v)$. The argument starts by thinking about the distribution of velocities in the individual x-, y-, and z-directions. We denote the distribution of velocities in the x-direction as $\phi(v_x)$. This means that the probability that a given atom has velocity in the x-direction between v_x and $v_x + dv_x$ is $\phi(v_x) \, dv_x$. Rotational symmetry means that the same distribution holds for the velocities in the y- and z-directions.

However, rotational invariance also requires that the full distribution can't depend on the direction of the velocity; it can only depend on the total speed $v = \sqrt{v_x^2 + v_y^2 + v_z^2}$. This means that the distribution function we're looking for must satisfy

$$f(v) \, d^3v = \phi(v_x)\phi(v_y)\phi(v_z) \, dv_x dv_y dv_z \; . \tag{6.8}$$

It doesn't look as if we have enough information to solve this equation for both f and ϕ. But, remarkably, there is a unique possibility! The only function which satisfies this equation is

$$\phi(v_x) = Ae^{-\beta mv_x^2/2} \tag{6.9}$$

for some constants A and β. This means that the distribution over speeds takes the form

$$f(v) = A^3 e^{-\beta mv^2/2} \; . \tag{6.10}$$

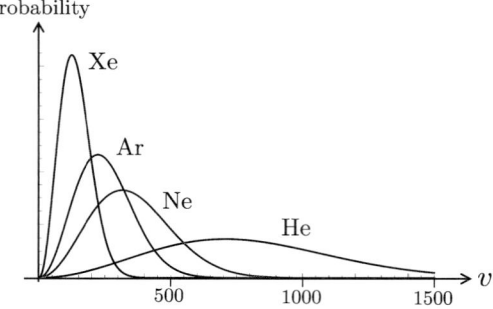

The Maxwell–Boltzmann distribution $4\pi v^2 f(v)$ for the inert gases He, Ne, Ar, and Xe. The lightest gas, helium, has the broadest spectrum of speeds, shown here in $\mathrm{m\,s}^{-1}$.

We can use the normalisation condition (6.7) to fix the coefficient A^3. We have

$$f(v) = \left(\frac{m\beta}{2\pi}\right)^{3/2} e^{-\beta mv^2/2} \ . \tag{6.11}$$

It remains to determine the other constant β. This we fix by computing the average kinetic energy of a gas particle

$$\frac{1}{2}m\langle v^2\rangle = \frac{1}{2}m \int d^3v \ v^2 f(v) = 2\pi m \int_0^\infty dv \ v^4 f(v) = \frac{3}{2\beta} \ . \tag{6.12}$$

Comparing this to our expected equipartition result (6.5) gives the rather simple relation

$$\beta = \frac{1}{k_B T} \ . \tag{6.13}$$

The result for the distribution of velocities in a gas is then

$$f(v) = \left(\frac{m}{2\pi k_B T}\right)^{3/2} e^{-mv^2/2k_B T} \ . \tag{6.14}$$

This lovely formula is the *Maxwell–Boltzmann distribution* for the velocities of particles in a gas. The factor of m in the exponent has the simple implication that, all other things being equal, heavier atoms travel slower than lighter atoms. The resulting probability distribution $4\pi v^2 f(v)$ is shown in Figure 6.1 for monatomic gases of different masses.

The Maxwell–Boltzmann distribution is the key, basic result of kinetic theory. We will derive it in a different way later in Section 6.3 using the Boltzmann equation and in yet a different way in Chapter 7 using the Fokker-Planck equation. (And in yet a different way again in the book on Statistical

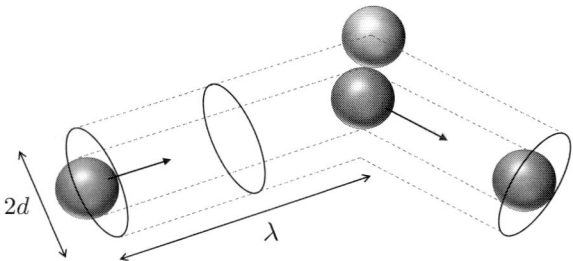

Fig. 6.2 A particle of radius $d/2$ travels, on average, a length λ between each collision. In this time it sweeps out a volume $\pi d^2 \lambda$.

Physics. A result as important as this one should be understood in as many different ways as possible!)

We can use the Maxwell–Boltzmann distribution to compute the average speed $\langle v \rangle$ as opposed to the average speed-squared $\langle v^2 \rangle$. (Note that $\langle v \rangle$ is the average *speed*, not the average velocity. The latter vanishes.) This is

$$\langle v \rangle = 4\pi \int_0^\infty dv \, v^3 \, f(v) = \sqrt{\frac{2}{\pi}} \sqrt{\frac{k_B T}{m}} \, . \tag{6.15}$$

The $\sqrt{k_B T/m}$ dependence follows purely on dimensional grounds; the numerical factor needs the Maxwell–Boltzmann distribution.

6.1.2 A First Look at Collisions

Now we turn to the collisions between particles. For the purposes of this section, we model our underlying atoms or molecules as hard spheres. We denote the diameter of these particles as d. The radius is $d/2$.

Viewed head on, the particle appears as a disc with area $\pi(d/2)^2$. However, more relevant for our purposes is the *effective cross-sectional area* of the particle which is given by πd^2. To see why, consider a single particle as it makes its way through the gas. If it travels a distance λ, it will effectively sweep out a volume $\pi d^2 \lambda$ as shown in Figure 6.2 and collide with any other particle whose centre lies within this volume.

The *mean free path* λ is defined to be the average distance travelled by the molecule between each collision. This is given by $\pi d^2 \lambda = V/N$, or

$$\lambda = \frac{V}{N} \frac{1}{\pi d^2} = \frac{1}{n \pi d^2} \tag{6.16}$$

where $n = N/V$ is the particle density.

In what follows, we'll assume that our gas is dilute, meaning $\lambda \gg d$, so the mean free path is much larger than the atomic size. For typical gases $d \approx 10^{-10}$ m while, at atmospheric pressure, $\lambda \approx 10^{-7}$ m. As we'll see, it's the hierarchy of scales $\lambda \gg d$ that will later allow us to make progress in deriving the Navier–Stokes equation.

The two, separated distance scales $\lambda \gg d$ can equivalently be viewed as two, separated time scales. The first of these is the average time τ that a particle travels before it collides with some other particle. This is known as the *scattering time* or *relaxation time*. It is roughly

$$\tau \sim \frac{\lambda}{\bar{v}_{\mathrm{rel}}} \tag{6.17}$$

with $\bar{v}_{\mathrm{rel}}$ the average relative speed of the particles. We can compute this average relative speed from the Maxwell–Boltzmann distribution. Assuming that the speeds of the two atoms are uncorrelated, we have

$$\bar{v}_{\mathrm{rel}}^2 = \int d^3v\, d^3v'\, (\mathbf{v} - \mathbf{v}')^2 f(v) f(v')$$

$$= \int d^3v\, d^3v'\, \left(v^2 + v'^2 - 2\mathbf{v} \cdot \mathbf{v}'\right) f(v) f(v')$$

$$= 2\langle v^2 \rangle - 2\langle \mathbf{v} \cdot \mathbf{v}' \rangle . \tag{6.18}$$

But the second term vanishes on rotational grounds $\langle \mathbf{v} \cdot \mathbf{v}' \rangle = 0$. (It's enough to show that the velocities are independent and each velocity component has $\langle v_x \rangle = 0$.) This means that we have $\bar{v}_{\mathrm{rel}}^2 = 2\langle v \rangle^2$ or, from (6.5)

$$\bar{v}_{\mathrm{rel}}^2 = \frac{6k_B T}{m} . \tag{6.19}$$

This means that the relaxation time scales as (ignoring numerical factors)

$$\tau \sim \frac{1}{n\pi d^2} \sqrt{\frac{m}{k_B T}} . \tag{6.20}$$

As the temperature increases, the mean free path stays the same but the relaxation time decreases.

There is a slightly different interpretation of the relaxation time that is useful to have in hand. Suppose that the probability that a molecule undergoes a collision between time t and time $t + dt$ is given by $w\,dt$, for some constant w, known as the *collision rate*. The fact that w is a constant means that we are assuming that no memory of previous collisions is kept: the chances of being hit again are not affected just because you already were hit a short time ago.

If $P(t)$ is the probability that the molecule makes it to time t unharmed, then the probability that it further makes it to time $t + dt$ without collision is

$$P(t + dt) = P(t)(1 - wdt) \ . \tag{6.21}$$

Writing this as a differential equation, we have

$$\frac{dP}{dt} = -wP \quad \Longrightarrow \quad P(t) = e^{-wt} \tag{6.22}$$

where we've chosen the normalisation so that $P(0) = 1$ and $P(\infty) = 0$. With this in hand, we can compute the average time between collisions, which coincides with the quantity that we called the relaxation time above. It is

$$\tau = \int_0^\infty P(t) \, dt = \frac{1}{w} \ . \tag{6.23}$$

We learn that $1/\tau$ is the same thing as the collision rate w.

The second time scale is the time τ_{coll} that it takes for a collision to occur. With d the size of an atom, this second time scale is roughly

$$\tau_{\text{coll}} \sim \frac{d}{\bar{v}_{\text{rel}}} \ . \tag{6.24}$$

This is called the *collision time*. If we are working with a dilute gas, with $\lambda \gg d$, then there is a clear ordering to these two different time scales, with $\tau \gg \tau_{\text{coll}}$. We will make use of this when we come to the Boltzmann equation.

6.1.3 Viscosity

We now turn our attention to the second question raised in the introduction to this chapter: How can we compute parameters like the viscosity from first principles? These parameters go by the name of *transport coefficients*.

As our operational definition of viscosity, we use Couette flow that we met in Section 2.2. Recall the basic set-up. The fluid is placed between two plates, a distance h apart in the z-direction, as shown in the figure. The lower plate is held stationary, while the upper plate is moved at a constant speed U in the

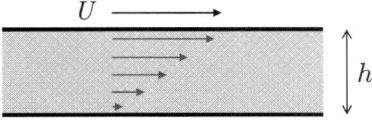

x-direction. The resulting velocity profile is linear in the z-direction

$$u(z) = \frac{Uz}{h} \ .$$

(6.25)

This satisfies the no-slip boundary conditions $u(0) = 0$ and $u(h) = U$ on the lower and upper plates, respectively. The viscosity of the fluid means that we must apply a constant force per unit area on the upper plate to keep it moving. This force was calculated in (2.41) and is given by the simple formula

$$\frac{F}{A} = \mu \frac{du}{dz} = \frac{\mu U}{h}$$

(6.26)

with μ the (dynamic) viscosity.

We would like to derive both the force law (6.26) and the viscosity μ from first principles. It's simple to get an intuition for what's happening on the atomic level: when the molecules collide with the upper plate, they pick up some x-momentum. They then collide with other molecules lower down, imparting some of this x-momentum to new molecules, which then collide with other molecules lower down, and so on. In this way, we set up the velocity gradient in the z-direction.

We'll think of a slab of gas at some fixed value of z. To figure out the force acting on this slab, we need to work out two things: the number of particles moving through the slab per unit of time; and the extra momentum in the x-direction that each particle imparts to the molecules in the slab.

Let's first deal with the number of particles. The density of particles in the fluid is $n = N/V$. How many of these pass through a slab in the z-direction in a given time period depends on how fast they're travelling in the z-direction. But we know how many particles there are with any given speed because this is described by the Maxwell–Boltzmann distribution (6.14). The net result is that the number of particles passing through the horizontal slab, per unit time per unit area, whose speed lies close to v (where "close" means in a region of size d^3v), is

$$\# \text{ of particles per unit time per unit area} = nv_z\, f(v)\, d^3v \ .$$

(6.27)

Now we can figure out the momentum that each of these molecules imparts. Consider a particle at some position z. It gets hit from below and it gets hit from above. The hits from above are likely to give it more momentum in the x-direction than those from below. Consider those arriving from above.

If they arrive from a position $z + \Delta z$, then they impart x-momentum

$$\Delta p = m(u_x(z + \Delta z) - u_x(z)) \approx m \frac{du_x}{dz} \Delta z \ . \tag{6.28}$$

What is the distance Δz here? Well, this depends on the angle the particles come in at. They have travelled the mean free path λ, so if they arrive at angle θ then, as shown below, we must have

$$\Delta z = \lambda \cos \theta \ . \tag{6.29}$$

Here $\theta \in [0, \pi/2)$ for particles arriving from above. But the same
argument also holds for particles
coming in from below. These have
$\theta \in (\pi/2, \pi]$ and, correspondingly,
$\Delta z < 0$ which, from (6.28), tells us
that these particles typically absorb
momentum from the layer at z.

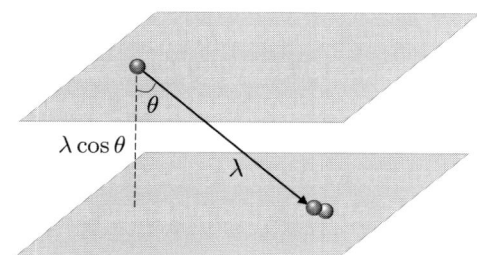

Our goal is to work out the force
per unit area acting on any z-slice.
This is given by the rate of change of momentum

$$\frac{F}{A} = -\frac{1}{A} \frac{\Delta p}{\Delta t} \tag{6.30}$$

where the minus sign arises because F defined in (6.26) is the force you
need to apply to keep the flow moving, while $\Delta p / \Delta t$ is the force of the fluid
pushing back. The rate of change of momentum per unit area is simply the
product of our two expressions (6.27) and (6.28). We have

$$\frac{F}{A} = -n \int d^3v \, \Delta p \, v_z \, f(v)$$

$$= -mn \frac{du_x}{dz} \int d^3v \, v_z \left(\frac{m}{2\pi k_B T} \right)^{3/2} e^{-mv^2/2k_B T} \lambda \cos \theta \ . \tag{6.31}$$

We've actually done something sly in this second line which is not really
justified. We're assuming that the fluid has an average velocity $\langle v_x \rangle = u_x$ in
the x-direction. Yet, at the same time we've used the Maxwell–Boltzmann
distribution for the velocity of the particles which has $\langle v_x \rangle = 0$. Presumably
this is not too bad if the speed of the flow $u \ll \langle v \rangle$, the average speed of the
particles in the fluid, but we really should be more careful in quantifying
this (and we will be later). Nonetheless, the spirit of this section is just to
get a heuristic feel for the physics, so let's push on regardless. Writing the

velocity integral in polar coordinates, we have

$$\frac{F}{A} = -mn\frac{du_x}{dz} \int dv\, v^2 \int_0^\pi d\theta\, \sin\theta$$
$$\times \int_0^{2\pi} d\phi\, (-v\cos\theta)\lambda \cos\theta \left(\frac{m}{2\pi k_B T}\right)^{3/2} e^{-mv^2/2k_B T} \,. \quad (6.32)$$

At this stage we can do the trivial integral $\int d\phi = 2\pi$ and the marginally less trivial integral $\int_0^\pi d\theta \cos^2\theta \sin\theta = 2/3$. We're left with

$$\frac{F}{A} = \frac{mn\lambda}{3}\frac{du_x}{dz} \int_0^\infty dv \left[4\pi\, v^3 \left(\frac{m}{2\pi k_B T}\right)^{3/2} e^{-\beta mv^2/2}\right] \,. \quad (6.33)$$

But the integrand of $\int dv$ is simply the expression (6.15) for the average speed $\langle v \rangle$ in the gas. We have our final expression

$$\frac{F}{A} = \frac{1}{3}mn\lambda\langle v\rangle \frac{du_x}{dz} \,. \quad (6.34)$$

Comparing with (6.26), our expression for the viscosity is

$$\mu = \frac{1}{3}mn\lambda\langle v\rangle \,. \quad (6.35)$$

We can replace the mean free path with the expression $\lambda = 1/n\pi d^2$ from (6.16), and average speed $\langle v \rangle$ with the temperature using (6.15). This gives the result for the viscosity

$$\mu = \sqrt{\frac{2}{\pi}}\frac{3}{\pi d^2}\sqrt{mk_B T} \,. \quad (6.36)$$

There is something surprising about this result: if we keep the temperature fixed, then the dynamic viscosity is independent of the density $n = N/V$ of the gas. That's somewhat counterintuitive since you might naively think that denser gases are more viscous.

The derivation above provides an explanation for what's going on. From (6.35), it's clear that the factor of the density n is cancelled by the mean free path which depends inversely on the density. If you halve the density, then there are half as many molecules coming from above. But the lower density means that each molecule travels twice as far and is therefore coming from a place with higher momentum, and is able to impart twice the momentum kick Δp when it finally hits.

The expression (6.35) holds a special place in the history of physics. It was first derived by Maxwell and is arguably the first novel prediction of the hypothesis that everything is made of atoms. (Previously, people were only able to derive known results, such as the ideal gas law, from an atomic perspective.) Indeed, Maxwell himself was surprised by the fact that μ is independent of the density of the gas, writing at the time:

> Such a consequence of the mathematical theory is very startling and the only experiment I have met with on the subject does not seem to confirm it.

Maxwell rose to the challenge, with the help of his wife Katherine. They built the necessary apparatus and performed the experiment, confirming Maxwell's own prediction.

6.1.4 Thermal Conductivity

The next transport process we will look at is the conduction of heat. Place a fluid between two plates, each held at a different temperature. Empirically, one finds a flow of energy in the fluid described by the *heat flow*, defined by the energy per unit time passing through an area. The heat flow $\mathbf{q}$ is proportional to the gradient of the temperature

$$\mathbf{q} = -\kappa \nabla T \tag{6.37}$$

where κ is called the *thermal conductivity*. Once again, we would like to derive this law, and an expression for κ.

Our calculation follows the same path that we took to determine the viscosity. We set up a temperature gradient in the z-direction. The number of particles with velocity $\mathbf{v}$ that pass through a slab at position z per unit time per unit area is again given by (6.27). We'll use equipartition and assume that the average kinetic energy of a particle at position z is given by

$$E(z) = \frac{3}{2} k_B T(z) \ . \tag{6.38}$$

We also need to know how particles deposit or gain energy when they reach the slab. If a particle came from a hot place with temperature $T(z + \Delta z)$, we'll assume the particle deposits the difference in energy. Similarly, if the particle arrives from a colder place, we'll assume it absorbs the difference. This means

$$\Delta E = E(z + \Delta z) - E(z) = \frac{3}{2} k_B \frac{dT}{dz} \Delta z \ . \tag{6.39}$$

Recall that the height Δz from which the particle arrives depends on both the mean free path and the angle at which it comes in: $\Delta z = \lambda \cos\theta$.

As in the derivation of the viscosity, there is something a little dodgy in what we've written above. We've equated the energy deposited or gained by a particle with the average energy. But this energy transfer will certainly depend on the velocity of the particle, which is dictated by the Maxwell–Boltzmann distribution in (6.27). As in the derivation of the viscosity, we will simply ignore this fact and proceed. We'll do better later in this chapter.

Modulo the concerns above, we now have enough information to compute the heat flow. It is

$$|\mathbf{q}| = n \int d^3v \; \Delta E \, v_z f(v) \;. \tag{6.40}$$

Doing the $\int d^3v$ integrals, and tracing the same steps that we took when computing the viscosity, we derive the law of heat flow (6.37)

$$|\mathbf{q}| = -\frac{1}{2} k_B n \lambda \langle v \rangle \frac{dT}{dz} \;. \tag{6.41}$$

The proportionality constant is the thermal conductivity,

$$\kappa = \frac{1}{2} k_B n \lambda \langle v \rangle = \frac{1}{3\pi d^2} \sqrt{\frac{k_B^3 T}{2\pi m}} \;. \tag{6.42}$$

We see that the viscosity and the thermal conductivity have the same basic structure: both are proportional to the mean free path λ multiplied by the average speed $\langle v \rangle$. Equivalently, both scale with temperature as $\sqrt{T}$.

6.2 From Liouville to BBGKY

In the rest of this chapter, we offer a more sophisticated approach to kinetic theory. The ultimate goal is to derive the Navier–Stokes, and other transport equations, from first principles. Along the way, we will also re-derive the results from the previous section, but with slightly less hand-waving.

Our starting point is simply the Hamiltonian dynamics for N identical point particles. The twist is that, as we mentioned above, the number of particles we're considering is ridiculously large, say $N \sim 10^{23}$ or something similar. We will not attempt the impossible task of following the trajectory of each individual particle, but will instead resort to probabilistic reasoning. We begin by reviewing the derivation of the Liouville equation which governs the evolution of probabilities in phase space. More details can be found in Volume 1 on Classical Mechanics.

We take our N particles to have positions $\mathbf{x}_a$ and momenta $\mathbf{p}_a$, where $a =$

$1, \ldots, N$ labels the particle. Their dynamics is governed by the Hamiltonian

$$H = \frac{1}{2m} \sum_{a=1}^{N} \mathbf{p}_a^2 + \sum_{a=1}^{N} V(\mathbf{x}_a) + \sum_{a<b} U(\mathbf{x}_a - \mathbf{x}_b) \ . \tag{6.43}$$

The Hamiltonian contains an external force $\mathbf{F} = -\nabla V$ that acts equally on all particles. There are also two-body interactions between particles, captured by the potential energy $U(\mathbf{x}_a - \mathbf{x}_b)$. At some point in our analysis (around Section 6.3.2) we will need to assume that this potential is short-ranged, meaning that $U(r) \approx 0$ for distances larger than some atomic distance scale.

Hamilton's equations are

$$\frac{\partial \mathbf{p}_a}{\partial t} = -\frac{\partial H}{\partial \mathbf{x}_a} \quad \text{and} \quad \frac{\partial \mathbf{x}_a}{\partial t} = \frac{\partial H}{\partial \mathbf{p}_a} \ . \tag{6.44}$$

Solving these equations for any $N > 2$ is typically tricky. Only a fool would try to solve them for $N \sim 10^{23}$. Instead, we work with probabilities. We consider a probability distribution $f(\mathbf{x}_a, \mathbf{p}_a; t)$ over the $6N$-dimensional phase space. The probability that the system sits in a state in the neighbourhood of the point $(\mathbf{x}_a, \mathbf{p}_a)$ is

$$f(\mathbf{x}_a, \mathbf{p}_a; t) \, dV \quad \text{with} \quad dV = \prod_{a=1}^{N} d^3 x_a \, d^3 p_a \ . \tag{6.45}$$

Because this is a probability distribution, it is necessarily normalised so that

$$\int dV \, f(\mathbf{x}_a, \mathbf{p}_a; t) = 1 \ . \tag{6.46}$$

Importantly, probability is *locally* conserved. This means that it must obey a continuity equation, so that any change of probability in one part of phase space must be compensated by a flow into neighbouring regions. There's something novel in this statement because we're not just talking about locality in position space, but also locality in momentum space. If one particle interacts with another through the potential $U(\mathbf{x}_a - \mathbf{x}_b)$, then this occurs only by one particle slowing down and the other speeding up, rather than an instantaneous exchange of momentum. This is what we mean by "locality in momentum space".

Mathematically, this means that the "∇" term in the continuity equation includes both $\partial/\partial \mathbf{x}_a$ and $\partial/\partial \mathbf{p}_a$ and, correspondingly, the velocity vector in phase space is $(\dot{\mathbf{x}}_a, \dot{\mathbf{p}}_a)$. The continuity equation of the probability distribution is then

$$\frac{\partial f}{\partial t} + \frac{\partial}{\partial \mathbf{x}_a} \cdot (f \dot{\mathbf{x}}_a) + \frac{\partial}{\partial \mathbf{p}_a} \cdot (f \dot{\mathbf{p}}_a) = 0 \tag{6.47}$$

where we implicitly sum over the $a = 1, \ldots, N$ index. Using Hamilton's equations (6.44), this equation becomes

$$\frac{\partial f}{\partial t} + \frac{\partial}{\partial \mathbf{x}_a} \cdot \left(f \frac{\partial H}{\partial \mathbf{p}_a} \right) - \frac{\partial}{\partial \mathbf{p}_a} \cdot \left(f \frac{\partial H}{\partial \mathbf{x}_a} \right) = 0 \ . \tag{6.48}$$

The $\partial^2 H / \partial x \partial p$ terms cancel and we're left with

$$\frac{\partial f}{\partial t} + \frac{\partial f}{\partial \mathbf{x}_a} \cdot \frac{\partial H}{\partial \mathbf{p}_a} - \frac{\partial f}{\partial \mathbf{p}_a} \cdot \frac{\partial H}{\partial \mathbf{x}_a} = 0 \ . \tag{6.49}$$

This is the *Liouville equation*. It can equivalently be written as a total derivative, instead of partial derivatives, in which case it becomes

$$\frac{df}{dt} = \frac{\partial f}{\partial t} + \frac{\partial f}{\partial \mathbf{x}_a} \cdot \dot{\mathbf{x}}_a + \frac{\partial f}{\partial \mathbf{p}_a} \cdot \dot{\mathbf{p}}_a = 0 \ . \tag{6.50}$$

Written this way, the Liouville equation is telling us that the probability doesn't change as you follow it along any trajectory in phase space.

To get a feel for how probability distributions evolve, one often evokes the closely related *Liouville's theorem*. This is the statement that if you follow some region of phase space under Hamiltonian evolution, then its shape can change but its volume remains the same. This means that the probability distribution on phase space acts like an incompressible fluid. Suppose, for example, that the probability is some constant, f, over some region of phase space and zero everywhere else. Then the distribution can't spread out over a larger volume, lowering its value. Instead, the probability must always be the same constant f over some region of phase space. The shape and position of this region can change, but not its volume.

The Liouville equation is often written using the *Poisson bracket*

$$\{A, B\} \equiv \frac{\partial A}{\partial \mathbf{x}_a} \cdot \frac{\partial B}{\partial \mathbf{p}_a} - \frac{\partial A}{\partial \mathbf{p}_a} \cdot \frac{\partial B}{\partial \mathbf{x}_a} \ . \tag{6.51}$$

With this notation, Liouville's equation becomes simply

$$\frac{\partial f}{\partial t} = \{H, f\} \ . \tag{6.52}$$

It's worth making a few simple comments about probability distributions. First, an *equilibrium distribution* is one which has no explicit time dependence, so

$$\frac{\partial f}{\partial t} = 0 \ . \tag{6.53}$$

This holds if $\{H, f\} = 0$. One way to satisfy this is if f is a function of H, and the most famous example is the Boltzmann distribution, $f \sim e^{-\beta H}$, which we will see emerging later in this chapter. Notice, however, that there is nothing (so far!) within the Hamiltonian framework that requires the equilibrium distribution to be Boltzmann: any function that Poisson commutes with H will do the job. We'll come back to this in Section 6.3.3.

Suppose that we have some function $A(\mathbf{x}_a, \mathbf{p}_a)$ on phase space that does not explicitly depend on time. Of course, $\mathbf{x}_a$ and $\mathbf{p}_a$ will evolve by Hamilton's equations, and so the lack of explicit time dependence of A does not mean that A is a constant. Typically it will change.

The expectation value of the function A is given by

$$\langle A \rangle = \int dV \, A(\mathbf{x}_a, \mathbf{p}_a) f(\mathbf{x}_a, \mathbf{p}_a; t) \ . \tag{6.54}$$

Because A has no explicit time dependence, this expectation value changes with time only if there is explicit time dependence in the distribution. So, for example, in equilibrium $\langle A \rangle$ is constant. In general, we have

$$\begin{aligned}
\frac{d\langle A \rangle}{dt} &= \int dV \, A \frac{\partial f}{\partial t} \\
&= \int dV \, A \left(\frac{\partial f}{\partial \mathbf{p}_a} \frac{\partial H}{\partial \mathbf{x}_a} - \frac{\partial f}{\partial \mathbf{x}_a} \frac{\partial H}{\partial \mathbf{p}_a} \right) \\
&= \int dV \left(-\frac{\partial A}{\partial \mathbf{p}_a} \frac{\partial H}{\partial \mathbf{x}_a} + \frac{\partial A}{\partial \mathbf{x}_a} \frac{\partial H}{\partial \mathbf{p}_a} \right) f
\end{aligned} \tag{6.55}$$

where we've used Liouville's equation (6.49) in going to the second line, and we have integrated by parts to get to the last line, throwing away boundary terms, which is justified in this context because f is normalised and so $f \to 0$ in asymptotic parts of phase space. We learn that expectation values evolve as

$$\frac{d\langle A \rangle}{dt} = \int dV \, \{A, H\} f = \langle \{A, H\} \rangle \ . \tag{6.56}$$

This should be ringing some bells. The evolution of expectation values in classical mechanics looks very much like the evolution of expectation values in quantum mechanics, with Poisson brackets in place of commutators. More discussion of this connection can be found in the book on Classical Mechanics.

6.2.1 The BBGKY Hierarchy

Humility comes in steps. In admitting ignorance, and embracing a probabilistic description of the system, we have taken the first step. But it hasn't
helped us much. This is because the probability distribution $f(\mathbf{x}_a, \mathbf{p}_a)$ is still
a function of $N \sim 10^{23}$ variables and that's not going to be easy to work
with. We need to take a second step. And this is to limit our ambition.

We will focus not on the probability distribution for all N particles, but
instead on the *one-particle distribution function*. This describes the expected
number of particles sitting at some point $\mathbf{x}$, with momentum $\mathbf{p}$. The one-
particle distribution function is defined by

$$f_1(\mathbf{x}, \mathbf{p}; t) = N \int \prod_{a=2}^{N} d^3 x_a \, d^3 p_a \, f(\mathbf{x}, \mathbf{x}_2, \dots, \mathbf{x}_N, \mathbf{p}, \mathbf{p}_2, \dots, \mathbf{p}_N; t) \, . \quad (6.57)$$

We have singled out the first particle for special treatment and integrated
over all the others. We will be interested in situations where all N particles
are identical, so we would have got the same function $f_1(\mathbf{x}, \mathbf{p}; t)$ had we
chosen to integrate over any other collection of $(N-1)$ particles. This is
also reflected in the factor N which sits out front, whose role is to ensure
that f_1 is normalised to

$$\int d^3 x \, d^3 p \, f_1(\mathbf{x}, \mathbf{p}; t) = N \, . \quad (6.58)$$

For many purposes, the function f_1 is all we really need. In particular, it
captures the most basic things that we might want to know about a system.
For example, the average density of particles in real space is simply

$$n(\mathbf{x}; t) = \int d^3 p \, f_1(\mathbf{x}, \mathbf{p}; t) \, . \quad (6.59)$$

There are a number of other important quantities, such as the average velocity and temperature of the gas, that we can construct from $f_1(\mathbf{x}, \mathbf{p}; t)$.
These will be described in detail in Section 6.5. For now, we will focus on
getting a handle on $f_1(\mathbf{x}, \mathbf{p}; t)$ itself.

We would like to derive an equation that tells us something about the
one-particle distribution function f_1. We can see how it changes with time
by looking at

$$\frac{\partial f_1}{\partial t} = N \int \prod_{a=2}^{N} d^3 x_a \, d^3 p_a \, \frac{\partial f}{\partial t} = N \int \prod_{a=2}^{N} d^3 x_a \, d^3 p_a \, \{H, f\} \, . \quad (6.60)$$

Using the Hamiltonian given in (6.43), this becomes

$$\frac{\partial f_1}{\partial t} = N \int \prod_{a=2}^{N} d^3 x_a \, d^3 p_a \left[-\sum_{b=1}^{N} \frac{\mathbf{p}_b}{m} \cdot \frac{\partial f}{\partial \mathbf{x}_b} + \sum_{b=1}^{N} \frac{\partial V}{\partial \mathbf{x}_b} \cdot \frac{\partial f}{\partial \mathbf{p}_b} \right.$$
$$\left. + \sum_{b=1}^{N} \sum_{c<d} \frac{\partial U(\mathbf{x}_c - \mathbf{x}_d)}{\partial \mathbf{x}_b} \cdot \frac{\partial f}{\partial \mathbf{p}_b} \right] . \quad (6.61)$$

Now, whenever $b = 2, \ldots, N$, we can always integrate by parts to move the derivatives away from f and onto the other terms. And, in each case, the result is simply zero because when we differentiate with respect to $\mathbf{x}_b$ the other terms depend only on momentum, and when we differentiate with respect to $\mathbf{p}_b$ the other terms depend only on position. This means that we're left only with the terms that involve derivatives with respect to $\mathbf{x}_1$ and $\mathbf{p}_1$ because we can't integrate these by parts. We revert to our previous notation and call $\mathbf{x}_1 = \mathbf{x}$ and $\mathbf{p}_1 = \mathbf{p}$. We have

$$\frac{\partial f_1}{\partial t} = N \int \prod_{a=2}^{N} d^3 x_a \, d^3 p_a \left[-\frac{\mathbf{p}}{m} \cdot \frac{\partial f}{\partial \mathbf{x}} + \frac{\partial V(\mathbf{x})}{\partial \mathbf{x}} \cdot \frac{\partial f}{\partial \mathbf{p}} + \sum_{c=2}^{N} \frac{\partial U(\mathbf{x} - \mathbf{x}_c)}{\partial \mathbf{x}} \cdot \frac{\partial f}{\partial \mathbf{p}} \right]$$
$$= \{H_1, f_1\} + N \int \prod_{a=2}^{N} d^3 x_a \, d^3 p_a \sum_{c=2}^{N} \frac{\partial U(\mathbf{x} - \mathbf{x}_c)}{\partial \mathbf{x}} \cdot \frac{\partial f}{\partial \mathbf{p}} . \quad (6.62)$$

where we have defined the one-particle Hamiltonian

$$H_1 = \frac{p^2}{2m} + V(\mathbf{x}) . \quad (6.63)$$

Notice that H_1 includes the external force V acting on the particle, but it knows nothing about the interaction with the other particles. All of that information is included in the last term in (6.62) with $U(\mathbf{x} - \mathbf{x}_c)$. The upshot is that the evolution of the one-particle distribution function is described by a Liouville-like equation, together with an extra term. We write

$$\frac{\partial f_1}{\partial t} = \{H_1, f_1\} + \left(\frac{\partial f_1}{\partial t} \right)_{\text{coll}} . \quad (6.64)$$

The first term is sometimes referred to as the *streaming term*, and sometimes just as the *Liouville term*. It tells you how the particles move in the absence of collisions. The second term, known as the *collision integral*, is given by the big integral in (6.62). We can simplify this slightly. Because all particles are the same, each of the $(N-1)$ terms in $\sum_{c=2}^{N}$ in (6.62) are identical and we can write

$$\left(\frac{\partial f_1}{\partial t} \right)_{\text{coll}} = N(N-1) \int d^3 x' \, d^3 p' \, \frac{\partial U(\mathbf{x} - \mathbf{x}')}{\partial \mathbf{x}} \cdot \frac{\partial}{\partial \mathbf{p}} \int \prod_{a=3}^{N} d^3 x_a \, d^3 p_a \, f$$

where $f = f(\mathbf{x}, \mathbf{x}', \ldots, \mathbf{p}, \mathbf{p}', \ldots; t)$. But now we've got something of a problem. The collision integral can't be expressed in terms of the one-particle distribution function. And that's not really surprising. As the name suggests, the collision integral captures the interactions, or collisions, of one particle with another. Yet f_1 contains no information about where any of the other particles are in relation to the first. The idea that we might get a self-contained equation for f_1 alone, without reference to the other particles, looks rather hopeless.

We might ask when this collision term is important. If we look back at the definition of particle density $n = \int d^3p \; f_1$ in (6.59), then we see that the collision term doesn't actually contribute to the time variation $\partial n / \partial t$. This is because, after integrating over $\int d^3p$, we can integrate the $\partial / \partial \mathbf{p}$ term in the collision integral by parts to see that it vanishes. So if we only care about the distribution of particles in space, then we can happily ignore the effects of the collision integral.

However, we will ultimately be interested in other quantities, such as the average velocity of the system. These involve averages over momentum, with terms like $\int d^3p \; \mathbf{p} f_1$, and that factor of $\mathbf{p}$ in the integrand means that our previous "just integrate by parts" argument fails and the collision integral is important. As you might imagine, if our goal is to understand the flow of fluids, then the average velocity is likely to be important. So we need to get some handle on the collision integral.

To move forwards, we define the *two-particle distribution function*,

$$f_2(\mathbf{x}_1, \mathbf{x}_2, \mathbf{p}_1, \mathbf{p}_2; t) = N(N-1) \int \prod_{a=3}^{N} d^3x_a \, d^3p_a \, f(\mathbf{x}_1, \mathbf{x}_2, \ldots, \mathbf{p}_1, \mathbf{p}_2, \ldots; t) \; .$$

With this definition, the collision integral is written simply as

$$\left(\frac{\partial f_1}{\partial t} \right)_{\text{coll}} = \int d^3x_2 \, d^3p_2 \, \frac{\partial U(\mathbf{x} - \mathbf{x}_2)}{\partial \mathbf{x}} \cdot \frac{\partial f_2}{\partial \mathbf{p}} \tag{6.65}$$

which is at least nicer in the sense that the equation doesn't almost spill over into the margin. But it's clear that if we want to understand how the one-particle distribution function f_1 evolves, then we need to get a handle on the two-particle distribution function f_2.

We can always figure out how f_2 evolves by repeating the same calculation that we did above for f_1. It's not hard to show that f_2 evolves by a Liouville-like equation, but with a corrected term that depends on the three-particle distribution function f_3. And f_3 evolves in a Liouville manner, but with a correction term that depends on f_4, and so on. In general, the n-particle distribution function, defined by

$$f_n(\mathbf{x}_1, \ldots, \mathbf{x}_n, \mathbf{p}_1, \ldots, \mathbf{p}_n; t)$$

$$= \frac{N!}{(N-n)!} \int \prod_{a=n+1}^{N} d^3x_a \, d^3p_a \, f(\mathbf{x}_1, \ldots, \mathbf{x}_N, \mathbf{p}_1, \ldots, \mathbf{p}_N; t)$$

obeys the equation

$$\frac{\partial f_n}{\partial t} = \{H_n, f_n\} + \sum_{a=1}^{n} \int d^3x_{n+1} \, d^3p_{n+1} \, \frac{\partial U(\mathbf{x}_a - \mathbf{x}_{n+1})}{\partial \mathbf{x}_a} \cdot \frac{\partial f_{n+1}}{\partial \mathbf{p}_a} \quad (6.66)$$

where the effective n-body Hamiltonian H_n includes the external force and any interactions between the n chosen particles, but neglects interactions with any particles outside of this set,

$$H_n = \sum_{a=1}^{n} \left(\frac{\mathbf{p}_a^2}{2m} + V(\mathbf{x}_a) \right) + \sum_{a<b\leq n} U(\mathbf{x}_a - \mathbf{x}_b) . \quad (6.67)$$

The equations (6.66) are known as the *BBGKY hierarchy*. (The initials stand for Bogoliubov, Born, Green, Kirkwood, and Yvon.) They are telling us that any group of n particles evolves in a Hamiltonian fashion, corrected by interactions with particles outside that group.

The hierarchy of equations means that there's no free lunch. If we want to understand everything in detail, then we're going to have to calculate everything. We started with the Liouville equation governing a complicated function f of $N \sim 10^{23}$ variables and it looks like all we've done is replace it with $\mathcal{O}(10^{23})$ coupled equations. This is reminiscent of the "closure problem" that we ran into in Section 5.1 when discussing turbulence.

However, there is an advantage in working with the hierarchy of equations (6.66) because they isolate the interesting, simple variables, namely f_1 and other lower f_n. This means that the equations are in a form that is ripe to start implementing various approximations. Given a particular problem, we can decide which terms are important and, ideally, which terms are so small that they can be ignored, truncating the hierarchy to something manageable. Exactly how you do this depends on the problem at hand. Here we explain the simplest, and most useful, of these truncations.

6.3 The Boltzmann Equation

Elegance should be left for tailors and shoemakers.

Ludwig Boltzmann

In this section, we explain how, despite the odds, we can write down a closed equation for f_1 alone. This will be the famous Boltzmann equation. The challenge is to find a way to approximate the collision integral so that it captures some aspects of the interactions, but without resorting to the two-particle distribution function.

To achieve this, we need to make an approximation. To this end, we assume that we are working with dilute gases rather than liquids. The advantage of this assumption is that it introduces two different time scales into the problem. One is the time between collisions, τ, known as the *scattering time* or *relaxation time*. The second is the *collision time*, τ_{coll}, which is roughly the time it takes for the process of collision between particles to occur. We will assume that there is a hierarchy of time scales,

$$\tau \gg \tau_{\text{coll}} \,. \tag{6.68}$$

In this situation, we should expect that, for much of the time, particles are just happily minding their own business, unaware of the other particles around them. Only occasionally is their path interrupted by a collision. This means that our one-particle distribution function f_1 spends most of its time evolving by the one-particle Hamiltonian H_1. Interactions with other particles happen only rarely, which gives us hope that we might somehow treat the collision integral in (6.64) perturbatively.

At this stage, there is a right way and a less-right way to proceed. The right way is to derive the Boltzmann equation starting from the BBGKY hierarchy. We will do this in Section 6.3.2. However, as is often the case, doing things the right way takes a little effort. So instead we're going to start by taking the less-right option. This has the advantage of getting the same answer in a much easier, but more hand-waving, fashion.

6.3.1 Motivating the Boltzmann Equation

We already got a glimpse of the basic form of the Boltzmann equation in (6.64). It is

$$\frac{\partial f_1}{\partial t} = \{H_1, f_1\} + \left(\frac{\partial f_1}{\partial t} \right)_{\text{coll}} \,. \tag{6.69}$$

But, of course, we don't yet have an expression for the collision integral $(\partial f_1/\partial t)_{\text{coll}}$ in terms of f_1. It's clear from the definition (6.65) that this term represents the change in momenta due to two-particle scattering. When $\tau \gg \tau_{\text{coll}}$, the collisions occur occasionally and abruptly. The collision integral should reflect the rate at which these collisions occur.

Suppose that our particle sits at $(\mathbf{x}, \mathbf{p})$ in phase space and collides with another particle at $(\mathbf{x}, \mathbf{p}_2)$. We're assuming here that collisions are local in space so that the two particles necessarily sit at the same point $\mathbf{x}$. These particles can collide and emerge with momenta $\mathbf{p}_1'$ and $\mathbf{p}_2'$. We will define the rate for this process to occur to be

$$\text{rate} = \omega(\mathbf{p}, \mathbf{p}_2 | \mathbf{p}_1', \mathbf{p}_2') \, f_2(\mathbf{x}, \mathbf{x}, \mathbf{p}, \mathbf{p}_2) \, d^3 p_2 \, d^3 p_1' \, d^3 p_2' \, . \qquad (6.70)$$

We will now unpick this formula. Roughly speaking, we're going to integrate this rate over $d^3 p_2$, $d^3 p_1'$, and $d^3 p_2'$ to get the collision integral. We don't integrate over our original one-particle momentum $\mathbf{p}$ because that's the independent variable in the one-particle distribution function, $f_1(\mathbf{x}, \mathbf{p})$, that we are trying to determine.

The rate (6.70) is proportional to the two-body distribution function f_2 because this tells us the chance that two particles originally sit at $(\mathbf{x}, \mathbf{p})$ and $(\mathbf{x}, \mathbf{p}_2)$. (We've dropped the explicit t dependence in f_2 only to keep the notation simple.)

In addition, the rate (6.70) also depends on $\omega(\mathbf{p}, \mathbf{p}_2 | \mathbf{p}_1', \mathbf{p}_2')$. This is a second function called the *scattering function*. This contains the information about the dynamics of the process. Ultimately, we will compute this scattering function from knowledge of the inter-atomic potential $U(\mathbf{r})$ in the Hamiltonian, but this is something that we're going to postpone to Section 6.3.2. For now, we will treat the scattering function ω as something that is given, and characterises the scattering between particles.

While collisions can deflect particles out of a state with momentum $\mathbf{p}$ and into a state with a different momentum, they can also deflect other particles into a state with momentum $\mathbf{p}$. This suggests that the collision integral should contain two terms,

$$\left(\frac{\partial f_1}{\partial t} \right)_{\text{coll}} = \int d^3 p_2 \, d^3 p_1' \, d^3 p_2' \, \left[\omega(\mathbf{p}_1', \mathbf{p}_2' | \mathbf{p}, \mathbf{p}_2) \, f_2(\mathbf{x}, \mathbf{x}, \mathbf{p}_1', \mathbf{p}_2') \right. \qquad (6.71)$$

$$\left. - \, \omega(\mathbf{p}, \mathbf{p}_2 | \mathbf{p}_1', \mathbf{p}_2') f_2(\mathbf{x}, \mathbf{x}, \mathbf{p}, \mathbf{p}_2) \right] \, .$$

The first term describes scattering into the state $\mathbf{p}$; the second scattering out of the state $\mathbf{p}$.

So far we've not made much progress. Our collision integral still depends on the two-particle distribution f_2 and, in addition, we have replaced the inter-atomic potential with a more abstract scattering function $\omega(\mathbf{p}, \mathbf{p}_2|\mathbf{p}_1', \mathbf{p}_2')$. Clearly we have more work to do.

The scattering function obeys a few simple requirements. First, there are conservation laws that must be obeyed. Two particles with momenta $\mathbf{p}$ and $\mathbf{p}_2$ can collide, and leave with momenta $\mathbf{p}_1'$ and $\mathbf{p}_2'$ only if the total momentum is conserved, so

$$\mathbf{p} + \mathbf{p}_2 = \mathbf{p}_1' + \mathbf{p}_2' \tag{6.72}$$

and total energy is conserved, so

$$p^2 + p_2^2 = p_1'^2 + p_2'^2 \ . \tag{6.73}$$

This means that the scattering function $\omega(\mathbf{p}, \mathbf{p}_2|\mathbf{p}_1', \mathbf{p}_2')$ is non-zero only if (6.72) and (6.73) are obeyed.

(An aside: Simple as they are, there is actually an assumption that is hiding behind (6.72) and (6.73). In general, we're considering particles in an external potential $V(\mathbf{x})$. This provides a force on the particles which, in principle, could mean that the momentum and kinetic energy of the particles is not the same before and after the collision. To eliminate this possibility, we will assume that the potential only varies appreciably over macroscopic distance scales, so that it can be neglected on the scale of atomic collisions. This, of course, is entirely reasonable for most external potentials such as gravity or electric fields. Then (6.72) and (6.73) continue to hold.)

In addition, there are various discrete spacetime symmetries that also give us important information about the scattering function. Under time reversal, $\mathbf{p} \to -\mathbf{p}$, and, of course, what was coming in is now going out. This means that any scattering which is invariant under time reversal (which is more or less anything of interest) must obey

$$\omega(\mathbf{p}, \mathbf{p}_2|\mathbf{p}_1', \mathbf{p}_2') = \omega(-\mathbf{p}_1', -\mathbf{p}_2'|-\mathbf{p}, -\mathbf{p}_2) \ . \tag{6.74}$$

Furthermore, under parity $(\mathbf{x}, \mathbf{p}) \to (-\mathbf{x}, -\mathbf{p})$. So any scattering process which is parity-invariant further obeys

$$\omega(\mathbf{p}, \mathbf{p}_2|\mathbf{p}_1', \mathbf{p}_2') = \omega(-\mathbf{p}, -\mathbf{p}_2|-\mathbf{p}_1', -\mathbf{p}_2') \ . \tag{6.75}$$

The combination of these two means that the scattering rate is invariant under exchange of ingoing and outgoing momenta,

$$\omega(\mathbf{p}, \mathbf{p}_2 | \mathbf{p}_1', \mathbf{p}_2') = \omega(\mathbf{p}_1', \mathbf{p}_2' | \mathbf{p}, \mathbf{p}_2) \ . \tag{6.76}$$

(There is actually a further assumption of translational invariance here, since the scattering rate at position $-\mathbf{x}$ should be equivalent to the scattering rate at position $+\mathbf{x}$.)

The symmetry property (6.76) allows us to simplify the collision integral (6.71) to

$$\left(\frac{\partial f_1}{\partial t}\right)_{\mathrm{coll}} = \int d^3 p_2 \, d^3 p_1' \, d^3 p_2' \ \omega(\mathbf{p}_1', \mathbf{p}_2' | \mathbf{p}, \mathbf{p}_2)$$
$$\times \left[f_2(\mathbf{x}, \mathbf{x}, \mathbf{p}_1', \mathbf{p}_2') - f_2(\mathbf{x}, \mathbf{x}, \mathbf{p}, \mathbf{p}_2) \right] \ . \tag{6.77}$$

To finish the derivation, we need to face up to our main goal of expressing the collision integral in terms of f_1 rather than f_2. To this end, we make the assumption that the velocities of the two incoming particles are uncorrelated. This means that

$$f_2(\mathbf{x}, \mathbf{x}, \mathbf{p}, \mathbf{p}_2) = f_1(\mathbf{x}, \mathbf{p}) f_1(\mathbf{x}, \mathbf{p}_2) \ . \tag{6.78}$$

This assumption goes by the name of *molecular chaos*. At first glance, it seems innocuous enough. But, it turns out, it is far from innocent!

To see why, let's look more closely at what we've actually assumed. Looking at (6.77), we can see that we have taken the rate of collisions to be proportional to $f_2(\mathbf{x}, \mathbf{x}, \mathbf{p}_1, \mathbf{p}_2)$ where $\mathbf{p}_1$ and $\mathbf{p}_2$ are the momenta of the particles before the collision. That means that if we substitute (6.78) into (6.77), then we are really assuming that the velocities are uncorrelated *before* the collision.

In contrast, after the collision the velocities are surely correlated. This is ensured by the conservation laws (6.72) and (6.73). However, although the velocities are correlated, the two particles then go in their separate directions and it is a long time τ before they collide again. Moreover, this next collision is typically with a completely different particle and so it seems entirely plausible that the velocity of this new particle has nothing to do with the velocity of the first, so that we may again assume that (6.78) holds.

All of these words suggest that the assumption of molecular chaos is eminently reasonable. And yet, the fact that we've assumed that velocities are uncorrelated *before* the collision rather than after has, rather slyly, introduced a preferred arrow of time into the game that was not there in our original Hamilton equations (6.44). As we will see in Section 6.4, where we derive the H-theorem, this arrow of time has dramatic and far reaching consequences.

With the assumption of molecular chaos (6.78) in hand, we may finally write down a closed expression for the evolution of the one-particle distribution function, given by

$$\frac{\partial f_1}{\partial t} = \{H_1, f_1\} + \left(\frac{\partial f_1}{\partial t}\right)_{\text{coll}} \tag{6.79}$$

with the collision integral

$$\left(\frac{\partial f_1}{\partial t}\right)_{\text{coll}} = \int d^3p_2 \, d^3p_1' \, d^3p_2' \, \omega(\mathbf{p}_1', \mathbf{p}_2'|\mathbf{p}, \mathbf{p}_2)$$
$$\times \left[f_1(\mathbf{x}, \mathbf{p}_1')f_1(\mathbf{x}, \mathbf{p}_2') - f_1(\mathbf{x}, \mathbf{p})f_1(\mathbf{x}, \mathbf{p}_2) \right] . \tag{6.80}$$

This is the *Boltzmann equation*. It is one of the more scary-looking equations of classical physics. The left-hand side of (6.79) is a differential equation, while the right-hand side is an integral equation and non-linear. All of which means that exact solutions are not that easy to come by. We'll see what we can do.

6.3.2 A Better Derivation of the Boltzmann Equation

Above, we derived an expression for the collision integral (6.80) using intuition for the scattering processes at play. The expression was a little heuristic in places and, most worryingly, depends on some mysterious scattering function $\omega(\mathbf{p}, \mathbf{p}_2|\mathbf{p}_1', \mathbf{p}_2')$.

That's a little unsatisfactory. After all, we've got a perfectly good expression for the collision integral in (6.65) involving the two-particle distribution function f_2 and the inter-particle potential $U(r)$. In this section, we do a better job. We'll see how we can derive the Boltzmann equation (6.80) directly from (6.65). This will help clarify some of the approximations that we need to use. At the same time, we will also review some basic classical mechanics that connects the scattering rate ω to the inter-particle potential $U(r)$.

We start by returning to the BBGKY hierarchy of equations. For simplicity, we'll turn off the external potential $V(\mathbf{x}) = 0$. We don't lose very much in doing this because most of the interesting physics is concerned with the scattering of atoms off each other, rather than the effect of an external force.

The first equation in the hierarchy is

$$\left(\frac{\partial}{\partial t} + \frac{\mathbf{p}_1}{m} \cdot \frac{\partial}{\partial \mathbf{x}_1}\right) f_1 = \int d^3x_2 \, d^3p_2 \, \frac{\partial U(\mathbf{x}_1 - \mathbf{x}_2)}{\partial \mathbf{x}_1} \cdot \frac{\partial f_2}{\partial \mathbf{p}_1} \, . \tag{6.81}$$

We've moved the streaming term to the left-hand side, and put the collision term, depending on f_2, on the right. The second equation in the hierarchy is

$$\left(\frac{\partial}{\partial t} + \frac{\mathbf{p}_1}{m} \cdot \frac{\partial}{\partial \mathbf{x}_1} + \frac{\mathbf{p}_2}{m} \cdot \frac{\partial}{\partial \mathbf{x}_2} - \frac{1}{2}\frac{\partial U(\mathbf{x}_1 - \mathbf{x}_2)}{\partial \mathbf{x}_1} \cdot \left[\frac{\partial}{\partial \mathbf{p}_1} - \frac{\partial}{\partial \mathbf{p}_2}\right]\right) f_2$$

$$= \int d^3x_3 \, d^3p_3 \left(\frac{\partial U(\mathbf{x}_1 - \mathbf{x}_3)}{\partial \mathbf{x}_1} \cdot \frac{\partial}{\partial \mathbf{p}_1} + \frac{\partial U(\mathbf{x}_2 - \mathbf{x}_3)}{\partial \mathbf{x}_2} \cdot \frac{\partial}{\partial \mathbf{p}_2}\right) f_3 \, . \tag{6.82}$$

Again, we've gathered the streaming terms onto the left-hand side, leaving only the higher distribution function on the right. To keep things clean, we've suppressed the arguments of the distribution functions: they are $f_1 = f_1(\mathbf{x}_1, \mathbf{p}_1; t)$ and $f_2 = f_2(\mathbf{x}_1, \mathbf{x}_2, \mathbf{p}_1, \mathbf{p}_2; t)$ and you can guess the arguments for f_3.

Our goal is to better understand the collision integral on the right-hand side of (6.81). It seems reasonable to assume that when particles are far separated, their distribution functions are uncorrelated. Here, "far separated" means that the distance between them is much farther than the atomic distance scale over which the potential $U(r)$ extends. We call this atomic distance scale d, and assume that

$$U(r) \approx 0 \quad \text{for} \quad r \gg d \, . \tag{6.83}$$

We then expect that, when $|\mathbf{x}_1 - \mathbf{x}_1| \gg d$, we have

$$f_2(\mathbf{x}_1, \mathbf{x}_2, \mathbf{p}_1, \mathbf{p}_2; t) \to f_1(\mathbf{x}_1, \mathbf{p}_1; t) f_1(\mathbf{x}_2, \mathbf{p}_2; t) \, . \tag{6.84}$$

But a glance at the right-hand-side of (6.81) tells us that $r \gg d$ isn't the regime of interest. Instead, f_2 is multiplied by $\partial U(r)/\partial r$ which is non-vanishing only over a region $r \le d$. This means that we need to understand f_2 when two particles get close to each other.

To proceed, we start by getting a feel for the order of magnitude of various terms in the hierarchy of equations. Dimensionally, each term in brackets in (6.81) and (6.82) is an inverse time scale. The terms involving the inter-atomic potential $U(r)$ are associated to the collision time τ_{coll}.

$$\frac{1}{\tau_{\text{coll}}} \sim \frac{\partial U}{\partial \mathbf{x}} \cdot \frac{\partial}{\partial \mathbf{p}} \, . \tag{6.85}$$

This is the time taken for a particle to cross the distance over which the potential $U(r)$ varies, which, for short-range potentials, is comparable to the atomic distance scale, d, itself and, as we saw in Section 6.1, is approximately $\tau_{\text{coll}} \sim d/\bar{v}_{\text{rel}}$, with $\bar{v}_{\text{rel}}$ the average relative speed between atoms. Our first approximation will be that this is the shortest time scale in the problem. This means that the terms involving $\partial U/\partial r$ are typically the largest terms in the equations above and determine how fast the distribution functions change.

With this in mind, we note that the equation for f_1 is special because it is the only one which does not include any collision terms on the left-hand side of the equation. (Compare the left-hand side of (6.81) to (6.82); only the latter has a $\partial U/\partial \mathbf{x}$ term.) This is because $U(r)$ appeared in all the Hamiltonians H_n defined in (6.67), but not in H_1 defined in (6.63). This means that the collision integral on the right-hand side of (6.81) will usually dominate the rate of change of f_1. (We'll meet some important exceptions to this statement in Section 6.5.)

In contrast, the equation that governs f_2 has collision terms on both the left- and the right-hand sides. But, importantly, for dilute gases, the term on the right is much smaller than the term on the left. To see why this is, we need to compare the f_3 term to the f_2 term. If we were to integrate f_3 over all space, we get

$$\int d^3x_3 \, d^3p_3 \, f_3 = N f_2 \qquad (6.86)$$

(where we've replaced $(N-2) \approx N$ in the above expression). However, although the right-hand side of (6.82) is integrated over all of space, it picks up a non-zero contribution only over an atomic scale $\sim d^3$. This means that the collision term on the right-hand side of (6.82) is suppressed compared to the one on the left by a factor of Nd^3/V where V is the volume of space. For gases that we live and breathe every day, $Nd^3/V \sim 10^{-3}$ to 10^{-4}. We make use of this small number to truncate the hierarchy of equations and replace (6.82) with

$$\left(\frac{\partial}{\partial t} + \frac{\mathbf{p}_1}{m} \cdot \frac{\partial}{\partial \mathbf{x}_1} + \frac{\mathbf{p}_2}{m} \cdot \frac{\partial}{\partial \mathbf{x}_2} - \frac{1}{2} \frac{\partial U(\mathbf{x}_1 - \mathbf{x}_2)}{\partial \mathbf{x}_1} \cdot \left[\frac{\partial}{\partial \mathbf{p}_1} - \frac{\partial}{\partial \mathbf{p}_2} \right] \right) f_2 \approx 0 \; .$$
$$(6.87)$$

This tells us that f_2 typically varies on a time scale of τ_{coll} and a length scale of d. Meanwhile, the variations of f_1 are governed by the right-hand side of (6.81) which, by the same arguments that we just made, are smaller than the variations of f_2 by a factor of Nd^3/V. In other words, f_1 varies on the larger time scale that we will call τ.

In fact, we can be a little more careful when we say that f_2 varies on a time scale τ_{coll}. We see that, as we would expect, only the relative position is affected by the collision term. For this reason, it's useful to change coordinates to the centre of mass and the relative positions of the two particles. We write

$$\mathbf{R} = \frac{1}{2}(\mathbf{x}_1 + \mathbf{x}_2) \quad \text{and} \quad \mathbf{r} = \mathbf{x}_1 - \mathbf{x}_2 \tag{6.88}$$

and similarly for the momentum

$$\mathbf{P} = \mathbf{p}_1 + \mathbf{p}_2 \quad \text{and} \quad \mathbf{p} = \frac{1}{2}(\mathbf{p}_1 - \mathbf{p}_2) \; . \tag{6.89}$$

We can think of $f_2 = f_2(\mathbf{R}, \mathbf{r}, \mathbf{P}, \mathbf{p}; t)$. The distribution function will depend on the centre-of-mass variables $\mathbf{R}$ and $\mathbf{P}$ in some slow fashion, much as f_1 depends on position and momentum. In contrast, the dependence of f_2 on the relative coordinates $\mathbf{r}$ and $\mathbf{p}$ is much faster. These vary over the short distance scale and can change on a time scale of order τ_{coll}.

Because the relative distributions in f_2 vary much more quickly than f_1, we will assume that f_2 reaches equilibrium and then feeds into the dynamics of f_1. This means that, ignoring the slow variations in $\mathbf{x}$ and $\mathbf{p}$, we will assume that $\partial f_2 / \partial t = 0$ and replace (6.87) with the equilibrium condition

$$\left(\frac{\mathbf{p}}{m} \cdot \frac{\partial}{\partial \mathbf{r}} - \frac{\partial U(\mathbf{r})}{\partial \mathbf{r}} \cdot \frac{\partial}{\partial \mathbf{p}} \right) f_2 \approx 0 \; . \tag{6.90}$$

This is now in a form that allows us to start manipulating the collision integral on the right-hand side of (6.81). We have

$$\left(\frac{\partial f_1}{\partial t} \right)_{\text{coll}} = \int d^3 x_2 \, d^3 p_2 \, \frac{\partial U(\mathbf{x}_1 - \mathbf{x}_2)}{\partial \mathbf{x}_1} \cdot \frac{\partial f_2}{\partial \mathbf{p}_1}$$

$$= \int d^3 x_2 \, d^3 p_2 \, \frac{\partial U(\mathbf{r})}{\partial \mathbf{r}} \cdot \left[\frac{\partial}{\partial \mathbf{p}_1} - \frac{\partial}{\partial \mathbf{p}_2} \right] f_2$$

$$= \frac{1}{m} \int_{|\mathbf{x}_1 - \mathbf{x}_2| \leq d} d^3 x_2 \, d^3 p_2 \, (\mathbf{p}_1 - \mathbf{p}_2) \cdot \frac{\partial f_2}{\partial \mathbf{r}} \tag{6.91}$$

where in the second line the extra term $\partial / \partial \mathbf{p}_2$ vanishes if we integrate by parts and, in the third line, we've used our equilibrium condition (6.90), with the limits on the integral in place to remind us that only the region $r \leq d$ contributes to the collision integral.

A Review of Scattering Cross-Sections

To complete the story, we still need to turn (6.91) into the collision integral (6.80). Most of the work involves clarifying how the scattering rate

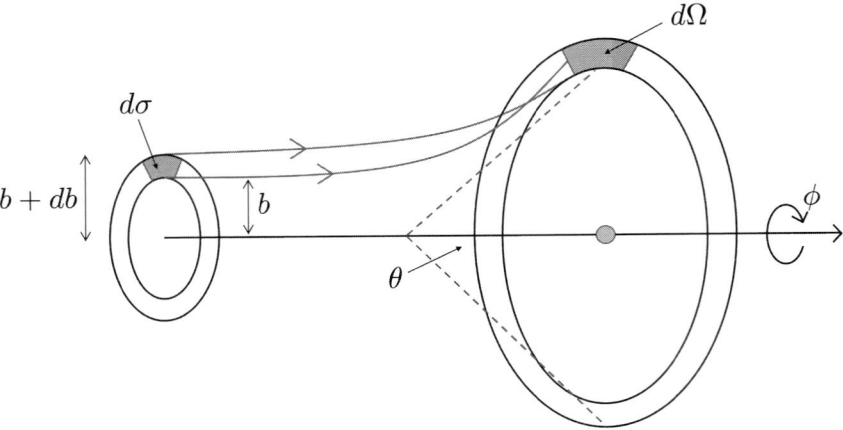

The differential cross-section.

$\omega(\mathbf{p}, \mathbf{p}_2|\mathbf{p}_1', \mathbf{p}_2')$ is defined for a given inter-atomic potential $U(\mathbf{x}_1 - \mathbf{x}_2)$. And, for this, we need to review the concept of the *differential cross-section*. This is an idea that we met briefly in Volume 1 on Classical Mechanics, and explored in much greater detail when discussing scattering in Volume 3 on Quantum Mechanics. Here we cover the basics.

We want to think about the collision between two particles. They start with momenta $\mathbf{p}_a = m\mathbf{v}_a$ and end with momenta $\mathbf{p}_a' = m\mathbf{v}_a'$ with $a = 1, 2$. Let's pick a favourite, say particle 1. We sit in its rest frame, which means that we will think of this particle as pinned in place. We then consider an onslaught of bombarding particles, each with velocity $\mathbf{v}_2 - \mathbf{v}_1$. The incoming particles in this beam do not all hit our favourite boy at the same point. Instead, they come in, randomly distributed over the plane perpendicular to $\mathbf{v}_2 - \mathbf{v}_1$. The flux, I, of these incoming particles is the number hitting this plane per unit area per second,

$$I = \frac{N}{V}|\mathbf{v}_2 - \mathbf{v}_1| \, . \qquad (6.92)$$

Now spend some time staring at Figure 6.3. Our chosen particle is the dot sitting in the centre of the two, large concentric rings. There are then a number of quantities defined in this picture.

First, the *impact parameter*, b, is the distance from the asymptotic trajectory to the centre line. This is the distance that the incoming particle would get to our preferred particle if there were no force between them. We will use b and ϕ as polar coordinates to parametrise the plane perpendicular to the incoming particle.

Next, the scattering angle, θ, is the angle by which the incoming particle is deflected. Finally, there are two solid angles, $d\sigma$ and $d\Omega$, depicted in the figure. Geometrically, we see that they are given by

$$d\sigma = b\,db\,d\phi \quad \text{and} \quad d\Omega = \sin\theta\,d\theta\,d\phi\ . \tag{6.93}$$

The number of particles scattered into $d\Omega$ in unit time is $I d\sigma$. We usually write this as

$$I\frac{d\sigma}{d\Omega}d\Omega = Ib\,db\,d\phi \tag{6.94}$$

where the *differential cross-section* is defined as

$$\left|\frac{d\sigma}{d\Omega}\right| = \frac{b}{\sin\theta}\left|\frac{db}{d\theta}\right| = \frac{1}{2}\left|\frac{d(b^2)}{d\cos\theta}\right|\ . \tag{6.95}$$

You should think of this in the following way: for a fixed $(\mathbf{v}_2 - \mathbf{v}_1)$, there is a unique relationship between the impact parameter b and the scattering angle θ and, for a given potential $U(r)$, you need to figure this out to get $|d\sigma/d\Omega|$ as a function of θ.

Now we can compare this to the notation that we used earlier in (6.70). There we talked about the rate of scattering into a small area $d^3p_1'\,d^3p_2'$ in momentum space. But this is the same thing as the differential cross-section

$$\omega(\mathbf{p},\mathbf{p}_2;\mathbf{p}_1',\mathbf{p}_2')\,d^3p_1'\,d^3p_2' = |\mathbf{v} - \mathbf{v}_2|\left|\frac{d\sigma}{d\Omega}\right|d\Omega\ . \tag{6.96}$$

If you're worried about the fact that $d^3p_1'\,d^3p_2'$ is a six-dimensional area while $d\Omega$ is a two-dimensional area, recall that conservation of energy and momenta provide four restrictions on the ability of particles to scatter. These are implicit on the left-hand side of (6.96), but explicit on the right.

An Example of a Cross-Section: Hard Spheres

A particularly simple picture of the microscopic constituents of a gas is to model the atoms as hard spheres. Here we derive the associated cross-section.

In fact, there are two different calculations that we can do. First, suppose that we throw point-like particles at a sphere of diameter d with an impact parameter $b \leq d/2$. From the left-hand diagram in Figure 6.4, we see that

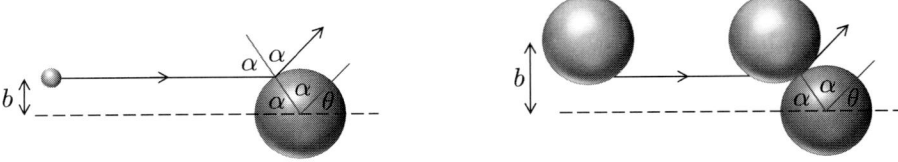

Fig. 6.4 On the left: a point particle scattering off a hard sphere. On the right: a hard sphere scattering off a hard sphere.

the scattering angle is $\theta = \pi - 2\alpha$, where

$$b = \frac{d}{2} \sin \alpha = \frac{d}{2} \sin\left(\frac{\pi}{2} - \frac{\theta}{2}\right) = \frac{d}{2} \cos \frac{\theta}{2} \qquad (6.97)$$

or

$$b^2 = \frac{d^2}{4} \cos^2 \frac{\theta}{2} = \frac{d^2}{8}(1 + \cos \theta) . \qquad (6.98)$$

From (6.95), we then find the differential cross-section

$$\left|\frac{d\sigma}{d\Omega}\right| = \frac{d^2}{16} . \qquad (6.99)$$

The total cross-section is defined as

$$\sigma_T = 2\pi \int_0^\pi d\theta \, \sin\theta \, \frac{d\sigma}{d\Omega} = \pi \left(\frac{d}{2}\right)^2 . \qquad (6.100)$$

This provides a nice justification for the name because this is indeed the cross-sectional area of a sphere of radius $d/2$.

Alternatively, we could consider two identical hard spheres, each of diameter d, one scattering off the other. Now the geometry changes a little, as shown in the right-hand diagram in Figure 6.4. The impact parameter is now the distance between the centres of the spheres, and given by

$$b = 2 \times \frac{d}{2} \sin \alpha . \qquad (6.101)$$

Clearly we now need $b \leq d$. The same calculation as above gives

$$\sigma_T = \pi d^2 . \qquad (6.102)$$

This is the effective cross-sectional area that a hard sphere presents to other hard spheres. It is larger than the geometrical cross-section (6.100). It is the same result that we got in Section 6.1 by thinking about the cylinder swept out by a hard sphere.

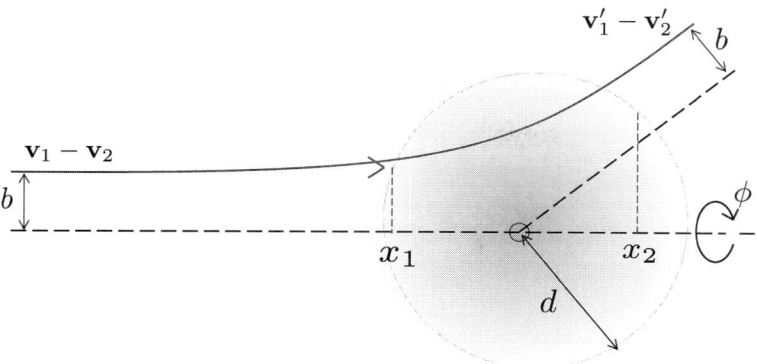

Fig. 6.5 Tw- particle scattering. The potential $U(r)$ has range d, forming the collision zone shown above.

Finally, the Collision Integral

With this refresher course on classical scattering, we can return to the collision integral (6.91) in the Boltzmann equation

$$\left(\frac{\partial f_1}{\partial t}\right)_{\text{coll}} = \int_{|\mathbf{x}_1 - \mathbf{x}_2| \leq d} d^3 x_2 \, d^3 p_2 \, (\mathbf{v}_1 - \mathbf{v}_2) \cdot \frac{\partial f_2}{\partial \mathbf{x}} . \qquad (6.103)$$

Here d is the range of the potential $U(r)$, in the sense of (6.80). We work in cylindrical polar coordinates, as shown in Figure 6.5. The direction parallel to $\mathbf{v}_2 - \mathbf{v}_1$ is parametrised by x, while the plane perpendicular is parametrised by ϕ and the impact parameter b. We've depicted the collision zone in this figure. Using the definitions (6.94) and (6.96), we have

$$\left(\frac{\partial f_1}{\partial t}\right)_{\text{coll}} = \int d^3 p_2 \, |\mathbf{v}_1 - \mathbf{v}_2| \int d\phi \, db \, b \int_{x_1}^{x_2} \frac{\partial f_2}{\partial x}$$

$$= \int d^3 p_2 d^3 p_1' d^3 p_2' \, \omega(\mathbf{p}_1', \mathbf{p}_2' | \mathbf{p}, \mathbf{p}_2) \, [f_2(x_2) - f_2(x_1)] . \qquad (6.104)$$

It remains only to decide what form the two-particle distribution function f_2 takes just before the collision at $x = x_1$ and just after the collision at $x = x_2$. At this point we invoke the assumption of molecular chaos. Just before we enter the collision, we assume that the two particles are uncorrelated. Moreover, we assume that the two particles are once again uncorrelated by the time they leave the collision zone, albeit now with their new momenta $\mathbf{v}_1' - \mathbf{v}_2'$

$$f_2(x_1) = f_1(\mathbf{x}, \mathbf{p}_1; t) f_1(\mathbf{x}, \mathbf{p}_2; t)$$
$$f_2(x_2) = f_1(\mathbf{x}, \mathbf{p}_1'; t) f_1(\mathbf{x}, \mathbf{p}_2'; t) . \qquad (6.105)$$

All functions f_1 are evaluated at the same point $\mathbf{x}$ in space since we've assumed that the single-particle distribution function is suitably coarse-grained so that it doesn't vary on scales of order d. With this final assumption, we get what we wanted: the collision integral is given by

$$\left(\frac{\partial f_1}{\partial t}\right)_{\text{coll}} = \int d^3 p_2 \, d^3 p_1' \, d^3 p_2' \; \omega(\mathbf{p}_1', \mathbf{p}_2' | \mathbf{p}, \mathbf{p}_2)$$
$$\times \left[f_1(\mathbf{x}, \mathbf{p}_1') f_1(\mathbf{x}, \mathbf{p}_2') - f_1(\mathbf{x}, \mathbf{p}) f_1(\mathbf{x}, \mathbf{p}_2) \right] . \quad (6.106)$$

This is the result that we previously derived in (6.80), now sitting on firmer footing.

6.3.3 Equilibrium

The simplest solutions to the Boltzmann equation are equilibrium distributions. Recall that equilibrium distributions have the property that there is no explicit time dependence, meaning

$$\frac{\partial f_1}{\partial t} = 0 . \quad (6.107)$$

The Boltzmann equation (6.79) tells us that there are two contributions to the time dependence of the distribution function f_1. The first is the Liouville contribution $\{f_1, H_1\}$. This vanishes if f_1 is any function of the energy, or any function that Poisson commutes with the energy. For simplicity, we'll restrict to situations with vanishing external force, so $V(\mathbf{x}) = 0$. Then, the Liouville contribution $\{f_1, H_1\}$ vanishes if $f_1(\mathbf{x}, \mathbf{p}) = f_1(\mathbf{p})$ is a function of momentum alone.

That leaves us with the contribution from the collision integral. One way to make the collision integral in (6.106) vanish is to find a distribution that obeys the *detailed balance* condition

$$f_1^{\text{eq}}(\mathbf{x}, \mathbf{p}_1') f_1^{\text{eq}}(\mathbf{x}, \mathbf{p}_2') = f_1^{\text{eq}}(\mathbf{x}, \mathbf{p}) f_1^{\text{eq}}(\mathbf{x}, \mathbf{p}_2) . \quad (6.108)$$

In fact, it's more useful to write this as

$$\log\left(f_1^{\text{eq}}(\mathbf{x}, \mathbf{p}_1')\right) + \log\left(f_1^{\text{eq}}(\mathbf{x}, \mathbf{p}_2')\right) = \log(f_1^{\text{eq}}(\mathbf{x}, \mathbf{p})) + \log(f_1^{\text{eq}}(\mathbf{x}, \mathbf{p}_2)) . \quad (6.109)$$

How can we ensure that this is true for all momenta? The momenta on the right are those before the collision and those on the left are after the collision. From the form of (6.109), it's clear that the sum of $\log f_1^{\text{eq}}$ must be the same before and after the collision. In other words, this sum must be conserved during the collision. But we know what things are conserved during collisions: momentum and energy as shown in (6.72) and (6.73) respectively. This means that we should take

$$\log(f_1^{\text{eq}}(\mathbf{x}, \mathbf{p})) = \beta \left(\mu_0 - E(\mathbf{p}) + \mathbf{u} \cdot \mathbf{p} \right) \qquad (6.110)$$

where $E(\mathbf{p}) = p^2/2m$ for non-relativistic particles and μ_0, β, and $\mathbf{u}$ are all constants. We'll adjust the constant μ to ensure that the overall normalisation of f_1 obeys (6.58). Then, writing $\mathbf{p} = m\mathbf{v}$, we have

$$f_1^{\text{eq}}(\mathbf{x}, \mathbf{p}) = \frac{N}{V} \left(\frac{\beta}{2\pi m} \right)^{3/2} e^{-\beta m (\mathbf{v} - \mathbf{u})^2 / 2} . \qquad (6.111)$$

This is the *Maxwell–Boltzmann distribution* that we met previously in (6.14). (It differs by a factor of $1/V$ because we need to integrate over d^3x, and also by a factor of m^3 because f_1^{eq} is a distribution over $\mathbf{p}$ while (6.14) is a distribution over $\mathbf{v}$.) We see that, while the Liouville term in the Boltzmann equations allows for many different equilibrium distributions, the collision term has forced upon us a rather special distribution (6.111).

The Maxwell–Boltzmann distribution depends on two constants, $\mathbf{u}$ and β. We've already seen in Section 6.1.1 that the coefficient β has the interpretation of inverse temperature

$$\beta = \frac{1}{k_B T} . \qquad (6.112)$$

As we saw previously, this identification ensures that the average kinetic energy of the gas is proportional to the temperature

$$\frac{1}{2} m \langle v^2 \rangle = \frac{3}{2} k_B T . \qquad (6.113)$$

The other constant $\mathbf{u}$ has a very simple meaning: it is just the average drift velocity of the atoms

$$\langle \mathbf{v} \rangle = \frac{1}{N} \int d^3x \, d^3p \, \mathbf{v} \, f_1^{\text{eq}}(\mathbf{x}, \mathbf{p}) = \mathbf{u} . \qquad (6.114)$$

This is a generalisation of the distribution that we introduced previously in (6.14), which had vanishing average drift velocity.

The Quantum Boltzmann Equation

Throughout this chapter (and, indeed, throughout this book), we have focussed only on classical systems. Here we break, very briefly, into the quantum world to make a quick observation.

One particular quantum novelty is that particles are indistinguishable and come in one of two types: bosons and fermions. This has an important implication for the scattering rate. We retain the assumption of molecular chaos, so $f_2 \sim f_1 f_1$ as in (6.78). But the scattering rate (6.70), for scattering $\mathbf{p}_1 + \mathbf{p}_2 \to \mathbf{p}_1' + \mathbf{p}_2'$, is replaced by

$$\text{rate} = \omega(\mathbf{p}, \mathbf{p}_2 | \mathbf{p}_1', \mathbf{p}_2') \, f_1(\mathbf{p}_1) f_1(\mathbf{p}_2)$$
$$\times \{1 \pm f_1(\mathbf{p}_1')\} \{1 \pm f_1(\mathbf{p}_2')\} \, d^3p_2 \, d^3p_1' \, d^3p_2' \ . \quad (6.115)$$

This expression now has two additional contributions, sitting in curly brackets, which depend on the distributions over the final momenta, $\mathbf{p}_1'$ and $\mathbf{p}_2'$. The $+$ sign is for bosons, and the $-$ sign for fermions.

The interpretation is particularly clear for fermions, where we can't have more than a single particle occupying any given quantum state. This means that it's not enough to know the probability that the initial state is filled. We also need to know the probability that the final state is free for the particle to scatter into, and that's what the $\{1 - f_1\}$ factors are telling us. It's a little harder to tell a similarly intuitive story for bosons, except to say that, if you're used to this game, then minus signs for fermions are very often replaced by plus signs for bosons. We'll give a proper derivation in the book on Statistical Physics.

The remaining arguments go forward as before, resulting in the quantum Boltzmann equation

$$\left(\frac{\partial f_1}{\partial t} \right)_{\text{coll}} = \int d^3p_2, d^3p_1' \, d^3p_2' \, \omega(\mathbf{p}_1', \mathbf{p}_2' | \mathbf{p}, \mathbf{p}_2)$$
$$\times \Big[f_1(\mathbf{p}_1') f_1(\mathbf{p}_2') \{1 \pm f_1(\mathbf{p})\} \{1 \pm f_1(\mathbf{p}_2)\}$$
$$- f_1(\mathbf{p}) f_1(\mathbf{p}_2) \{1 \pm f_1(\mathbf{p}_1')\} \{1 \pm f_1(\mathbf{x}, \mathbf{p}_2')\} \Big] \ . \quad (6.116)$$

We can again look for equilibrium solutions. The condition of detailed balance now becomes

$$\log \left(\frac{f_1^{\text{eq}}(\mathbf{p}_1')}{1 \pm f_1^{\text{eq}}(\mathbf{p}_1')} \right) + \log \left(\frac{f_1^{\text{eq}}(\mathbf{p}_2')}{1 \pm f_1^{\text{eq}}(\mathbf{p}_2')} \right)$$
$$= \log \left(\frac{f_1^{\text{eq}}(\mathbf{p})}{1 \pm f_1^{\text{eq}}(\mathbf{p})} \right) + \log \left(\frac{f_1^{\text{eq}}(\mathbf{p}_2)}{1 \pm f_1^{\text{eq}}(\mathbf{p}_2)} \right) \quad (6.117)$$

which is again solved by relating each log to a linear combination of the energy and momentum. We find

$$f_1^{\text{eq}}(\mathbf{p}) = \frac{1}{e^{-\beta(\mu - E(\mathbf{p}) + \mathbf{u} \cdot \mathbf{p})} \mp 1} \ . \quad (6.118)$$

These are famous distribution functions. For bosons, with a plus sign, it is called the *Bose–Einstein distribution*. For fermions, with a minus sign, it is called the *Fermi–Dirac distribution*. We won't discuss their properties here, and flag them up only to show how they arise from the Boltzmann equation. We will look at both distributions in detail in the book on Statistical Physics.

6.4 The H-Theorem

One of the key ideas that underpins thermodynamics and fluid mechanics is the concept of *equilibrium*. We assume that, if we have 10^{23} atoms flying around and we wait long enough, then, regardless of their starting point, they will eventually settle down to equilibrium. But how do we know this? Moreover, it seems that it would be rather tricky to prove because settling down to equilibrium clearly involves an arrow of time that distinguishes the future from the past. Yet the underlying classical mechanics is invariant under time reversal.

The purpose of this section is to demonstrate that, within the framework of the Boltzmann equation, systems do indeed settle down to equilibrium. As we described above, we have introduced an arrow of time into the Boltzmann equation. We didn't do this in any crude way like adding friction to the system. Instead, we merely assumed that particle velocities were uncorrelated before collisions. That would seem to be a rather minor input but, as we will now show, it's enough to demonstrate the approach to equilibrium.

Specifically, we will prove the *H-theorem*, named after a quantity H introduced by Boltzmann. (H is not to be confused with the Hamiltonian. Boltzmann originally called this quantity something like a German $\mathcal{E}$, but the letter was somehow lost in translation and the name H has stuck.) This quantity is

$$H(t) = \int d^3x \, d^3p \ f_1(\mathbf{x}, \mathbf{p}; t) \, \log(f_1(\mathbf{x}, \mathbf{p}; t)) \ . \tag{6.119}$$

In 1872, Boltzmann proved that H always decreases with time.

These days, the quantity H is rarely used. Instead, we multiply by a minus sign and a factor of Boltzmann's constant and use the quantity

$$S = -k_B H \ . \tag{6.120}$$

This is much more famous: it is the *entropy*, a definition that is due to Gibbs, who showed that the entropy for any probability distribution p_n can be defined as $S = -k_B \sum_n p_n \log p_n$. We will explore this definition further in the book on Statistical Physics. Here we prove Boltzmann's H-theorem: H always decreases. In other words, entropy always increases.

We start by differentiating H with respect to time

$$\frac{dH}{dt} = \int d^3x \, d^3p \, (\log f_1 + 1) \frac{\partial f_1}{\partial t} = \int d^3x \, d^3p \, \log f_1 \frac{\partial f_1}{\partial t} \, . \quad (6.121)$$

We can drop the $+1$ because $\int f_1 = N$ is unchanging, ensuring that $\int \partial f_1/\partial t = 0$. For the other terms, we replace $\partial f_1/\partial t$ with its expression in the Boltzmann equation (6.79). We have

$$\frac{dH}{dt} = \int d^3x \, d^3p \, \log f_1 \left(\frac{\partial V}{\partial \mathbf{x}} \cdot \frac{\partial f_1}{\partial \mathbf{p}} - \frac{\mathbf{p}}{m} \cdot \frac{\partial f_1}{\partial \mathbf{x}} + \left(\frac{\partial f_1}{\partial t} \right)_{\text{coll}} \right) \, . \quad (6.122)$$

But the first two terms in this expression both vanish. You can see this by integrating by parts twice, first moving the derivative away from f_1 and onto $\log f_1$, and then moving it back. So, for example

$$\int d^3x \, d^3p \, \log f_1 \frac{\partial V}{\partial \mathbf{x}} \cdot \frac{\partial f_1}{\partial \mathbf{p}} = - \int d^3x \, d^3p \, \frac{1}{f_1} \frac{\partial f_1}{\partial \mathbf{p}} \cdot \frac{\partial V}{\partial \mathbf{x}} f_1$$

$$= + \int d^3x \, d^3p \, f_1 \frac{\partial}{\partial \mathbf{p}} \cdot \frac{\partial V}{\partial \mathbf{x}} = 0 \, . \quad (6.123)$$

We learn that the change in H is governed entirely by the collision terms

$$\frac{dH}{dt} = \int d^3x \, d^3p \, \log f_1 \left(\frac{\partial f_1}{\partial t} \right)_{\text{coll}}$$

$$= \int d^3x \, d^3p_1 \, d^3p_2 \, d^3p_1' \, d^3p_2' \, \omega(\mathbf{p}_1', \mathbf{p}_2' | \mathbf{p}_1, \mathbf{p}_2)$$

$$\times \, \log f_1(\mathbf{p}_1) \left[f_1(\mathbf{p}_1') f_1(\mathbf{p}_2') - f_1(\mathbf{p}_1) f_1(\mathbf{p}_2) \right] \quad (6.124)$$

where I've suppressed $\mathbf{x}$ and t arguments of f_1 to keep things looking vaguely reasonable, but kept the momentum arguments as they differ in different terms. I've also relabelled the integration variable $\mathbf{p} \to \mathbf{p}_1$. At this stage, all momenta are integrated over so they are really nothing but dummy variables. Let's relabel $1 \leftrightarrow 2$ on the momenta. All the terms remain unchanged except the log. So we can also write

$$\frac{dH}{dt} = \int d^3x \, d^3p_1 \, d^3p_2 \, d^3p_1' \, d^3p_2' \, \omega(\mathbf{p}_1', \mathbf{p}_2' | \mathbf{p}_1, \mathbf{p}_2)$$

$$\times \, \log f_1(\mathbf{p}_2) \left[f_1(\mathbf{p}_1') f_1(\mathbf{p}_2') - f_1(\mathbf{p}_1) f_1(\mathbf{p}_2) \right] \, . \quad (6.125)$$

Adding (6.124) and (6.125), we have the more symmetric looking expression

$$\frac{dH}{dt} = \frac{1}{2} \int d^3x \, d^3p_1 \, d^3p_2 \, d^3p_1' \, d^3p_2' \, \omega(\mathbf{p}_1', \mathbf{p}_2' | \mathbf{p}_1, \mathbf{p}_2)$$

$$\times \, \log\left[f_1(\mathbf{p}_1) \, f_1(\mathbf{p}_2) \right] \left[f_1(\mathbf{p}_1') f_1(\mathbf{p}_2') - f_1(\mathbf{p}_1) f_1(\mathbf{p}_2) \right] . \quad (6.126)$$

With all momenta integrated over, we're allowed to just flip the dummy indices again. This time we swap $\mathbf{p} \leftrightarrow \mathbf{p}'$ in the above expression. But, the symmetry property (6.76) tells us that the scattering function remains unchanged

$$\omega(\mathbf{p}, \mathbf{p}_2 | \mathbf{p}_1', \mathbf{p}_2') = \omega(\mathbf{p}_1', \mathbf{p}_2' | \mathbf{p}, \mathbf{p}_2) . \quad (6.127)$$

(As an aside: we can actually get away with a weaker assumption at this step. Rather than assuming (6.127), it's enough if

$$\int d^3p_1' \, d^3p_2' \, \omega(\mathbf{p}_1', \mathbf{p}_2' | \mathbf{p}_1, \mathbf{p}_2) = \int d^3p_1' \, d^3p_2' \, \omega(\mathbf{p}_1, \mathbf{p}_2 | \mathbf{p}_1', \mathbf{p}_2') . \quad (6.128)$$

This follows from unitarity of the scattering matrix in quantum mechanics.)

Making the switch $\mathbf{p} \leftrightarrow \mathbf{p}'$ in the expression (6.126) gives

$$\frac{dH}{dt} = -\frac{1}{2} \int d^3x \, d^3p_1 \, d^3p_2 \, d^3p_1' \, d^3p_2' \, \omega(\mathbf{p}_1', \mathbf{p}_2' | \mathbf{p}_1, \mathbf{p}_2)$$

$$\times \log\left[f_1(\mathbf{p}_1') \, f_1(\mathbf{p}_2') \right] \left[f_1(\mathbf{p}_1') f_1(\mathbf{p}_2') - f_1(\mathbf{p}_1) f_1(\mathbf{p}_2) \right] . (6.129)$$

Finally, we add (6.126) and (6.129). We get the final expression

$$\frac{dH}{dt} = -\frac{1}{2} \int d^3x \, d^3p_1 \, d^3p_2 \, d^3p_1' \, d^3p_2' \, \omega(\mathbf{p}_1', \mathbf{p}_2' | \mathbf{p}_1, \mathbf{p}_2)$$

$$\times \left[\log\left[f_1(\mathbf{p}_1') \, f_1(\mathbf{p}_2') \right] - \log\left[f_1(\mathbf{p}_1) \, f_1(\mathbf{p}_2) \right] \right]$$

$$\times \left[f_1(\mathbf{p}_1') f_1(\mathbf{p}_2') - f_1(\mathbf{p}_1) f_1(\mathbf{p}_2) \right] . \quad (6.130)$$

It looks rather messy, but it's straightforward to extract the information that we need. The scattering rate on the first line is positive. Meanwhile, the rest of the integrand is a function of the form $(\log x - \log y)(x - y)$ and this too is positive for all values of x and y. This is all we need to prove the H-theorem

$$\frac{dH}{dt} \leq 0 . \quad (6.131)$$

And there we see the arrow of time seemingly emerging from time-invariant Hamiltonian mechanics!

The H-theorem is more familiar when expressed in terms of the entropy

$S = -k_B H$. If H can only decrease, the entropy can only increase. This is intimately related to the second law of thermodynamics.

Clearly, something is awry here. It should be impossible to start with time-invariant equations of motion and derive an arrow of time, a point first made by Loschmidt soon after Boltzmann's original derivation. But, as we saw earlier, everything hinges on the seemingly innocent assumption of molecular chaos (6.78). This was where we broke time-reversal symmetry, ultimately ensuring that entropy increases only in the future. Had we instead decided in (6.77) that the rate of scattering was proportional to f_2 *after* the collision, again assuming $f_2 \sim f_1 f_1$, then we would find that entropy always decreases as we move into the future.

The H-theorem is not a strict inequality. For some distributions, the entropy remains unchanged. From (6.130), we see that these obey

$$f_1(\mathbf{p}_1')f_1(\mathbf{p}_2') = f_1(\mathbf{p}_1)f_1(\mathbf{p}_2) . \tag{6.132}$$

But this is simply the requirement of detailed balance that we met previously in (6.108).

6.5 Towards the Equations of Fluid Mechanics

The narrative arc of this chapter takes us from the dynamics of 10^{23} interacting particles, through to the equations of fluid mechanics. So far, we have reached the Boltzmann equation. In this section, we take the next steps. We will complete our journey in Section 6.6 when we reach the Navier–Stokes equation.

Fluid mechanics emerges from the Boltzmann equation when we focus on quantities that vary slowly in space and time. As we will see, in the simplest case of a gas, there are only three such quantities. They are the following:

- The density $\rho(\mathbf{x}, t) = mn(\mathbf{x}, t)$. Here m is the mass of the atom or molecule and $n(\mathbf{x}, t)$ is the number density of particles that we already met briefly in (6.59). It is defined by

$$n(\mathbf{x}, t) = \int d^3p \, f_1(\mathbf{x}, \mathbf{p}, t) \tag{6.133}$$

and is normalised so that $\int d^3x \, n(\mathbf{x}, t) = N$, the total number of particles.

- The velocity $\mathbf{u}(\mathbf{x}, t)$. This is defined as the average velocity of the particles,

$$n(\mathbf{x}, t)\mathbf{u}(\mathbf{x}, t) = \int d^3p \, \frac{\mathbf{p}}{m} f_1(\mathbf{x}, \mathbf{p}; t) \ . \tag{6.134}$$

- The temperature $T(\mathbf{x}, t)$. This is defined as the average kinetic energy of the particles, *relative* to the background fluid flow

$$n(\mathbf{x}, t)T(\mathbf{x}, t) = \frac{1}{3k_B m} \int d^3p \, \mathbf{p}^2 f_1(\mathbf{x}, \mathbf{p}; t) \ . \tag{6.135}$$

We will see why these are the appropriate definitions shortly.

These three fields $n(\mathbf{x}, t)$, $\mathbf{u}(\mathbf{x}, t)$, and $T(\mathbf{x}, t)$ will play the role of our dynamical variables. Our goals in this section are to define these quantities more precisely, to understand why they are the relevant variables, and finally to derive the equations that govern their dynamics.

6.5.1 Conserved Quantities

We start by explaining why these are the variables of interest. The answer is that these are quantities which don't relax back down to their equilibrium value in an atomic blink of an eye, but instead change on a much slower, domestic time scale. At heart, this is because all three are associated to conserved quantities. Let's see why.

Consider a general function $A(\mathbf{x}, \mathbf{p})$ over the single-particle phase space. Because we live in real space instead of momentum space, the question of how things vary with $\mathbf{x}$ is more immediately interesting. For this reason, we integrate over momentum and define the average of a quantity $A(\mathbf{x}, \mathbf{p})$ to be

$$\langle A \rangle (\mathbf{x}, t) = \frac{\int d^3p \, A(\mathbf{x}, \mathbf{p}) f_1(\mathbf{x}, \mathbf{p}; t)}{\int d^3p' \, f_1(\mathbf{x}, \mathbf{p}'; t)}$$

$$= \frac{1}{n(\mathbf{x}, t)} \int d^3p \, A(\mathbf{x}, \mathbf{p}) f_1(\mathbf{x}, \mathbf{p}; t) \ . \tag{6.136}$$

This is different from the average that we defined earlier in (6.54) when discussing Liouville evolution. Here we're integrating only over momenta and the resulting average is a function of space.

We're interested in how the average of A changes with time. We looked at this kind of question for Liouville evolution earlier in this chapter and found the answer (6.56). Now we want to ask the same question when evolution is governed by the Boltzmann equation. Before we actually write down the answer, you can guess what it will look like: there will be a streaming term and a term due to the collision integral. Moreover, we know from our previous discussion that the term involving the collision integral will vary much faster than the streaming term.

Because we're ultimately interested in quantities that vary slowly, this motivates looking at functions A which vanish when integrated against the collision integral. The relevant criterion is

$$\int d^3p \, A(\mathbf{x}, \mathbf{p}) \left(\frac{\partial f_1}{\partial t} \right)_{\text{coll}} = 0 \ . \tag{6.137}$$

We'd like to find quantities A which have this property for any distribution f_1. Using our expression (6.79) for the collision integral, we want

$$\int d^3p_1 \, d^3p_2 \, d^3p_1' \, d^3p_2' \, \omega(\mathbf{p}_1', \mathbf{p}_2' | \mathbf{p}, \mathbf{p}_2) \, A(\mathbf{x}, \mathbf{p}_1)$$
$$\times \left[f_1(\mathbf{x}, \mathbf{p}_1') f_1(\mathbf{x}, \mathbf{p}_2') - f_1(\mathbf{x}, \mathbf{p}) f_1(\mathbf{x}, \mathbf{p}_2) \right] = 0 \ . \tag{6.138}$$

This now looks rather similar to equation (6.124), just with the log f replaced by A. Indeed, we can follow the steps between (6.124) and (6.129), using the symmetry properties of the scattering rate ω, to massage this into the form

$$\int d^3p_1 \, d^3p_2 \, d^3p_1' \, d^3p_2' \, \omega(\mathbf{p}_1', \mathbf{p}_2' | \mathbf{p}_1, \mathbf{p}_2) \left[f_1(\mathbf{p}_1') f_1(\mathbf{p}_2') - f_1(\mathbf{p}_1) f_1(\mathbf{p}_2) \right]$$
$$\times \left[A(\mathbf{x}, \mathbf{p}_1) + A(\mathbf{x}, \mathbf{p}_2) - A(\mathbf{x}, \mathbf{p}_1') - A(\mathbf{x}, \mathbf{p}_2') \right] = 0 \ . \tag{6.139}$$

Now it's clear that if we want this to vanish for all distributions, then A itself must have the property that it remains unchanged before and after the collision,

$$A(\mathbf{x}, \mathbf{p}_1) + A(\mathbf{x}, \mathbf{p}_2) = A(\mathbf{x}, \mathbf{p}_1') + A(\mathbf{x}, \mathbf{p}_2') \ . \tag{6.140}$$

Quantities which obey this are sometimes called *collisional invariants*. Of course, in the simplest situation we already know what they are: momentum (6.72) and energy (6.73) and, not forgetting, the trivial solution $A = 1$. We will turn to each of these in turn shortly. But first we derive an expression for the time evolution of any quantity obeying (6.140).

Take the Boltzmann equation (6.79), multiply by a collisional invariant $A(\mathbf{x}, \mathbf{p})$, and integrate over $\int d^3p$. Because the collision term vanishes, we have

$$\int d^3p \, A(\mathbf{x}, \mathbf{p}) \left(\frac{\partial}{\partial t} + \frac{\mathbf{p}}{m} \cdot \frac{\partial}{\partial \mathbf{x}} + \mathbf{F} \cdot \frac{\partial}{\partial \mathbf{p}} \right) f_1(\mathbf{x}, \mathbf{p}, t) = 0 \tag{6.141}$$

where we've put back a general external force $\mathbf{F} = -\nabla V$. We integrate the last term by parts, remembering that the force $\mathbf{F}$ can depend on position but not on momentum. We can't integrate the middle term by parts because we're not integrating over space, but nonetheless, we can rewrite it. Finally, because A has no explicit time dependence, we can take it inside the time derivative. The upshot is that we have

$$\frac{\partial}{\partial t} \int d^3 p \; Af + \frac{\partial}{\partial \mathbf{x}} \cdot \int d^3 p \; \frac{\mathbf{p}}{m} Af - \int d^3 p \; \frac{\mathbf{p}}{m} \cdot \frac{\partial A}{\partial \mathbf{x}} f - \int d^3 p \; \mathbf{F} \cdot \frac{\partial A}{\partial \mathbf{p}} f = 0 \; .$$

(6.142)

Although this doesn't really look like an improvement, the advantage of writing it in this way is apparent when we remember our expression for the average (6.136). Using this notation, we can write the evolution of A as

$$\frac{\partial}{\partial t} \left(n\langle A\rangle \right) + \frac{\partial}{\partial \mathbf{x}} \cdot \left(n\langle \mathbf{v}A\rangle \right) - n \left\langle \mathbf{v} \cdot \frac{\partial A}{\partial \mathbf{x}} \right\rangle - n \left\langle \mathbf{F} \cdot \frac{\partial A}{\partial \mathbf{p}} \right\rangle = 0 \quad (6.143)$$

where $\mathbf{v} = \mathbf{p}/m$. This is our master equation. It tells us how any collisional invariant changes. The next step is to look at specific quantities. There are three and we take each in turn.

Density

Our first collisional invariant is the trivial one, with $A = 1$. If we plug this into (6.143) we get the equation for the particle density $n(\mathbf{x}, t)$ or, multiplying by the mass m of the atom, the density $\rho(\mathbf{x}, t) = m \, n(\mathbf{x}, t)$

$$\frac{\partial \rho}{\partial t} + \nabla \cdot (\rho \mathbf{u}) = 0 \; .$$

(6.144)

We've returned to using $\nabla = \partial/\partial \mathbf{x}$, and the average velocity $\mathbf{u}(\mathbf{x}, t)$ of the particles is defined by

$$\mathbf{u}(\mathbf{x}, t) = \langle \mathbf{v} \rangle \; .$$

(6.145)

This, of course, is the continuity equation (1.9) that captures the conservation of mass. It's comforting to know that, almost 350 pages into the book, we can finally confirm the simplest equation that we kicked things off with!

Momentum

Our next collisional invariant is the momentum. We substitute $A = m\mathbf{v}$ into (6.143) to find

$$\frac{\partial}{\partial t}(mnu_i) + \frac{\partial}{\partial x_j} \left(mn\langle v_j v_i \rangle \right) - n\langle F_i \rangle = 0$$

(6.146)

where we've resorted to index notation, with $i = 1, 2, 3$ labelling the three spatial directions. We can play around with the middle term a little. We write

$$\langle v_j v_i \rangle = \langle (v_j - u_j)(v_i - u_i) \rangle + u_i \langle v_j \rangle + u_j \langle v_i \rangle - u_i u_j$$
$$= \langle (v_j - u_j)(v_i - u_i) \rangle + u_i u_j \qquad (6.147)$$

where, to go to the second line, we have used (6.145). We define the symmetric stress tensor

$$\sigma_{ij} = \sigma_{ji} = -\rho \langle (v_j - u_j)(v_i - u_i) \rangle . \qquad (6.148)$$

This is the same stress tensor that we met previously in Chapter 2 and we'll develop these connections as we go along. This tensor is computing the flux of i-momentum in the j-direction. With this definition, (6.146) becomes

$$\rho \left(\frac{\partial}{\partial t} + \mathbf{u} \cdot \nabla \right) u_i = \frac{\rho}{m} F_i + \frac{\partial}{\partial x_j} \sigma_{ij} . \qquad (6.149)$$

This is again a continuity equation, this time capturing the conservation of momentum. Indeed, if we use the mass conservation equation (6.144), we can rewrite (6.149) as

$$\frac{\partial \rho u_i}{\partial t} + \frac{\partial \Pi_{ij}}{\partial x_j} = \frac{\rho}{m} F_i \quad \text{with} \quad \Pi_{ij} = \rho u_i u_j - \sigma_{ij} . \qquad (6.150)$$

which coincides with the form of the Navier–Stokes equation that we saw in (2.25), albeit with a driving force $\mathbf{F}$ on the right-hand side. Of course, we know from Chapter 2 that the stress tensor σ_{ij} should include a pressure term and a viscosity term to reproduce the Navier–Stokes equation. They will both come shortly.

Kinetic Energy, also known as Temperature

Our final collisional invariant is the kinetic energy of the particles. However, rather than take the absolute kinetic energy, it is more convenient to work with the relative kinetic energy,

$$A = \frac{1}{2} m \left(\mathbf{v} - \mathbf{u} \right)^2 . \qquad (6.151)$$

We want to substitute this into the master equation (6.143). There is actually a subtlety here. In deriving the master equation (6.143), we assumed that A has no explicit time dependence, but the A defined in (6.151) does have explicit time dependence through $\mathbf{u}(\mathbf{x}, t)$. Nonetheless, you can check that (6.143) still holds, essentially because the extra term that you get is $\sim \langle (\mathbf{v} - \mathbf{u}) \cdot \partial \mathbf{u} / \partial t \rangle = \langle \mathbf{v} - \mathbf{u} \rangle \cdot \partial \mathbf{u} / \partial t = 0$.

The term involving the force in the master equation also vanishes (because $\langle \mathbf{v} - \mathbf{u} \rangle = 0$). However, the term that involves ∇A is not zero because the average velocity $\mathbf{u}$ depends on $\mathbf{x}$. We have

$$\frac{1}{2}\frac{\partial}{\partial t}\left(\rho\langle(\mathbf{v}-\mathbf{u})^2\rangle\right) + \frac{1}{2}\frac{\partial}{\partial x_i}\left(\rho\langle v_i(\mathbf{v}-\mathbf{u})^2\rangle\right) - \rho\left\langle v_i\frac{\partial u_j}{\partial x_i}(v_j-u_j)\right\rangle = 0 \ .$$

(6.152)

At this point, we define the *temperature* $T(\mathbf{x}, t)$ of our non-equilibrium system as proportional to the average, relative kinetic energy

$$\frac{3}{2}k_B T(\mathbf{x}, t) = \frac{1}{2}m\langle(\mathbf{v}-\mathbf{u}(\mathbf{x},t))^2\rangle \ .$$

(6.153)

This coincides with the temperature of an ideal gas (6.5) when $\mathbf{u} = 0$ so the background velocity vanishes. Note, also, that the temperature is something that is captured by the stress tensor (6.148), with $\operatorname{Tr}\sigma = -3\rho k_B T/m$.

We also define a new quantity which is cubic in velocities. This is the *heat flux*

$$q_i = \frac{1}{2}\rho\langle(v_i - u_i)\,(\mathbf{v}-\mathbf{u})^2\rangle \ .$$

(6.154)

The utility of both of these definitions becomes apparent if we play around with the middle term in (6.152). We can write

$$\frac{1}{2}\rho\langle v_i(\mathbf{v}-\mathbf{u})^2\rangle = \frac{1}{2}\rho\langle(v_i-u_i)\,(\mathbf{v}-\mathbf{u})^2\rangle + \frac{1}{2}\rho u_i\langle(\mathbf{v}-\mathbf{u})^2\rangle$$

$$= q_i + \frac{3}{2}nu_i k_B T \ .$$

(6.155)

Invoking the definition of the stress tensor (6.148), we can now rewrite (6.152) as

$$\frac{3}{2}\frac{\partial}{\partial t}(\rho k_B T) + \frac{\partial}{\partial x_i}\left(mq_i + \frac{3}{2}\rho u_i k_B T\right) - m\sigma_{ij}\frac{\partial u_j}{\partial x_i} = 0 \ .$$

(6.156)

Because $\sigma_{ij} = \sigma_{ji}$, we can replace $\partial u_j/\partial x_i$ in the last term with the symmetric rate of strain tensor that we previously met in (2.18)

$$E_{ij} = \frac{1}{2}\left(\frac{\partial u_i}{\partial x_j} + \frac{\partial u_j}{\partial x_i}\right) \ .$$

(6.157)

Finally, with a little help from the continuity equation (6.144), our expression for the conservation of energy becomes

$$\frac{\rho}{m}\left(\frac{\partial}{\partial t} + \mathbf{u}\cdot\nabla\right)k_B T + \frac{2}{3}\nabla\cdot\mathbf{q} - \frac{2}{3}E_{ij}\sigma_{ij} = 0 \ .$$

(6.158)

It's been a bit of a slog, but finally we have three equations describing how the particle density n (6.144), the velocity $\mathbf{u}$ (6.149), and the temperature T (6.158) change with time. It's worth stressing that these equations hold for any distribution f_1. We made no assumption that f_1 was something close to equilibrium. Our three equations are capturing nothing more than the underlying conservation laws for mass, momentum, and energy.

However, importantly, the set of three equations (6.144), (6.149), and (6.158) are not closed. The equation for n depends on $\mathbf{u}$; the equation for $\mathbf{u}$ depends on σ_{ij}, and the equation for T (which is related to the trace of σ_{ij}) depends on a new quantity, the heat flux $\mathbf{q}$. And to determine any of these, we need to solve the Boltzmann equation and compute the distribution f_1. But the Boltzmann equation is hard! How to proceed?

6.5.2 Local Equilibrium and the Euler Equation

We previously derived the Maxwell–Boltzmann distribution (6.111), describing the equilibrium state of a gas. It is

$$f_1^{\mathrm{eq}}(\mathbf{x}, \mathbf{p}) = n \left(\frac{1}{2\pi m k_B T} \right)^{3/2} e^{-m(\mathbf{v}-\mathbf{u})^2/2k_B T} \ . \tag{6.159}$$

Important as this distribution is, it's a little bit boring when it comes to describing fluid motion. That's because the distribution depends on neither time nor space, and so our three dynamical variables $n(\mathbf{x}, t)$, $\mathbf{u}(\mathbf{x}, t)$, and $T(\mathbf{x}, t)$ are all constant when evaluated on this distribution. Indeed, you can see their constant values n, $\mathbf{u}$, and T sitting as parameters of the distribution.

To get to equations that describe the motion of fluids, we want to look at distributions that solve the Boltzmann equation and are, in some sense, close to equilibrium. Recall that there are two terms in the Boltzmann equation: the streaming term and the collision term. But the time scales of these two terms are rather different, with the collision term inducing a fast change to something near equilibrium, and the streaming term then giving a more relaxed evolution. This suggests that if we're looking for a distribution that is close to equilibrium, we should work with something that vanishes when evaluated on the collision term, so

$$\left(\frac{\partial f_1}{\partial t} \right)_{\mathrm{coll}} = 0 \ . \tag{6.160}$$

We already saw that we can find solutions to this equation whenever the detailed balance condition (6.108) is met. Previously, requiring that both the streaming and collision terms vanish led us to the equilibrium distribution (6.159). But if we're only after the weaker condition of a vanishing collision term, then there is a much larger class of solutions to detailed balance. These solutions are again of the form (6.159), but now with the constants n, $\mathbf{u}$, and T promoted to functions of space and time,

$$f_1^{(0)}(\mathbf{x}, \mathbf{p}; t) = n(\mathbf{x}, t) \left(\frac{1}{2\pi m k_B T(\mathbf{x}, t)} \right)^{3/2}$$

$$\times \exp\left(-\frac{m}{2k_B T(\mathbf{x}, t)} (\mathbf{v} - \mathbf{u}(\mathbf{x}, t))^2 \right) . \qquad (6.161)$$

Distributions of this kind are said to be in *local equilibrium*. It's simple to check that, if you compute the average values using the definitions (6.133), (6.145), and (6.153), then they do indeed give the functions $n(\mathbf{x}, t)$, $\mathbf{u}(\mathbf{x}, t)$, and $T(\mathbf{x}, t)$ appearing in the distribution (6.161).

The distribution (6.161) is *not* a solution to the Boltzmann equation because, while the collision integral vanishes, the streaming term does not. Nonetheless, we will take it as our first approximation to the true solution and later see what we're missing.

As a first step, we can use the distribution (6.161) to compute the stress tensor σ_{ij} and heat flux $\mathbf{q}$. It's straightforward to see that it gives $\mathbf{q} = 0$ and

$$\sigma_{ij} = -k_B n(\mathbf{x}, t) T(\mathbf{x}, t) \, \delta_{ij} . \qquad (6.162)$$

If we write this stress tensor as

$$\sigma_{ij} = -P(\mathbf{x}, t) \, \delta_{ij} \qquad (6.163)$$

then we see that we get a local equation of state for the gas, with

$$P(\mathbf{x}, t) = k_B n(\mathbf{x}, t) T(\mathbf{x}, t) . \qquad (6.164)$$

This, of course, is the ideal gas equation of state, but now where the pressure P, the number density n, and the temperature T are all dynamical fields. We saw this same form previously in (3.153).

We can substitute the expressions for $\mathbf{q}$ and the stress tensor into the dynamical equations. The continuity equation (6.144) remains unchanged

$$\left(\frac{\partial}{\partial t} + \mathbf{u} \cdot \nabla \right) \rho + \rho \nabla \cdot \mathbf{u} = 0 . \qquad (6.165)$$

Meanwhile, the equation (6.149) governing the velocity flow becomes the Euler equation (1.29) describing fluid motion

$$\left(\frac{\partial}{\partial t} + \mathbf{u} \cdot \nabla \right) \mathbf{u} = \frac{1}{\rho} \nabla P + \frac{1}{m} \mathbf{F} . \qquad (6.166)$$

The final equation (6.158) describing the flow of heat reduces to

$$\left(\frac{\partial}{\partial t} + \mathbf{u} \cdot \nabla\right) T + \frac{2T}{3}\nabla \cdot \mathbf{u} = 0 \ . \tag{6.167}$$

This is the equation (3.199) that we promised long ago when discussing sound waves. (To compare, you should set $\gamma = 5/3$ in (3.199), reflecting the fact that we're dealing with a monatomic gas, with no internal degrees of freedom.) We don't yet have the term involving heat diffusivity. That will come soon.

The kinetic theory approach gives us the equations (6.165), (6.166), and (6.167) for a dilute gas, which is very much not incompressible. That's why we're seeing the equation governing temperature sitting on the same footing as the Euler equation while, for much of the earlier chapters of this book, we could largely ignore temperature and just focus on $\mathbf{u}(\mathbf{x}, t)$ and $P(\mathbf{x}, t)$. But, as we've seen in those earlier chapters, there is one thing that these equations are missing: dissipation. There is no irreversibility sown into these equations, no mechanism for the fluid to return to equilibrium.

In fact, we may have anticipated that these equations lack dissipation. The starting point was the local equilibrium distribution (6.161) but these obey detailed balance and, as we mentioned in (6.132), that's sufficient to show that the Boltzmann H-function is constant when evaluated on this distribution. That means that entropy does not increase. We can show this statement directly from the equations above. We can combine (6.165) and (6.166) to find

$$\left(\frac{\partial}{\partial t} + \mathbf{u} \cdot \nabla\right)(\rho T^{-3/2}) = 0 \tag{6.168}$$

which tells us that the quantity $\rho T^{-3/2}$ is constant along streamlines. But this is the requirement that motion along streamlines is adiabatic, not increasing the entropy. This too is a result that we anticipated previously in (3.200).

6.6 From 10^{23} Atoms to Navier–Stokes

We're making progress. We've succeeded in deriving the equations for an ideal fluid, including the Euler equation. But we're still missing dissipation.

In fact, it's clear what we need to do to find this. Our first guess for the distribution function took the form of local equilibrium

$$f_1^{(0)}(\mathbf{x}, \mathbf{p}; t) = n(\mathbf{x}, t) \left(\frac{1}{2\pi m k_B T(\mathbf{x}, t)} \right)^{3/2}$$
$$\times \exp \left(-\frac{m}{2k_B T(\mathbf{x}, t)} [\mathbf{v} - \mathbf{u}(\mathbf{x}, t)]^2 \right) . \qquad (6.169)$$

We chose this on the grounds that it gives a vanishing contribution to the collision integral. But we never worried about the fact that it doesn't actually solve the Boltzmann equation because it doesn't vanish when evaluated on the streaming term. Now we worry.

Using the definition of the Poisson bracket and the one-particle Hamiltonian H_1 (6.63), the Liouville evolution of the distribution is given by

$$\frac{\partial f_1^{(0)}}{\partial t} - \{H_1, f_1^{(0)}\} = \frac{\partial f_1^{(0)}}{\partial t} + \mathbf{F} \cdot \frac{\partial f_1^{(0)}}{\partial \mathbf{p}} + \mathbf{v} \cdot \frac{\partial f_1^{(0)}}{\partial \mathbf{x}} . \qquad (6.170)$$

Now the dependence on $\mathbf{p} = m\mathbf{v}$ in local equilibrium is easy: it is simply

$$\frac{\partial f_1^{(0)}}{\partial \mathbf{p}} = -\frac{1}{k_B T}(\mathbf{v} - \mathbf{u}) f_1^{(0)} . \qquad (6.171)$$

Meanwhile all $\mathbf{x}$ dependence and t dependence of $f_1^{(0)}$ lies in the functions $n(\mathbf{x}, t)$, $T(\mathbf{x}, t)$, and $\mathbf{u}(\mathbf{x}, t)$. From (6.169) we have

$$\frac{\partial f_1^{(0)}}{\partial n} = \frac{f_1^{(0)}}{n}$$
$$\frac{\partial f_1^{(0)}}{\partial T} = -\frac{3}{2} \frac{f_1^{(0)}}{T} + \frac{m}{2k_B T^2}(\mathbf{v} - \mathbf{u})^2 f_1^{(0)} \qquad (6.172)$$
$$\frac{\partial f_1^{(0)}}{\partial \mathbf{u}} = \frac{m}{k_B T}(\mathbf{v} - \mathbf{u}) f_1^{(0)} .$$

Using all these relations, we get

$$\frac{\partial f_1^{(0)}}{\partial t} - \{H_1, f_1^{(0)}\} = \left[\frac{1}{n} \tilde{D}_t n + \left(\frac{m(\mathbf{v} - \mathbf{u})^2}{2k_B T^2} - \frac{3}{2T} \right) \tilde{D}_t T \right.$$
$$\left. + \frac{m}{k_B T}(\mathbf{v} - \mathbf{u}) \cdot \tilde{D}_t \mathbf{u} - \frac{1}{k_B T} \mathbf{F} \cdot (\mathbf{v} - \mathbf{u}) \right] f_1^{(0)} \quad (6.173)$$

where we've introduced the notation $\tilde{D}_t$ which differs from the more familiar material derivative D_t in that it depends on the microscopic velocity $\mathbf{v}$ rather than the average velocity $\mathbf{u}$,

$$\tilde{D}_t = \frac{\partial}{\partial t} + \mathbf{v} \cdot \frac{\partial}{\partial \mathbf{x}} = D_t + (\mathbf{v} - \mathbf{u}) \cdot \frac{\partial}{\partial \mathbf{x}} . \qquad (6.174)$$

Our first attempt at deriving the equations of fluid mechanics gave us three equations describing how n (6.165), $\mathbf{u}$ (6.166), and T (6.167) change with time. We substitute these into (6.173). You need a couple of lines of algebra, cancelling some terms, using the relationship $P = nk_BT$, and the definition of E_{ij} in (6.157), but it's not hard to show that we ultimately get

$$\frac{\partial f_1^{(0)}}{\partial t} - \{H_1, f_1^{(0)}\} = \left[\frac{1}{T}\left(\frac{m}{2k_BT}(\mathbf{v} - \mathbf{u})^2 - \frac{5}{2}\right)(\mathbf{v} - \mathbf{u})\cdot\nabla T \right.$$
$$\left. + \frac{m}{k_BT}\left((v_i - u_i)(v_j - u_j) - \frac{1}{3}(\mathbf{v} - \mathbf{u})^2\delta_{ij}\right)E_{ij}\right]f_1^{(0)} .$$

$$(6.175)$$

If the distribution $f_1^{(0)}$ solved the Boltzmann equation, then the right-hand side would vanish. But there's no reason for this to be the case.

Nonetheless, there's reason to think that all is not lost. The terms on the right-hand side depend on ∇T and $\partial\mathbf{u}/\partial\mathbf{x}$ (the latter through E_{ij}). That means that if we stick to long wavelength variations in the temperature and velocity then we almost have a solution. And it opens up the possibility of building an actual solution to the Boltzmann equation perturbatively, where we first add the additional term

$$f_1 = f_1^{(0)} + \delta f_1 . \qquad (6.176)$$

We would next like to see how this correction term changes things.

6.6.1 Relaxation Time Approximation

The correction term δf_1 gives a contribution to the Boltzmann equation from the collision integral (6.80). Dropping the $\mathbf{x}$ argument due to space limitations, we have

$$\left(\frac{\partial f_1}{\partial t}\right)_{\text{coll}} = \int d^3p_2\, d^3p_1'\, d^3p_2'\, \omega(\mathbf{p}_1', \mathbf{p}_2'|\mathbf{p}_1, \mathbf{p}_2)\left[f_1(\mathbf{p}_1')f_1(\mathbf{p}_2') - f_1(\mathbf{p}_1)f_1(\mathbf{p}_2)\right]$$
$$= \int d^3p_2\, d^3p_1'\, d^3p_2'\, \omega(\mathbf{p}_1', \mathbf{p}_2'|\mathbf{p}_1, \mathbf{p}_2)\left[f_1^{(0)}(\mathbf{p}_1')\delta f_1(\mathbf{p}_2')\right.$$
$$\left. + \delta f(\mathbf{p}_1')f_1^{(0)}(\mathbf{p}_2') - f_1^{(0)}(\mathbf{p}_1)\delta f_1(\mathbf{p}_2) - \delta f(\mathbf{p}_1)f_1^{(0)}(\mathbf{p}_2)\right]$$

where, in the second line, we have used the fact that $f_1^{(0)}$ vanishes in the collision integral and ignored quadratic terms $\sim \delta f_1^2$. The resulting collision integral is a linear function of δf_1. But it's still kind of a mess and not easy to play with.

At this point, there are a couple of different ways in which we could proceed. This first involves taking more care in the expansion of δf_1 (using what is known as the Chapman–Enskog expansion) and then treating the linear operator above correctly.

The second is just to make something up. We're going to take this route on the grounds that it's significantly easier.

We replace the complicated expression above with something much simpler that captures the relevant physics. We take

$$\left(\frac{\partial f_1}{\partial t}\right)_{\text{coll}} = -\frac{\delta f_1}{\tau} \tag{6.177}$$

where τ is the *relaxation time*, which we met previously in (6.68). This time scale governs the rate of change of f_1.

The choice of operator (6.177) is called the *relaxation time approximation*. It is most certainly not exact, but rather a slightly cheap model which will help us build some intuition for what's going on in an otherwise complicated setting. The key idea stems from the solution to the simple differential equation

$$\frac{\partial \delta f_1}{\partial t} = -\frac{\delta f_1}{\tau} \quad \Longrightarrow \quad \delta f_1 \sim e^{-t/\tau} . \tag{6.178}$$

This suggests that the right-hand side in (6.177) will force the distribution towards equilibrium in something close to the relaxation time τ.

The relaxation time approximation is not without its flaws. Most importantly, it does not preserve the collisional invariants that are the basis for our conserved quantities. Nonetheless, we will see that it gives physically reasonable answers.

In the relaxation time approximation, the Boltzmann equation becomes

$$\frac{\partial (f_1^{(0)} + \delta f_1)}{\partial t} - \{H_1, f_1^{(0)} + \delta f_1\} = -\frac{\delta f_1}{\tau} . \tag{6.179}$$

The correction to the distribution $\delta f_1 \ll f_1^{(0)}$, so we replace $f_1^{(0)} + \delta f_1 \approx f_1^{(0)}$ on the left-hand side. But we don't have this option on the right-hand side. Then, using (6.175), we find a simple expression for the correction to the distribution,

$$\delta f_1 = -\tau \left[\frac{1}{T} \left(\frac{m}{2k_BT} (\mathbf{v} - \mathbf{u})^2 - \frac{5}{2} \right) (\mathbf{v} - \mathbf{u}) \cdot \frac{\partial T}{\partial \mathbf{x}} \right.$$
$$\left. + \frac{m}{k_BT} \left((v_i - u_i)(v_j - u_j) - \frac{1}{3}(\mathbf{v} - \mathbf{u})^2 \delta_{ij} \right) E_{ij} \right] f_1^{(0)} \; . (6.180)$$

By construction, this correction is proportional to the relaxation time τ. We can now use this small correction to revisit our fluid equations of motion.

6.6.2 Thermal Conductivity

We start by computing the heat flux, defined by

$$q_i = \frac{1}{2} \rho \langle (v_i - u_i) \, (\mathbf{v} - \mathbf{u})^2 \rangle \; . \tag{6.181}$$

We evaluate this on the corrected distribution (6.176). We've already seen that the local equilibrium distribution $f_1^{(0)}$ gives a vanishing contribution so we can focus on δf_1. As shown in (6.180), δf_1 has two terms, the first odd in velocities, the second even. Meanwhile, the expression for $\mathbf{q}$ is itself odd in velocities. When we compute the average, we integrate over momenta (or, equivalently, over velocities), so only terms that are even survive. This means that only the first term in (6.180) contributes to the heat flux. We have

$$\mathbf{q} = -\kappa \nabla T \; . \tag{6.182}$$

The coefficient κ is known as the *thermal conductivity* and is given by

$$\kappa = \frac{\tau \rho}{2T} \int d^3p \, (\mathbf{v}_i - \mathbf{u}_i)^2 (\mathbf{v} - \mathbf{u})^2 \left[\frac{m}{2k_BT}(\mathbf{v} - \mathbf{u})^2 - \frac{5}{2} \right] f_1^{(0)}$$
$$= \frac{\tau \rho}{6T} \left[\frac{m}{2k_BT} \langle v^6 \rangle_0 - \frac{5}{2} \langle v^4 \rangle_0 \right] \; . \tag{6.183}$$

In the second line, we've replaced all $(v - u)$ factors with v by performing an ($\mathbf{x}$-dependent) shift of the integration variable. The subscript on $\langle \cdot \rangle_0$ means that these averages are to be taken in the local Maxwell–Boltzmann distribution $f_1^{(0)}$ with $u = 0$. These integrals are simple to perform. We have $\langle v^4 \rangle_0 = 15k_B^2 T^2/m^2$ and $\langle v^6 \rangle_0 = 105k_B^3 T^3/m^3$, giving

$$\kappa = \frac{5}{2} \frac{\tau n}{m} k_B^2 T \; . \tag{6.184}$$

The factor of $5/2$ here has followed us throughout the calculation. Its presence in this formula can be traced to the fact that, for a monatomic gas, the specific heat at constant pressure is $c_P = \frac{5}{2}k_B$.

This result is parametrically the same as we found earlier in (6.42). (To see this, you need to replace $\tau \sim l/\bar{v}_{\rm rel}$ and $\bar{v}_{\rm rel} = \sqrt{6k_BT/m}$.) While the parametric dependence coincides, the overall numerical factors do not, but it is not really to be trusted here, not least because the only definition of τ that we have is in the implementation of the relaxation time approximation.

We can substitute the expression for the heat flux $\mathbf{q}$ into the equation (6.158) governing the evolution of temperature. To highlight some physics, we can look at this equation for a static fluid, with $\mathbf{u} = 0$, which means that we turn off advection. We assume that we can ignore variations in the heat conductivity, so that $\nabla\kappa \approx 0$. We then have

$$\frac{\rho k_B}{m}\frac{\partial T}{\partial t} = \frac{2}{3}\kappa\nabla^2 T \ . \tag{6.185}$$

This is the *heat equation*.

We can compare this to our earlier incarnations of the heat equation from Sections 3.4 and 4.4. There we worked with the thermal diffusivity, α, defined as

$$\frac{\partial T}{\partial t} = \alpha\nabla^2 T \ . \tag{6.186}$$

We see that this thermal diffusivity coefficient can be expressed in terms of the heat conductivity κ

$$\alpha = \frac{2}{3}\frac{\kappa m}{\rho k_B} = \frac{5}{3}\frac{\kappa}{n c_P} \ . \tag{6.187}$$

6.6.3 Viscosity Revisited

Next, we turn to the viscosity. We know from our discussion of Couette flow in Section 2.2 that the relevant experimental set-up to measure viscosity is a fluid with a velocity gradient, $\partial u_x/\partial z \neq 0$. The shear viscosity is associated to the flux of x-momentum in the z-direction. But this is precisely what is computed by the off-diagonal component of the stress tensor,

$$\sigma_{xz} = -\rho\langle(v_x - u_x)(v_z - u_z)\rangle \ . \tag{6.188}$$

We've already seen that the local equilibrium distribution gives a diagonal stress tensor (6.163), corresponding to vanishing viscosity. What happens if we use the corrected distribution (6.176)?

Now only the second term in (6.180) contributes (because the first term is an odd function of $(\mathbf{v} - \mathbf{u})$). We write the stress tensor as

$$\sigma_{ij} = -P\,\delta_{ij} + \sigma'_{ij} \tag{6.189}$$

where the extra term is given by

$$
\sigma'_{ij} = -\frac{m\tau\rho}{k_B T} E_{kl} \int d^3p \, (v_i - u_i)(v_j - u_j)
$$

$$
\times \left((v_k - u_k)(v_l - u_l) - \frac{1}{3}(\mathbf{v} - \mathbf{u})^2 \delta_{kl} \right) f_1^{(0)}
$$

$$
= -\frac{m\tau\rho}{k_B T} E_{kl} \left[\langle v_i v_j v_k v_l \rangle_0 - \frac{1}{3} \delta_{kl} \langle v_i v_j v^2 \rangle_0 \right] . \tag{6.190}
$$

Before we compute σ'_{ij}, note that it is a traceless tensor. This is because the first term above becomes $\langle v^2 v_k v_l \rangle_0 = \delta_{kl} \langle v^2 v_x^2 \rangle_0$ which is easily calculated to be $\langle v^2 v_x^2 \rangle_0 = 5 k_B^2 T^2 / m^2 = \frac{1}{3} \langle v^4 \rangle_0$. Moreover, σ'_{ij} depends linearly on the rate of strain tensor E_{ij} defined in (6.157). These two facts mean that the stress tensor σ_{ij} must be of the form

$$
\sigma_{ij} = -P\delta_{ij} + 2\mu \left(E_{ij} - \frac{1}{3} \delta_{ij} \nabla \cdot \mathbf{u} \right) \tag{6.191}
$$

for some coefficient μ that we recognise as the viscosity. This coincides with the form of the stress tensor that we met previously in (2.16). Note, however, that unlike in (2.16), there is no bulk viscosity term $\zeta \nabla \cdot \mathbf{u} \, \delta_{ij}$. This term didn't play a role for much of this book because we were dealing with incompressible fluids. Now we're in the world of compressible gases, it could certainly arise. But we learn that, for dilute gases, we have $\zeta = 0$, at least within the relaxation time approximation.

To compute the viscosity μ, we evaluate the off-diagonal term in the stress tensor

$$
\sigma_{xz} = -\frac{m\tau\rho}{k_B T} E_{kl} \left[\langle v_x v_z v_k v_l \rangle_0 - \frac{1}{3} \delta_{kl} \langle v_x v_z v^2 \rangle_0 \right]
$$

$$
= -\frac{m\tau\rho}{k_B T} (E_{xz} + E_{zx}) \langle v_x v_z v_x v_z \rangle_0
$$

$$
= -\frac{2m\tau\rho}{k_B T} E_{xz} \langle v_x^2 \rangle_0 \langle v_z^2 \rangle_0 . \tag{6.192}
$$

But $\langle v_x^2 \rangle_0 = \langle v_z^2 \rangle_0 = \frac{1}{3} \langle v^2 \rangle_0 = k_B T / m$. Comparing to (6.191), we can read off the viscosity in terms of the relaxation time

$$
\mu = n k_B T \tau . \tag{6.193}
$$

This, again, has the same parametric dependence as the result (6.35) that we derived previously using more down-to-earth arguments.

We can now put all the pieces together. Our previous equation (6.149) describing momentum conservation now has an extra contribution from the stress tensor (6.189). As with the heat conductivity, we assume that spatial variations in the viscosity can be neglected, so $\nabla\mu \approx 0$. We're then left with

$$\left(\frac{\partial}{\partial t} + \mathbf{u}\cdot\nabla\right)\mathbf{u} = \frac{\mathbf{F}}{m} - \frac{1}{\rho}\nabla P + \frac{\mu}{\rho}\nabla^2\mathbf{u} + \frac{\mu}{3\rho}\nabla(\nabla\cdot\mathbf{u}) \ . \qquad (6.194)$$

This, of course, is the *Navier–Stokes* equation. Note that we include an extra term proportional to $\nabla\cdot\mathbf{u}$ that we largely ignored in earlier chapters because we were dealing with incompressible fluids.

In addition, we have the heat conduction equation (6.158). We again drop some terms on the grounds that they are small. This time, we set $\nabla\kappa \approx 0$ and $E_{ij}\sigma'_{ij} \approx 0$ since both are small at the order we are working to. We're left with

$$\frac{\rho}{m}\left(\frac{\partial}{\partial t} + \mathbf{u}\cdot\nabla\right)T - \frac{2}{3}\kappa\nabla^2 T + \frac{2}{3k_B}P\,\nabla\cdot\mathbf{u} = 0 \ . \qquad (6.195)$$

This is the equation (3.202) that we saw previously when discussing sound waves. (To compare the two equations, you need to use the ideal gas law $mP = \rho k_B T$.)

Briefly, Viscosity of Liquids

Our discussion in this chapter has focussed entirely on dilute gases, where the separation of scales between the mean free path λ and the atomic distance $d \ll \lambda$ is, ultimately, what allows us to make analytic progress.

For liquids, there is no such simplification, and the mean free path is comparable to the atomic scale. We can be bold and use our previous formulae in this case, setting $\lambda \approx d$. This suggests that the viscosity of a liquid will be roughly

$$\nu \sim c_s d \qquad (6.196)$$

with c_s the speed of sound, and this indeed gives a ballpark figure for the viscosity of most liquids, as we noted previously in Section 2.3.

There is also an interesting story with the temperature dependence. As we've seen, for gases the viscosity increases with temperature, which can be traced to the fact that the molecules move faster at higher temperature, and hence exchange more momentum.

In contrast, for liquids the viscosity decreases with temperature. This is because the molecules are much more closely packed, held in place by intermolecular forces which give rise to the friction between different layers. As the temperature increases, the molecules again move faster, but this time can more easily wriggle free from the forces that bind them. This means that the viscosity will typically exhibit a minimum value.

The data for the kinematic viscosity of oxygen O_2, which is fairly typical, is shown in the figure. This data is taken at the high pressure of 30 MPa, which is above the critical point of oxygen (which sits at 155 K and 5 MPa) to ensure that there is no phase transition between the liquid and the gas (see the book on Statistical Physics for an explanation of this point).

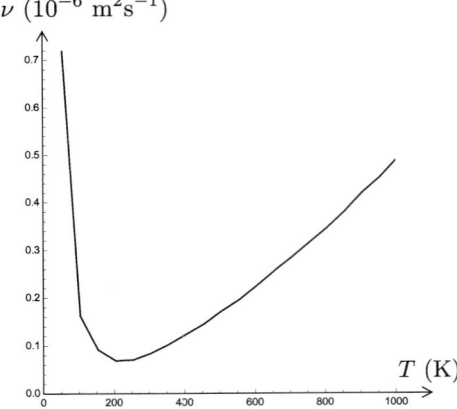

7   Diffusion

In this chapter, we continue our study of fluids from the microscopic perspective. We're going to ask how objects move in a fluid. Of course, we've already done this in earlier chapters where we looked at macroscopic phenomena, such as the lift and drag forces. But now we go smaller. We're interested in motion where the constant bombardment by the underlying atoms or molecules is important.

We will be interested in questions like how a blob of ink spreads in a glass of water. Or, more generally, how a particular kind of particle makes its way through a background liquid or gas. These kinds of phenomena are governed by a process known as *diffusion*.

As we will see, key to diffusion is an appreciation that the underlying dynamics has a random element. The particle doesn't just go directly from point X to point Y. Instead, it goes via many other points in between. This means that diffusive processes are slow. In Newtonian mechanics, in time t a particle travelling with constant velocity will travel a distance

$$L \sim t \ . \tag{7.1}$$

In the present context, such motion is sometimes referred to as *ballistic*. (Always with the military analogies in physics!) In contrast, diffusive processes spread out at a characteristic rate

$$L \sim \sqrt{t} \ . \tag{7.2}$$

It's like the difference between a student learning physics by reading a well-crafted set of textbooks and a student reading Wikipedia.

We've mentioned aspects of diffusion previously in this book. For example, in Chapter 2 we stressed that viscosity is a measure of how quickly momentum diffuses. Our purpose in this chapter is to put these comments on a firmer footing. We start by describing a few basic features of diffusion, deriving the diffusion equation, also known as the heat equation, as a consequence of the underlying randomness. We then look more closely at this randomness by introducing the idea of a *stochastic differential equation*, in which the driving force is not a known function, but a random variable.

Diffusion as a Random Walk

A random walk provides a particularly simple model that gives rise to motion proceeding at a rate $\sqrt{t}$.

Consider a lattice which, for now, we take to be one-dimensional. To make contact with our discussion in Chapter 6, we could think of the spacing between the lattice sites as set by the mean free path, λ. After a time, τ, the particle jumps either left or right. The direction of the jump is entirely random: 50% of the time it goes left, 50% right. This model is known as a *random walk*.

The particle starts at the origin and we want to know the probability $P(x, t)$ that it sits at $x = p\lambda$ at time $t = N\tau$, with $p, N \in \mathbb{Z}$. We start by giving a simple combinatoric derivation of the answer. For simplicity, we take $N + p$ to be even and we'll look at $p \ll N$.

To get to $x = p\lambda$, the particle must have made $\frac{1}{2}(N + p)$ forward jumps and $\frac{1}{2}(N - p)$ backward jumps. The probability is just the number of different ways we can do this, divided by 2^N, the total number of possible combinations, or

$$P(x, t) = \frac{2^{-N} N!}{[\frac{1}{2}(N + p)]! [\frac{1}{2}(N - p)]!} \ . \tag{7.3}$$

We replace the factorials with Stirling's approximation which, for large N, says

$$\log N! = N \log N - N + \frac{1}{2} \log 2\pi N + \mathcal{O}(1/N) \ . \tag{7.4}$$

If we further expand to leading order in p/N, we have

$$P(x, t) \approx \sqrt{\frac{2}{\pi N}} e^{-p^2/2N} = \sqrt{\frac{2\tau}{\pi t}} e^{-x^2 \tau/2\lambda^2 t} \ . \tag{7.5}$$

We see that the probability distribution of the particle is an ever-spreading Gaussian ensemble. The mean is simply $\langle x \rangle = 0$, reflecting the fact that the particle is equally likely to travel forwards as backwards. The variance is

$$\langle x^2 \rangle = \frac{\lambda^2}{\tau} t \ . \tag{7.6}$$

The root-mean-square (rms) distance travelled by the particle grows as $\sqrt{\langle x^2 \rangle} \sim \sqrt{t}$. This is the characteristic behaviour of random walks.

It is simple to repeat our analysis for a random walk in three dimensions.

For a cubic lattice, we assume that the motion in each of the directions is independent and equally likely. On average, the particle moves in the x-direction only every 3τ, so (7.6) should be replaced by $\langle x^2 \rangle = \lambda^2 t / 3\tau$. But this means that

$$\langle \mathbf{x}^2 \rangle = \langle x^2 \rangle + \langle y^2 \rangle + \langle z^2 \rangle = \frac{\lambda^2}{\tau} t \qquad (7.7)$$

and the total rms distance covered remains unchanged.

7.1 The Heat Equation

We now repeat the random walk experiment but with many particles – say, N – rather than one, moving in a volume V. After some time, we can talk about a density of particles $n = N/V$, that is a function of both time and space. With this formulation, we can recast the random walk in terms of a differential equation for $n(\mathbf{x}, t)$.

We again start with the one-dimensional situation. Consider the density $n(x, t)$ at some fixed time. We would like to understand what the density is at the same point x a short time Δt later. Some particles will leave, while others will come in to replace them. Those particles that come in must have been sitting at a nearby position $x + \Delta x$ at time t. Here we will think of Δx as a random variable that may be positive or negative but with mean

$$\langle \Delta x \rangle = 0 \;. \qquad (7.8)$$

We can then write an expression for the density at time $t + \Delta t$ as an average over all the different Δx,

$$n(t + \Delta t, x) = \langle n(t, x + \Delta x) \rangle$$
$$= n(t, x) + \frac{\partial n}{\partial x} \langle \Delta x \rangle + \frac{1}{2} \frac{\partial^2 n}{\partial x^2} \langle \Delta x^2 \rangle + \dots \;. \qquad (7.9)$$

The first order derivative vanishes, by virtue of (7.8), leaving the second order derivative as the leading term. We can similarly Taylor expand the left-hand side, but this time we keep the first order time derivative. The upshot is that we get the *heat equation*

$$\frac{\partial n}{\partial t} = D \frac{\partial^2 n}{\partial x^2} \qquad (7.10)$$

where the diffusion constant is $D = \langle \Delta x^2 \rangle / 2\Delta t$. We will also refer to this as the *diffusion equation*.

Returning to our dilute gas, with mean free path λ and scattering time τ, on dimensional grounds we must have

$$D \sim \frac{\lambda^2}{\tau} \,. \tag{7.11}$$

We will shortly look at solutions to the diffusion equation where we will show that they evolve so as to iron out any inhomogeneities in particle density. As an example, suppose that all N particles start out life sitting at the origin, giving us the initial condition $n(x, t = 0) = N\delta(x)$. The solution to the diffusion equation with this initial condition is an ever-spreading Gaussian

$$n(x,t) = N\sqrt{\frac{1}{4\pi Dt}}e^{-x^2/4Dt} \,. \tag{7.12}$$

This reproduces the discretised result (7.5). If we think of the rms distance travelled as the width of the spreading cloud of particles, we again have the $\sqrt{t}$ scaling, with $\langle x^2 \rangle = 2Dt$.

It is simple to extend the derivation above to three dimensions. Going through the same steps, we now find the 3d diffusion equation,

$$\frac{\partial n}{\partial t} = D\nabla^2 n \,. \tag{7.13}$$

We again expect that $D \sim \lambda^2/\tau$, although the overall numerical factor is not necessarily the same as in the 1d case. (A simple analysis suggests that it is a factor of 3 less.) The Gaussian again provides a solution, now with

$$\langle \mathbf{x}^2 \rangle = 6Dt \,. \tag{7.14}$$

We again see the characteristic $\sqrt{t}$ scaling.

7.1.1 Conserved Quantities Diffuse

There was a hidden assumption in our discussion above. In the random walk, the particle moved either to the left or to the right. It didn't just disappear. In other words, the number of particles is conserved.

This conservation law underpins diffusion. While there's no law of physics that says that the amount of ink in the universe has been the same since the Big Bang, if you drop a blob of ink in a glass of water then the amount of ink doesn't change. It just spreads out.

But we have other, more fundamental conservation laws in nature. In particular, both energy and momentum are conserved and, moreover, they are conserved *locally*. This means that if the energy density $\mathcal{E}(\mathbf{x}, t)$ in some region of space changes, then it's because the energy has moved to a nearby region of space. This is captured by a continuity equation

$$\frac{\partial \mathcal{E}}{\partial t} + \nabla \cdot \mathbf{q} = 0 \ . \tag{7.15}$$

with $\mathbf{q}$ the *heat flow* which, itself, is determined by temperature differences

$$\mathbf{q} = -\kappa \nabla T \ . \tag{7.16}$$

Here κ the *thermal conductivity*. At constant pressure, the temperature and energy are related by $\mathcal{E} = c_P T$, with c_P the relevant heat capacity. Putting it together, we see that these two equations imply diffusion of temperature

$$\frac{\partial T}{\partial t} = -\frac{1}{c_P} \nabla \cdot \mathbf{q} = \alpha \nabla^2 T \tag{7.17}$$

where, in the second equality, we assumed that the thermal conductivity κ is constant, and so escapes the clutches of ∇ to sit inside the diffusion constant $\alpha = \kappa/c_P$. In Chapter 6 (see, for example, (6.42)) we saw that the thermal conductivity, and hence the diffusion constant, scales as $\alpha \sim n\lambda \langle v \rangle \sim n\lambda^2/\tau$ in agreement with (7.11).

The name "heat equation" arises because, as we see above, it describes the evolution of temperature. In addition, equation (7.16) is known as *Fick's first law* and (7.17) as *Fick's second law*.

There is a similar story for momentum density $\mathbf{p} = \rho \mathbf{u}$, with ρ the density and $\mathbf{u}$ the velocity. In index notation, the continuity equation reads

$$\frac{\partial p_i}{\partial t} + \frac{\partial \Pi_{ji}}{\partial x_j} = 0 \tag{7.18}$$

where Π_{ji} is the momentum current, describing the flux of the p_i momentum in the jth direction. We saw this form of the continuity equation previously in (1.34) and (2.25) and, when discussing kinetic theory, in (6.150), each with increasingly complicated expressions for Π_{ji}. For a fluid, the full expression for Π_{ji} yields the Navier–Stokes equation.

For our purposes, we focus just on the viscosity term (see (2.25)) which gives

$$\Pi_{ij} = -\mu \left(\frac{\partial u_i}{\partial x_j} + \frac{\partial u_j}{\partial x_i} \right) \ . \tag{7.19}$$

For an incompressible fluid, with ρ constant and $\nabla \cdot \mathbf{u} = 0$, the continuity equation becomes

$$\rho \frac{\partial \mathbf{u}}{\partial t} = \mu \nabla^2 \mathbf{u} \ . \tag{7.20}$$

Again, we see the diffusion equation, this time for momentum or, equivalently, the fluid velocity. The viscosity plays the role of the diffusion constant. In Chapter 6 (see, for example, (6.35)) we saw that the viscosity scales as $\mu/mn \sim \lambda\langle v\rangle \sim \lambda^2/\tau$, again in agreement with our expectation (7.11).

7.1.2 Diffusion on a Finite Interval

We can look more closely at the solutions to the 1d diffusion equation,

$$\frac{\partial n}{\partial t} = D\frac{\partial^2 n}{\partial x^2} \ . \tag{7.21}$$

These are particularly straightforward if we consider the system on a finite interval $x \in [0, L]$. We will impose boundary conditions on both ends

$$n(0, t) = n_0 \quad \text{and} \quad n(L, t) = n_1 \tag{7.22}$$

with n_0 and n_1 both constant. To start, we can look for a steady-state solution with no time dependence. This is straightforward. We have

$$\frac{\partial^2 n}{\partial x^2} = 0 \quad \implies \quad n(x, t) = n^\star(x) = n_0 + (n_1 - n_0)\frac{x}{L} \tag{7.23}$$

where we've implemented the boundary conditions (7.22).

What happens when we deviate from the steady state? Now we need to specify the initial value of the field at $t = 0$. The resulting solution takes the form

$$n(x, 0) = n^\star(x) + c(x, t) \tag{7.24}$$

where $c(x, t)$ also solves the diffusion equation

$$\frac{\partial c}{\partial t} = D\frac{\partial^2 c}{\partial x^2} \tag{7.25}$$

now with boundary conditions $c(0, t) = c(L, t) = 0$. We can look for separable solutions of the form

$$c(x, t) = f(x)\,g(t) \ . \tag{7.26}$$

Substituting this into the diffusion equation, we see that the two functions must obey

$$f\dot{g} = Dgf'' \tag{7.27}$$

where the dot means differentiation with respect to time, and the prime

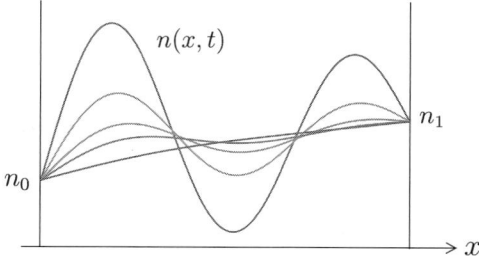

Fig. 7.1 Diffusion of an initial wiggly profile quickly settles down to the linear, steady state.

means differentiation with respect to space. Dividing through by fg, we have

$$\frac{\dot{g}}{g} = D\frac{f''}{f} \ . \tag{7.28}$$

But the left-hand side is a function only of t, and the right-hand side is a function only of x, which means that actually both sides must be constant. The solution for $f(x)$ is constrained by the boundary conditions $f(0) = f(L) = 0$ and so the solutions must be of the form

$$f''(x) \propto f(x) \quad \Longrightarrow \quad f(x) = \sin\left(\frac{\pi n x}{L}\right) \text{ with } n = 1, 2 \ldots . \tag{7.29}$$

Now (7.28) fixes the form of $g(t)$,

$$\dot{g} = -\lambda_n g \quad \Longrightarrow \quad g = e^{-\lambda_n t} \text{ with } \lambda_n = \frac{D\pi^2 n^2}{L^2} \ . \tag{7.30}$$

Because the heat equation is linear, we can simply add together separable solutions for different values of n. Moreover, the most general solution can be constructed in this way and takes the form

$$c(x, t) = \sum_{n=1}^{\infty} c_n e^{-\lambda_n t} \sin\left(\frac{\pi n x}{L}\right) \ . \tag{7.31}$$

Here the c_n are determined by the initial conditions where they are essentially Fourier components of the initial profile at time $t = 0$. We now see the key feature of the heat equation: all the Fourier modes die off exponentially quickly, tending towards the steady-state solution. The higher the Fourier mode, so the more wiggly the profile, the faster it decays away. This is the characteristic behaviour of the heat equation: it smooths things out. An example of the evolution of $n(x, t)$, plotted for increasing values of t, is shown in Figure 7.1.

7.1.3 How to Cook a Turkey

If you buy a turkey, the instructions will typically tell you to cook it for 20 minutes per kg, and then another 70 minutes for a 2-4 kg turkey, or another 90 minutes for a 4-10 kg turkey. These slightly convoluted rules arise because the relationship between the cooking time and the weight is not linear. The correct relationship was suggested by the particle physicist Pief Panofsky who pointed out that the time τ depends on the mass M by

$$\tau \sim M^{2/3} . \tag{7.32}$$

We can derive this formula using the results that we've reviewed above.

The temperature $T(\mathbf{x}, t)$ of the turkey is described by the 3d heat equation,

$$\frac{\partial T}{\partial t} = D\nabla^2 T . \tag{7.33}$$

In fine tradition, we will assume that our turkey is spherical. (The analysis below also holds for spherical cows.) Then, further assuming full spherical symmetry, we can think of $T = T(r, t)$ with r the radial coordinate, and the heat equation takes the form

$$\frac{\partial T}{\partial t} = \frac{D}{r^2}\frac{\partial}{\partial r}\left(r^2\frac{\partial T}{\partial r}\right) . \tag{7.34}$$

At this point, we use a trick and write $V(r, t) = rT(r, t)$. Then we have $T' = V'/r - V/r^2$ and the heat equation becomes

$$\frac{\partial V}{\partial t} = D\frac{\partial^2 V}{\partial r^2} . \tag{7.35}$$

We see that in this new variable V, we're back solving the 1d diffusion equation. And we know how to do that! A separable solution takes the form

$$V(r, t) = e^{-\lambda D t}\left[A\cos(\sqrt{\lambda}r) + B\sin(\sqrt{\lambda}r)\right] \tag{7.36}$$

for some $\lambda > 0$. The temperature T is given by $T = V/r$ so if we want to avoid a divergence at $r = 0$ then we need to set $A = 0$. Our solution will involve only $\sin(\sqrt{\lambda}r)$.

We can determine the allowed values of λ, together with the constant B, by looking at the boundary conditions. If the turkey has radius R then we have the boundary condition

$$T(r, t) = T_{\text{hot}} \quad \text{for all } t \text{ and } r \geq R . \tag{7.37}$$

Here T_{hot} is the temperature of the oven. This is telling us that we can deviate from the uniform temperature only inside the turkey, $r < R$. We do this by taking $\sqrt{\lambda} = n\pi/R$ with $n \in \mathbb{Z}$ and writing down the general solution

$$T(r,t) = T_{\text{hot}} + \frac{1}{r} \sum_{n=1}^{\infty} \left[V_n \sin \left(\frac{n\pi r}{R} \right) e^{-n^2\pi^2 Dt/R^2} \right] . \qquad (7.38)$$

The coefficients V_n are set by the initial condition. We'll take this to be

$$T(r,0) = T_0 \ll T_{\text{hot}} \quad \text{for } 0 \leq r < R \qquad (7.39)$$

This initial data is discontinuous at $r = R$ where the temperature jumps from T_0 to T_{hot}, but it's straightforward to implement this. We just need to pick the coefficients V_n so that it gives the Fourier decomposition of a linear function r, cancelling the $1/r$ factor in front of the sum. This is achieved by the solution

$$T(r,t) = T_{\text{hot}} - \frac{2R}{\pi^2} \frac{(T_{\text{hot}} - T_0)}{r} \sum_{n=1}^{\infty} \left[\frac{(-1)^n}{n} \sin \left(\frac{n\pi r}{R} \right) e^{-n^2\pi^2 Dt/R^2} \right] .$$

The initial condition decays away in characteristic time

$$\tau = \frac{R^2}{n^2\pi^2 D} . \qquad (7.40)$$

When the lowest $n = 1$ mode has decayed away, the turkey has reached temperature $T = T_{\text{hot}}$ all the way through and is cooked. Importantly, $\tau \sim R^2$. This is the origin of the Panofsky turkey rule (7.32), since the mass is proportional to volume $M \sim R^3$.

We can put some numbers on this. We need to know the diffusivity for heat in a turkey. That can be easily measured and turns out to be $D \approx 2 \times 10^{-3}$ cm^2 s^{-1}. Suppose that our turkey has radius $R \approx 10$ cm, then we find $\tau \approx 5000$ seconds, or about 80 minutes. You might want to wait for, say $2 \times \tau$, to be convinced that you're not going to get salmonella, so pop it in for three hours and voilà. Don't let anyone tell you that maths isn't useful.

7.1.4 Diffusion on the Line

The boundary conditions played a crucial role in constructing the solutions above. What happens if we want to solve the heat equation

$$\frac{\partial n}{\partial t} = D \frac{\partial^2 n}{\partial x^2} \qquad (7.41)$$

on a line?

We will insist that our density is localised somewhere (say, near the origin) and, moreover, that $dn/dx \to 0$ as $x \to \pm\infty$. This ensures that the total amount of stuff

$$N = \int_{-\infty}^{+\infty} dx \; n(x,t) \tag{7.42}$$

is constant, with

$$\frac{dN}{dt} = \int_{-\infty}^{+\infty} dx \; \frac{\partial n}{\partial t} = D \int_{-\infty}^{+\infty} dx \; \frac{\partial^2 n}{\partial x^2} = D \left[\frac{\partial n}{\partial x} \right]_{-\infty}^{+\infty} = 0 \; . \tag{7.43}$$

We won't give the most general solution to the heat equation. Instead, we will find a particular solution that is "self-similar", meaning that after scaling space and time in a certain way, it looks the same.

The essence of these self-similar solutions is that we can replace the partial differential equation (7.41) with an ordinary differential equation, where the relevant variable is a suitable combination of x and t. To figure out what linear combination works, we do a little dimensional analysis.

We have two variables x and t and two constants with dimension $[D] = L^2 T^{-1}$ and $[N] = L$. (Here we're assuming that $n(x,t)$ itself is dimensionless; you could assign it a dimension but this is carried only by N and it doesn't change the conclusions.) We then introduce the dimensionless combination

$$\xi = \frac{x}{\sqrt{Dt}} \; . \tag{7.44}$$

Furthermore, we look for solutions of the form

$$n(\mathbf{x}, t) = \frac{N}{\sqrt{Dt}} \, f(\xi) \; . \tag{7.45}$$

The idea here is that the constant N sets the overall scale of the solution, and the overall factor of $(Dt)^{-1/2}$ ensures that the function $f(\xi)$ is dimensionless. At this point we have to figure out what the heat equation looks like when written in terms of ξ. We have

$$\frac{\partial \xi}{\partial t} = -\frac{1}{2} \frac{\xi}{t} \quad \text{and} \quad \frac{\partial \xi}{\partial x} = \frac{1}{\sqrt{Dt}} = \frac{\xi}{x} \; . \tag{7.46}$$

The time derivative of $n(x,t)$ is then

$$\begin{aligned}
\frac{\partial n}{\partial t} &= -\frac{1}{2t} \frac{N}{\sqrt{Dt}} f + \frac{N}{\sqrt{Dt}} f'(\xi) \frac{\partial \xi}{\partial t} \\
&= -\frac{1}{2t} \frac{N}{\sqrt{Dt}} (f + \xi f') \\
&= -\frac{1}{2t} \frac{N}{\sqrt{Dt}} \frac{d}{d\xi} (\xi f) \; .
\end{aligned} \tag{7.47}$$

Meanwhile, the spatial derivatives are

$$\frac{\partial}{\partial x} = \frac{1}{\sqrt{Dt}}\frac{\partial}{\partial \xi} \quad \text{and} \quad \frac{\partial^2}{\partial x^2} = \frac{1}{Dt}\frac{\partial^2}{\partial \xi^2} \ . \tag{7.48}$$

Putting this together, the heat equation (7.41) becomes the ordinary differential equation

$$\frac{d^2 f}{d\xi^2} + \frac{1}{2}\frac{d}{d\xi}(\xi f) = 0 \ . \tag{7.49}$$

It's simple to integrate this once7

$$\frac{df}{d\xi} + \frac{1}{2}\xi f = \text{constant} \ . \tag{7.50}$$

If we want a localised solution, with $f, f' \to 0$ as $\xi \to \infty$, then this constant must vanish. We learn that we must solve

$$\frac{df}{d\xi} = -\frac{1}{2}\xi f \quad \Longrightarrow \quad f(\xi) = A e^{-\xi^2/4} \ . \tag{7.51}$$

The normalisation condition (7.42) translates to the requirement

$$\int_{-\infty}^{+\infty} d\xi \ f(\xi) = 1 \quad \Longrightarrow \quad A = \frac{1}{\sqrt{4\pi}} \ . \tag{7.52}$$

The upshot of this analysis is that we have a self-similar solution to the heat equation given by

$$n(x,t) = \frac{N}{\sqrt{4\pi Dt}} e^{-x^2/4Dt} \ . \tag{7.53}$$

This is a Gaussian of ever-spreading width. If we trace it back to $t \to 0^-$, it becomes a delta function localised at the origin. Again, we see the tendency of the heat equation to take a solution and spread it out. The resulting profile for various values of t is shown on the left of Figure 7.2.

Changing Boundary Conditions at Infinity

We can get solutions with different boundary conditions using a slight variation of this argument. Suppose that we want a solution to the heat equation such that

$$n(x,t) \to \begin{cases} +1 & \text{as } x \to +\infty \\ -1 & \text{as } x \to -\infty \end{cases} \ . \tag{7.54}$$

Now there's no analogue of the conserved quantity N because the spatial

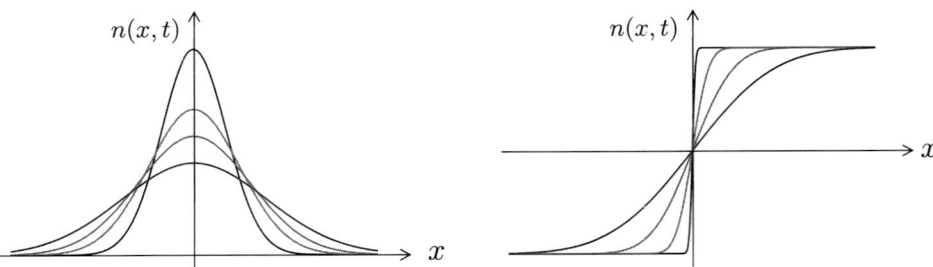

On the left: diffusion of a Gaussian wavepacket. On the right: diffusion of the error function. In both cases, diffusion takes the edge off.

integral over $n(x,t)$ diverges. But, inspired by the approach above, we could look for solutions of the form

$$n(x,t) = t^\alpha f(\xi) \quad \text{with} \quad \xi = \frac{x}{\sqrt{Dt}} \tag{7.55}$$

and some constant α that we need to determine. The two sides of the heat equation then become

$$\frac{\partial n}{\partial t} = t^{\alpha-1}\left(\alpha - \frac{1}{2}\xi f'\right) \quad \text{and} \quad \frac{\partial^2 n}{\partial x^2} = \frac{t^{\alpha-1}}{D} f'' . \tag{7.56}$$

If we want to impose the boundary condition $n(x,t) \to \pm 1$ on both sides, then we had better take $\alpha = 0$. The heat equation becomes

$$f'' + \frac{1}{2}\xi f' = 0 . \tag{7.57}$$

Again, we can integrate to get

$$f'(\xi) = A e^{-\xi^2/4} \quad \Longrightarrow \quad f(\xi) = B + A \int_0^\xi d\eta \, e^{-\eta^2/4} \tag{7.58}$$

with A and B both integration constants. This definite integral defines the so-called *error function*

$$\mathrm{Erf}(x) = \frac{2}{\sqrt{\pi}} \int_0^x dy \, e^{-y^2} . \tag{7.59}$$

It has the property that $\mathrm{Erf}(x) \approx 2x/\sqrt{\pi}$ for $|x| \ll 1$ and $\mathrm{Erf}(x) \to \pm 1$ as $x \to \pm\infty$. The integration constants A and B are then fixed by the boundary conditions (7.54), and we have the solution

$$n(x,t) = \mathrm{Erf}\left(\frac{x}{\sqrt{4Dt}}\right) . \tag{7.60}$$

The evolution of this function with t is shown on the right-hand side of Figure 7.2.

7.2 The Langevin Equation

The diffusion equation describes the evolution of a large number of particles immersed in a fluid. Here we return to the more basic question of what happens to an individual particle.

A particularly vivid demonstration arises in the phenomenon of *Brownian motion*. A large particle, like a pollen grain, suspended in a fluid will jitter around like a child in desperate need of a wee. Famously, this jittering is caused by the constant bombardment from molecules in the surrounding fluid. Our purpose in this section is to introduce a mathematical formalism that allows us to model such random behaviour.

Any particle obeys Newton's law of motion,

$$m\ddot{\mathbf{x}} = -\gamma\dot{\mathbf{x}} + \mathbf{F} \ . \tag{7.61}$$

Here $\mathbf{F}$ is the external force acting on the particle. In addition, we have included a friction term with coefficient γ. If we model the particle as a sphere of radius a, then the Stokes formula (2.105) gives us an expression for the friction coefficient in terms of the viscosity μ of the surrounding fluid,

$$\gamma = 6\pi\mu a \ . \tag{7.62}$$

In the presence of a time-independent force, the steady-state solution with $\ddot{\mathbf{x}} = 0$ is

$$\dot{\mathbf{x}} = \frac{1}{\gamma}\mathbf{F} \ . \tag{7.63}$$

For this reason, the quantity $1/\gamma$ is sometimes referred to as the *mobility*.

Returning to (7.61), for any specified force $\mathbf{F}$, the path of the particle is fully determined. This is seemingly at odds with the random behaviour observed in Brownian motion. The way in which we reconcile these two points is, hopefully, obvious: in Brownian motion the force $\mathbf{F}$ that the particle feels is itself random. In fact, we will split the force into two pieces,

$$\mathbf{F} = -\nabla V + \mathbf{f}(t) \ . \tag{7.64}$$

Here $V(\mathbf{x})$ is a fixed background potential in which the particle is moving. Perhaps V arises because the particle is moving in gravity; perhaps because it is attached to a spring. But, either way, there is nothing random about V. In contrast, $\mathbf{f}(t)$ is the random force that the particle experiences due to the constant bombardment by all the other atoms in the fluid. It is sometimes referred to as *noise*. The resulting equation is called the *Langevin equation*

$$m\ddot{\mathbf{x}} = -\gamma \dot{\mathbf{x}} - \nabla V + \mathbf{f}(t) \ . \tag{7.65}$$

Although this looks just like an ordinary differential equation, it is, in fact, a different beast, known as a *stochastic differential equation*. The reason that it's different is that we don't actually know what $\mathbf{f}(t)$ is. Yet, somehow, we must solve this equation anyway!

A few words may help to clarify what we're trying to achieve. Suppose that you did know the microscopic force $\mathbf{f}(t)$ that is experienced by a given particle. Then you could, in principle, go ahead and solve (7.65). But the next particle that you look at will experience a different force $\mathbf{f}(t)$ so you'll have to solve (7.65) again. And for the third particle, you'll have to solve it yet again. Clearly, this is going to become tedious. What's more, it's unrealistic to think that we will actually know $\mathbf{f}(t)$ in any specific case. Instead, we admit that we only know certain crude features of the force $\mathbf{f}(t)$ such as, for example, its average value. Then we might hope that this is sufficient information to figure out, say, the average value of $\mathbf{x}(t)$. That is the goal when solving the Langevin equation.

Throughout this section, we will only solve the Langevin equation with vanishing potential, so $V = 0$. This is because things get a little messy when a potential is involved as the equation is, typically, non-linear. We will, however, discuss some of the physics that arises when we include a potential in Section 7.3 where we discuss a different formalism known as the Fokker–Planck equation.

7.2.1 Diffusion in a Very Viscous Fluid

We can simplify the problem even further by considering Brownian motion in a very viscous liquid. In this case, motion is entirely dominated by the friction term in the Langevin equation and we ignore the inertial term, which is tantamount to setting $m = 0$.

When $m = 0$, we're left with a first order equation,

$$\dot{\mathbf{x}}(t) = \frac{1}{\gamma} \mathbf{f}(t) \ . \tag{7.66}$$

For any $\mathbf{f}(t)$, this can be trivially integrated to give

$$\mathbf{x}(t) = \mathbf{x}(0) + \frac{1}{\gamma} \int_0^t dt' \ \mathbf{f}(t') \ . \tag{7.67}$$

At this point, we can't go any further until we specify some of the properties of the noise $\mathbf{f}(t)$. Our first assumption is that, on average, the noise vanishes at any given time. We will denote averages by $\langle \cdot \rangle$, so this assumption reads

$$\langle \mathbf{f}(t) \rangle = 0 \ . \tag{7.68}$$

Taking the average of (7.67) then gives us the result:

$$\langle \mathbf{x}(t) \rangle = \mathbf{x}(0) \ . \tag{7.69}$$

This is deeply unsurprising: if the average noise vanishes, the average position of the particle is simply where we left it to begin with. Nonetheless, it's worth stressing that this doesn't mean that all particles sit where you leave them. It means that if you drop many identical particles at the origin, $\mathbf{x}(0) = \mathbf{0}$, then they will all move but their average position (or, equivalently, their centre of mass) will remain at the origin.

We can get more information by looking at the variance of the position

$$\langle \, (\mathbf{x}(t) - \mathbf{x}(0))^2 \, \rangle \ . \tag{7.70}$$

This will tell us the average spread of the particles. We derive an expression for the variance by first squaring (7.67) and then taking the average

$$\langle \, (\mathbf{x}(t) - \mathbf{x}(0))^2 \, \rangle = \frac{1}{\gamma^2} \int_0^t dt_1' \int_0^t dt_2' \, \langle \mathbf{f}(t_1') \cdot \mathbf{f}(t_2') \rangle \ . \tag{7.71}$$

In order to compute this, we need to specify more information about the noise, namely its correlation function $\langle \, f_i(t_1) f_j(t_2) \, \rangle$ where we have resorted to index notation, $i, j = 1, 2, 3$, to denote the direction of the force. This is specifying how likely it is that the particle will receive a kick f_j at time t_2 given that it received a kick f_i at time t_1.

In many cases of interest, including that of Brownian motion, the kicks imparted by the noise are both fast and uncorrelated. I should explain what this means. Suppose that a given collision between our particle and an atom takes time τ_{coll}. If we focus on time scales less than τ_{coll} then there will clearly be a correlation between the forces imparted on our particle because these forces are due to the same process that's already taking place. (If an atom is coming in from the left, then it's still coming in from the left at a time $t \ll \tau_{\mathrm{coll}}$ later.) However, if we look on time scales $t \gg \tau_{\mathrm{coll}}$, the force will be due to a different collision with a different atom. The statement that the noise is uncorrelated means that the force imparted by later collisions knows nothing about earlier collisions. Mathematically, this means

$$\langle \, f_i(t_1) f_j(t_2) \, \rangle = 0 \quad \text{when} \quad t_2 - t_1 \gg \tau_{\mathrm{coll}} \ . \tag{7.72}$$

The statement that the collisions are fast means that we *only* care about time scales $t_2 - t_1 \gg \tau_{\text{coll}}$ and so can effectively take the limit $\tau_{\text{coll}} \to 0$. However, that doesn't quite mean that we can just ignore the short time correlations. Instead, when we take the limit $\tau_{\text{coll}} \to 0$, we're left with a delta function contribution,

$$\langle\, f_i(t_1) f_j(t_2)\,\rangle = 2D\gamma^2\, \delta_{ij}\, \delta(t_2 - t_1)\;. \tag{7.73}$$

Here the coefficient D measures the strength of the correlation, and the factor of γ^2 has been put in for convenience. Random forces which obey (7.68) and (7.73) are referred to as *white noise*. It is valid whenever the environment relaxes back down to equilibrium much faster than the system of interest. This guarantees that, although the system is still reeling from the previous kick, the environment remembers nothing of what went before and kicks again, as fresh and random as the first time.

Using this expression for white noise, the variance (7.71) of the position of the particles is

$$\langle\, (\mathbf{x}(t) - \mathbf{x}(0))^2\,\rangle = 6D\,t\;. \tag{7.74}$$

This is the characteristic $\sqrt{t}$ diffusive behaviour, with the coefficient $6D$ that matches our previous result (7.14). We see that, from this perspective, the diffusion constant D appears as the strength of the white noise (7.73). (We put the factor of γ^2 in the correlation function (7.73) so that this equation would come out nicely.)

7.2.2 Diffusion in a Less Viscous Liquid

We now return to the Langevin equation (7.65) and repeat our analysis, this time retaining the inertia term, so $m \neq 0$.

As before, computing average quantities $\langle\, \dot{\mathbf{x}}(t)\,\rangle$ and $\langle\, \mathbf{x}(t)\,\rangle$ is straightforward and relatively uninteresting. For a given $\mathbf{f}(t)$, it is not difficult to solve (7.65). After multiplying by an integrating factor $e^{\gamma t/m}$, the equation becomes

$$\frac{d}{dt}\left(\dot{\mathbf{x}} e^{\gamma t/m}\right) = \frac{1}{m}\mathbf{f}(t) e^{\gamma t/m} \tag{7.75}$$

which can be happily integrated to give

$$\dot{\mathbf{x}}(t) = \dot{\mathbf{x}}(0) e^{-\gamma t/m} + \frac{1}{m}\int_0^t dt'\, \mathbf{f}(t')\, e^{\gamma(t'-t)/m} \tag{7.76}$$

with $\dot{\mathbf{x}}(0)$ the initial velocity of the particle. We now use the fact that the average noise vanishes (7.68) to find that the average velocity is simply that of a damped particle in the absence of any noise,

$$\langle \dot{\mathbf{x}}(t) \rangle = \dot{\mathbf{x}}(0) e^{-\gamma t/m} \ . \tag{7.77}$$

Similarly, to determine the average position we have

$$\mathbf{x}(t) = \mathbf{x}(0) + \int_0^t dt' \ \dot{\mathbf{x}}(t') \ . \tag{7.78}$$

From which we get

$$\begin{aligned} \langle \mathbf{x}(t) \rangle &= \mathbf{x}(0) + \int_0^t dt' \ \langle \dot{\mathbf{x}}(t') \rangle \\ &= \mathbf{x}(0) + \frac{m}{\gamma} \dot{\mathbf{x}}(0) \left(1 - e^{-\gamma t/m} \right) \ . \end{aligned} \tag{7.79}$$

Again, this is unsurprising: when the average noise vanishes, the average position of the particle coincides with that of a particle that didn't experience any noise.

Things get more interesting when we look at the expectation values of quadratic quantities. This includes the variance in position $\langle \mathbf{x}(t) \cdot \mathbf{x}(t) \rangle$ and velocity $\langle \dot{\mathbf{x}}(t) \cdot \dot{\mathbf{x}}(t) \rangle$, but also more general correlation functions in which the two quantities are evaluated at different times. For example, the correlation function $\langle \dot{x}_i(t_1) \dot{x}_j(t_2) \rangle$ tells us information about the velocity of the particle at time t_2 given its velocity at time t_1. From (7.76), we have the expression,

$$\begin{aligned} \langle \dot{x}_i(t_1) \dot{x}_j(t_2) \rangle = \langle \dot{x}_i(t_1) \rangle \langle \dot{x}_j(t_2) \rangle \\ + \frac{1}{m^2} \int_0^{t_1} dt_1' \int_0^{t_2} dt_2' \ \langle f_i(t_1') f_j(t_2') \rangle \ e^{\gamma(t_1' + t_2' - t_1 - t_2)/m} \end{aligned}$$

where the average velocities $\langle \dot{\mathbf{x}} \rangle$ in the first term are given by (7.77), and we made use of the fact that $\langle \mathbf{f}(t) \rangle = 0$ to drop the terms linear in the noise $\mathbf{f}$. If we use the white noise correlation function (7.73), and assume $t_2 \geq t_1 > 0$, then we can rewrite this as

$$\begin{aligned} \langle \dot{x}_i(t_1) \dot{x}_j(t_2) \rangle &= \langle \dot{x}_i(t_1) \rangle \langle \dot{x}_j(t_2) \rangle + \frac{2D\gamma^2}{m^2} \delta_{ij} \ e^{-\gamma(t_1 + t_2)/m} \int_0^{t_1} dt' \ e^{2\gamma t'/m} \\ &= \langle \dot{x}_i(t_1) \rangle \langle \dot{x}_j(t_2) \rangle + \frac{D\gamma}{m} \delta_{ij} \left(e^{-\gamma(t_2 - t_1)/m} - e^{-\gamma(t_1 + t_2)/m} \right) \ . \end{aligned}$$

For very large times, $t_1, t_2 \gg m/\gamma$, we can drop the last term as well as the average velocities since, from (7.77), $\langle \dot{\mathbf{x}}(t) \rangle \to 0$ as $t \to \infty$. We learn that the correlation between velocities decays exponentially as

$$\langle \dot{x}_i(t_1) \dot{x}_j(t_2) \rangle \ \to \ \frac{D\gamma}{m} \delta_{ij} \ e^{-\gamma(t_2 - t_1)/m} \ . \tag{7.80}$$

This means that if you know the velocity of the particle at some time t_1, then you can be fairly confident that it will have a similar velocity at a time $t_2 \lesssim t_1 + m/\gamma$ later. But if you wait much longer than time m/γ then you would be a fool to make any bets on the velocity based only on your knowledge at time t_1.

Finally, we can also use this result to compute the average velocity-squared, which, of course, is the kinetic energy of the system. At late times, any initial velocity has died away and the resulting kinetic energy is due entirely to the bombardment by the environment. It is independent of time and given by

$$\langle \dot{\mathbf{x}}(t) \cdot \dot{\mathbf{x}}(t) \rangle = \frac{3D\gamma}{m} . \tag{7.81}$$

One can compute similar correlation functions for position $\langle x_i(t_1)x_j(t_2) \rangle$. The expressions are a little more tricky but still quite manageable. (Combining equations (7.78) and (7.76), you can see that you will have a quadruple integral to perform and figuring out the limits is a little fiddly.) At late times, it turns out that the variance of the position is given by the same expression that we saw for the viscous liquid (7.74),

$$\langle (\mathbf{x}(t) - \mathbf{x}(0))^2 \rangle = 6D\,t \tag{7.82}$$

again exhibiting the now-familiar $\sqrt{t}$ behaviour for the root-mean-square distance.

7.2.3 The Einstein Relation

We brushed over something important and lovely in the previous discussion. We computed the average kinetic energy of a particle in (7.81). It is

$$E = \frac{1}{2}m\langle \dot{\mathbf{x}} \cdot \dot{\mathbf{x}} \rangle = \frac{3}{2}D\gamma . \tag{7.83}$$

But we already know what the average energy of a particle is when it's bombarded by its environment: it is given by the equipartition theorem and, crucially, depends only on the temperature of the surroundings

$$E = \frac{3}{2}k_B T . \tag{7.84}$$

This means that, for a fixed temperature, the diffusion constant D is related to the mobility $1/\gamma$ by

$$D = \frac{k_B T}{\gamma} . \tag{7.85}$$

That's rather surprising! The diffusion constant captures the amount a particle is kicked around due to the background medium; the mobility expresses how hard it is for a particle to plough through the background medium. And yet they are related. This equation is telling us that diffusion and viscosity both have their microscopic origin in the random bombardment of molecules. Notice that D is inversely proportional to γ which, in turn, is linear in viscosity μ as shown by the Stokes formula (7.62). Yet you might have thought that the amount the particle is kicked increases as the viscosity increases. Indeed, looking back at (7.73), you can see that the amount the particle is kicked is actually proportional to $D\gamma^2 \sim T\gamma$. Which is more in line with our intuition.

Equation (7.85) is known as the *Einstein relation*. It is an important example of a more general result called the *fluctuation-dissipation theorem*. The fluctuations of the particle as it undergoes its random walk are related to the drag force (or dissipation of momentum) that the particle feels as it moves through the fluid.

The Einstein relation gives a way to determine Boltzmann's constant k_B experimentally. Watch a particle perform a Brownian jitter. After time t, the distance travelled by the particle (7.74) should be

$$\langle \mathbf{x}^2 \rangle = \frac{k_B T}{\pi \eta a} t \tag{7.86}$$

where we have used the Stokes formula (7.62) to relate the mobility to the viscosity μ and radius a of the particle. This experiment was done in 1909 by the French physicist Jean Baptiste Perrin and won him the 1926 Nobel prize.

7.2.4 Noise Probability Distributions

So far, we've only needed to use two pieces of information about the noise, namely the one- and two-point functions

$$\langle \mathbf{f}(t) \rangle = 0 \tag{7.87}$$

$$\langle f_i(t_1) f_j(t_2) \rangle = 2D\gamma^2 \delta_{ij} \delta(t_1 - t_2) \ . \tag{7.88}$$

However, if we wanted to compute correlation functions involving more than two velocities or positions, it should be clear from the calculation that we would need to know higher moments of the probability distribution for $\mathbf{f}(t)$.

In fact, the definition of *white noise* is that there are no non-trivial correlations other than $\langle f_i(t_1) f_j(t_2) \rangle$. This doesn't mean that the higher correlation functions are vanishing, just that they can be reduced to the two-point correlator. This means that, for N even

$$\langle\, f_{i_1}(t_1)\dots f_{i_N}(t_N)\,\rangle = \langle f_{i_1}(t)f_{i_2}(t_2)\rangle \dots \langle f_{i_{N-1}}(t_{N-1})f_{i_N}(t_N)\rangle$$
$$+ \text{ permutations} \qquad (7.89)$$

while, for N odd

$$\langle\, f_{i_1}(t_1)\dots f_{i_N}(t_N)\rangle = 0\ . \qquad (7.90)$$

Another way of saying this is that all but the second cumulant of the probability distribution vanish.

Instead of specifying all these moments of the distribution, it is often much more useful to specify the probability distribution for $\mathbf{f}(t)$ directly. However, this is a slightly subtle object because we want to specify the probability for an entire function $\mathbf{f}(t)$, rather than a single random variable. This means that the probability distribution must be a functional: you give it a function $\mathbf{f}(t)$ and it spits back a number which, because we're talking about probabilities, should be between zero and one.

The good news is that, among the class of probability distributions over functions, the white noise distribution is by far the easiest. If we were dealing with a single random variable, the distribution that has only two-point correlators, but no higher, is the Gaussian. And, suitably generalised, this also works for our functional probability distribution. The probability distribution that gives white noise is

$$\text{Prob}[f(t)] = \mathcal{N}\exp\left(-\int_{-\infty}^{+\infty} dt\ \frac{\mathbf{f}(t)\cdot\mathbf{f}(t)}{4D\gamma^2}\right) \qquad (7.91)$$

where $\mathcal{N}$ is a normalisation factor which is needed to ensure that the sum over all probabilities gives unity. This "sum" is really a sum over all functions $\mathbf{f}(t)$ or, in other words, a functional integral. The normalisation condition which fixes $\mathcal{N}$ is then

$$\int \mathcal{D}f(t)\ \text{Prob}[f(t)] = 1\ . \qquad (7.92)$$

Here the measure factor is written as $\mathcal{D}f(t)$, rather than the more usual df, to highlight the fact that we're integrating over all functions $f(t)$, rather than over a single variable. This kind of functional integral also goes by the name of a path integral: roughly speaking, it is an infinite number of ordinary integrals, one for each value of t. A much fuller discussion of path integrals can be found in Volume 3 on Quantum Mechanics.

With this probability distribution, all averaging over the noise can now be computed as a functional integral. If you have any function $g(x(t))$, then its average is

$$\langle\, g(x)\,\rangle = \mathcal{N} \int \mathcal{D}f(t)\; g(x_f)\, e^{-\int dt\; \mathbf{f}(t)^2/4D\gamma^2} \tag{7.93}$$

where the notation x_f means that $g(x(t))$ is evaluated on the solution to the Langevin equation (7.65) in the presence of a fixed source $\mathbf{f}$.

We now show that the Gaussian probability distribution indeed reproduces the white noise correlations as claimed. To do this, we first introduce an object $Z[\mathbf{J}(t)]$ known as a *generating function*. We can introduce a generating function for any probability distribution, so let's keep things general for now and later specialise to the Gaussian distribution. The generating function is defined by

$$Z[\mathbf{J}(t)] = \int \mathcal{D}f(t)\; \mathrm{Prob}[f(t)] \exp\left(\int_{-\infty}^{+\infty} dt\; \mathbf{J}(t)\cdot \mathbf{f}(t)\right)\,. \tag{7.94}$$

This generating function is a functional: it is a function of any function $\mathbf{J}(t)$ that we care to feed it. By construction, $Z[0] = 1$, courtesy of (7.92).

The generating function has much in common with the partition function that has a starring role in the book on Statistical Physics. These objects are also important in quantum field theory, where the names "partition function" and "generating function" are often used synonymously.

The function $\mathbf{J}(t)$ that we have introduced is, in this context, really little more than a trick that allows us to encode all the correlation functions in one object, $Z[\mathbf{J}(t)]$. To see how this works, suppose that we differentiate Z with respect to $\mathbf{J}(t)$, evaluated at some time $t = t_1$, and then set $\mathbf{J} = 0$. We have

$$\left.\frac{\delta Z}{\delta J_i(t_1)}\right|_{\mathbf{J}=0} = \int \mathcal{D}f(t)\; f_i(t_1)\, \mathrm{Prob}[f(t)] = \langle\, f_i(t_1)\,\rangle\,. \tag{7.95}$$

Playing the same game, but now taking n derivatives, gives

$$\left.\frac{\delta^n Z}{\delta J_{i_1}(t_1)\delta J_{i_2}(t_2)\dots\delta J_{i_n}(t_n)}\right|_{\mathbf{J}=0}$$
$$= \int \mathcal{D}f(t)\; f_{i_1}(t_1) f_{i_2}(t_2)\dots f_{i_n}(t_n)\, \mathrm{Prob}[f(t)]$$
$$= \langle\, f_{i_1}(t_1) f_{i_2}(2)\dots f_{i_n}(t_n)\,\rangle\,. \tag{7.96}$$

So we see that if we can compute $Z[\mathbf{J}]$, then successive correlation functions are simply the coefficients of a Taylor expansion in $\mathbf{J}$. This is particularly useful for the Gaussian distribution (7.91) for the simple, but expedient, reason that we can actually do the integral and compute $Z[\mathbf{J}]$. If we substitute the Gaussian distribution, the generating function is

$$Z[\mathbf{J}(t)] = \mathcal{N} \int \mathcal{D}f(t) \, \exp\left(-\int_{-\infty}^{+\infty} dt \, \frac{\mathbf{f}(t) \cdot \mathbf{f}(t)}{4D\gamma^2} - \mathbf{J}(t) \cdot \mathbf{f}(t)\right) \, . \quad (7.97)$$

But this is nothing more than a Gaussian integral. (Ok, it's an infinite number of Gaussian integrals because it's a functional integral. But the trick is to be brave and not let that bother us.) We first complete the square

$$Z[\mathbf{J}(t)] = \mathcal{N} \int \mathcal{D}f(t) \, \exp\left(-\frac{1}{4D\gamma^2} \int_{-\infty}^{+\infty} dt \, \left[\mathbf{f}(t) - 2D\gamma^2 \mathbf{J}(t)\right]^2 \right.$$
$$\left. - \, 4D^2\gamma^4 \mathbf{J}(t) \cdot \mathbf{J}(t)\right) \quad (7.98)$$

and then shift the integration variable to $\mathbf{f} \to \mathbf{f} - 2D\gamma^2\mathbf{J}$. If we're cavalier about the measure, then the integral reduces to (7.92), leaving behind

$$Z[\mathbf{J}(t)] = \exp\left(D\gamma^2 \int_{-\infty}^{+\infty} dt \, \mathbf{J}(t) \cdot \mathbf{J}(t)\right) \, . \quad (7.99)$$

Now it is an easy matter to compute correlation functions. Taking one derivative, we have

$$\frac{\delta Z}{\delta J_i(t_1)} = 2D\gamma^2 \, J_i(t_1) \, Z[\mathbf{J}] \, . \quad (7.100)$$

But this vanishes when we set $J = 0$, in agreement with our requirement (7.87) that the average noise vanishes. Taking a second derivative gives

$$\frac{\delta^2 Z}{\delta J_i(t_1)\delta J_j(t_2)} = 2D\gamma^2\delta_{ij}\delta(t_1 - t_2)Z[\mathbf{J}] + 4D^2\gamma^4 J_i(t_1)J_j(t_2)Z[\mathbf{J}] \, . \quad (7.101)$$

Now setting $\mathbf{J} = 0$, only the first term survives and reproduces the white noise correlation (7.88). One can continue the process to see that all higher correlation functions are entirely determined by $\langle f_i \, f_j \rangle$.

7.3 The Fokker–Planck Equation

Drop a particle at some position, say $\mathbf{x}_0$ at time t_0. If the subsequent evolution is noisy, so that it is governed by a stochastic Langevin equation, then we've got no way to know for sure where the particle will be. The best that we can do is talk about probabilities. We will denote the probability that the particle sits at $\mathbf{x}$ at time t as $P(\mathbf{x}, t; \mathbf{x}_0, t_0)$.

In the previous section we expressed our uncertainty in the position of the particle in terms of correlation functions. Here we shift perspective a little. We would like to ask: What probability distribution $P(\mathbf{x}, t; \mathbf{x}_0, t_0)$ would give rise to the same correlation functions that arose from the Langevin equation?

We should stress that we care nothing about the particular path $\mathbf{x}(t)$ that the particle took. The probability distribution over paths would be rather complicated. Instead we will ask the much simpler question of the probability that the particle sits at $\mathbf{x}$ at time t, regardless of how it got there.

It is simple to write down a formal expression for the probability distribution. We denote the solution to the Langevin equation for a given noise function $\mathbf{f}$ as $\mathbf{x}_f(t)$. Of course, if we know the noise, then there is no uncertainty in the probability distribution for $\mathbf{x}$. It is simply $P(\mathbf{x}, t) = \delta(\mathbf{x} - \mathbf{x}_f(t))$. Now averaging over all possible noise functions, we can write the probability distribution as

$$P(\mathbf{x}, t) = \langle\, \delta(\mathbf{x} - \mathbf{x}_f(t)) \,\rangle \ . \tag{7.102}$$

In this section, we will show that $P(\mathbf{x}, t)$ obeys a simple partial differential equation known as the Fokker–Planck equation.

7.3.1 The Diffusion Equation for Probabilities

The simplest stochastic process we studied was a particle subject to random forces in a very viscous fluid. The Langevin equation is

$$\dot{\mathbf{x}}(t) = \frac{1}{\gamma}\, \mathbf{f}(t) \ . \tag{7.103}$$

In Section 7.2.1 we showed that the average position of the particle remains unchanged, meaning that if $\mathbf{x}(t = 0) = \mathbf{0}$, then $\langle \mathbf{x}(t) \rangle = \mathbf{0}$ for all t. But the variance of the position grows linearly in time (7.74)

$$\langle\, \mathbf{x}(t)^2 \,\rangle = 6Dt \ . \tag{7.104}$$

For this simple case, we won't derive the associated probability distribution: we'll just write it down. This variance can be reproduced by a Gaussian probability distribution

$$P(\mathbf{x}, t) = \left(\frac{1}{4\pi Dt}\right)^{3/2} e^{-\mathbf{x}^2/4Dt} \tag{7.105}$$

where the factor out front is determined by the normalisation requirement
that

$$\int d^3x\, P(x,t) = 1 \qquad (7.106)$$

for all time t. Note that there is more information contained in this probability distribution than just the variance (7.104). Specifically, all higher cu-
mulants vanish. This means, for example, that $\langle \mathbf{x}^3 \rangle = 0$ and $\langle \mathbf{x}^4 \rangle = 3\langle \mathbf{x}^2 \rangle$
and so on. But it is simple to check that this is indeed what arises from the
Langevin equation with white noise, as described in Section 7.2.4.

Interestingly, the probability distribution (7.105) obeys the *diffusion equation*

$$\frac{\partial P}{\partial t} = D\nabla^2 P \;. \qquad (7.107)$$

We said previously that diffusion occurs for conserved quantities. Since probability is certainly conserved, it is perhaps not surprising that it, too, obeys
the diffusion equation.

The diffusion equation is the simplest example of a Fokker–Planck equation. Here our Langevin equation was simple enough for us to just write
down the associated probability distribution, and then reconstruct the associated equation. For more complicated versions of the Langevin equation,
we will have to work harder to derive the analogous equation governing the
probability distribution P.

7.3.2 Meet the Fokker–Planck Equation

We now turn to a general stochastic process. We'll still work in the viscous
limit, which means that we set $m = 0$ in the Langevin equation. But, in
contrast to Section 7.2, we will keep the potential term $V(\mathbf{x})$ in Newton's
force law. This means that we have the first order Langevin equation,

$$\gamma\dot{\mathbf{x}} = -\nabla V + \mathbf{f} \;. \qquad (7.108)$$

A quadratic $V(\mathbf{x})$ corresponds to a harmonic oscillator potential and the
Langevin equation is not difficult to solve. (In fact, mathematically it is
the same problem that we solved in Section 7.2.2. You just have to replace
$\dot{\mathbf{x}} = \mathbf{v} \rightarrow \mathbf{x}$.) Any other $V(\mathbf{x})$ gives rise to a non-linear stochastic equation
and no general solution is available. Nonetheless, we will still be able to
massage the Langevin equation (7.108) into the form of a Fokker–Planck
equation.

We begin by extracting some information from the Langevin equation. Consider a particle sitting at some point $\mathbf{x}$ at time t. If we look again a short time δt later, the particle will have moved a small amount

$$\delta \mathbf{x} = \dot{\mathbf{x}}\,\delta t = -\frac{1}{\gamma}\nabla V\,\delta t + \frac{1}{\gamma}\int_t^{t+\delta t} dt'\,\mathbf{f}(t') \; . \tag{7.109}$$

Here we've done something a little slippery. We've taken the average value of the noise function $\mathbf{f}(t)$ over the small time interval. However, we've assumed that the displacement of the particle $\delta \mathbf{x}$ is small enough so that we can evaluate the force ∇V at the original position $\mathbf{x}$. It turns out that this is an appropriate approximation in the present context, but there are often pitfalls in making such assumptions in general in the theory of stochastic processes. We'll comment on one such pitfall in Section 7.3.5.

We can now compute the average. Because $\langle \mathbf{f}(t) \rangle = 0$, we have

$$\langle\, \delta \mathbf{x} \,\rangle = -\frac{1}{\gamma}\nabla V\,\delta t \; . \tag{7.110}$$

The computation of the two-point function $\langle\, \delta x_i\,\delta x_j \,\rangle$ is also straightforward

$$\gamma^2 \langle\, \delta x_i\,\delta x_j \,\rangle = \langle \partial_i V \partial_j V \rangle \delta t^2 - \delta t \int_t^{t+\delta t} dt'\, \langle \partial_i V\, f_j(t') + \partial_j V\, f_i(t') \rangle$$

$$+ \int_t^{t+\delta t} dt' \int_t^{t+\delta t} dt''\langle\, f_i(t')\, f_j(t'') \,\rangle \; . \tag{7.111}$$

Both the first two terms are of order δt^2. At first glance it looks like the third term is also of order δt^2 because there are two integrals, both over an interval $\mathcal{O}(\delta t)$. However, we saw in (7.73) that the noise two-point function includes a delta function, $\langle f(t_1)f(t_2)\rangle \sim \delta(t_1 - t_2)$, and this kills one of the integrals, leaving just a single integral standing. This means that the leading contribution comes from the third term and is actually proportional to δt,

$$\langle\, \delta x_i\,\delta x_j \,\rangle = 2\delta_{ij} D\,\delta t + \mathcal{O}(\delta t^2) \; . \tag{7.112}$$

We will ignore the terms of order δt^2. It is simple to check that all higher correlation functions are higher order in δt. For example, $\langle \mathbf{x}^4 \rangle \sim \delta t^2$. These too will be ignored.

Our strategy now is to construct a probability distribution that reproduces (7.110) and (7.112). We start by considering the conditional probability $P(\mathbf{x}, t + \delta t; \mathbf{x}', t)$ that the particle sits at $\mathbf{x}$ at time $t + \delta t$ given that, a moment earlier, it was sitting at $\mathbf{x}'$. From the definition (7.102) we can write this as

$$P(\mathbf{x}, t + \delta t; \mathbf{x}', t) = \langle\, \delta(\mathbf{x} - \mathbf{x}' - \delta \mathbf{x}) \,\rangle \tag{7.113}$$

where $\delta\mathbf{x}$ is the random variable here; it is the distance moved in time δt. Next, we're going to do something that may make your heart race. We Taylor expand the delta function. If you're (rightly) nervous about treating a distribution in such a cavalier fashion, then you could always regulate the delta function in your favourite manner to turn it into a well-behaved function. Alternatively, from a happy-go-lucky physicist's perspective, these kinds of manipulations are harmless because any potentially offensive expression ends up sitting inside an integral where it makes perfect sense. For now, we just proceed naively

$$P(\mathbf{x}, t + \delta t; \mathbf{x}', t) = \Big(1 + \langle\, \delta x_i\,\rangle \frac{\partial}{\partial x_i'}$$

$$+ \frac{1}{2}\langle\, \delta x_i\, \delta x_j\,\rangle \frac{\partial^2}{\partial x_i' \partial x_j'} + \ldots \Big)\delta(\mathbf{x} - \mathbf{x}') . \quad (7.114)$$

We have truncated at second order because we want to compare this to (7.118) and, as we saw above, $\langle\, \delta\mathbf{x}\,\rangle$ and $\langle\, \delta\mathbf{x}^2\,\rangle$ are the only terms that are of order δt.

We now have all the information that we need. We just have to compare (7.118) and (7.114) and figure out how to deal with those delta functions. To do this, we need one more trick. First recall that our real interest is in the evolution of the probability $P(\mathbf{x}, t; \mathbf{x}_0, t_0)$, given some initial, arbitrary starting position $\mathbf{x}(t = t_0) = \mathbf{x}_0$. There is an obvious property that this probability must satisfy: if you look at some intermediate time $t_0 < t' < t$, then the particle has to be somewhere. Written as an equation, this "has to be somewhere" property is called the *Chapman–Kolmogorov equation* and reads

$$P(\mathbf{x}, t; \mathbf{x}_0, t_0) = \int_{-\infty}^{+\infty} d^3x' \; P(\mathbf{x}, t; \mathbf{x}', t') \, P(\mathbf{x}', t'; \mathbf{x}_0, t_0) . \quad (7.115)$$

Replacing t by $t + \delta t$, we can substitute our expression (7.114) into the Chapman–Kolmogorov equation, and then integrate by parts so that the derivatives on the delta function turn and hit $P(\mathbf{x}', t'; \mathbf{x}_0, t_0)$. The delta function, now unattended by derivatives, kills the integral, leaving

$$P(\mathbf{x}, t + \delta t; \mathbf{x}_0, t_0) = P(\mathbf{x}, t; \mathbf{x}_0, t_0) - \frac{\partial}{\partial x_i}\Big(\langle\, \delta x_i\,\rangle P(\mathbf{x}, t; \mathbf{x}_0, t_0)\Big)$$

$$+ \frac{1}{2}\langle\, \delta x_i\, \delta x_j\,\rangle \frac{\partial^2}{\partial x_i \partial x_j} P(\mathbf{x}, t; \mathbf{x}_0, t_0) + \ldots . \quad (7.116)$$

Using our expressions for $\langle\, \delta x\,\rangle$ and $\langle\, \delta x\, \delta x\,\rangle$ given in (7.110) and (7.112), this becomes

$$P(\mathbf{x}, t + \delta t; \mathbf{x}_0, t_0) = P(\mathbf{x}, t; \mathbf{x}_0, t_0) + \frac{1}{\gamma} \frac{\partial}{\partial x_i} \left(\frac{\partial V}{\partial x_i} P(\mathbf{x}, t; \mathbf{x}_0, t_0) \right) \delta t$$

$$+ D \frac{\partial^2}{\partial x^2} P(\mathbf{x}, t; \mathbf{x}_0, t_0) \, \delta t + \dots . \qquad (7.117)$$

But we can also get a much simpler expression for the left-hand side simply by Taylor expanding with respect to time

$$P(\mathbf{x}, t + \delta t; \mathbf{x}_0, t_0) = P(\mathbf{x}, t; \mathbf{x}_0, t_0) + \frac{\partial}{\partial t} P(\mathbf{x}, t; \mathbf{x}_0, t_0) \, \delta t + \dots . \quad (7.118)$$

Equating (7.118) with (7.117) gives us our final result

$$\frac{\partial P}{\partial t} = \frac{1}{\gamma} \nabla \cdot (P \nabla V) + D \nabla^2 P . \qquad (7.119)$$

This is the *Fokker–Planck* equation. This form also goes by the name of the *Smoluchowski equation* or, for probabilists, *Kolmogorov's forward equation*.

Properties of the Fokker–Planck Equation

It is useful to write the Fokker–Planck equation as a continuity equation

$$\frac{\partial P}{\partial t} = -\nabla \cdot \mathbf{J} \qquad (7.120)$$

where the *probability current* is

$$\mathbf{J} = -\frac{1}{\gamma} P \nabla V + D \nabla P . \qquad (7.121)$$

The second term is clearly due to diffusion because there's a big capital D in front of it. The first term is due to the potential and is referred to as the *drift*, meaning the overall motion of the particle due to background forces that we understand.

One advantage of writing the Fokker–Planck equation in terms of a current is that we see immediately that probability is conserved, meaning that if $\int d^3x \, P = 1$ at some point in time then it will remain so for all later times. This follows by a standard argument,

$$\frac{\partial}{\partial t} \int d^3x \, P = \int d^3x \, \frac{\partial P}{\partial t} = -\int d^3x \, \nabla \cdot \mathbf{J} = 0 \qquad (7.122)$$

where the last equality follows because we have a total derivative and we are implicitly assuming that there's no chance that the particle escapes to infinity so we can drop the boundary term.

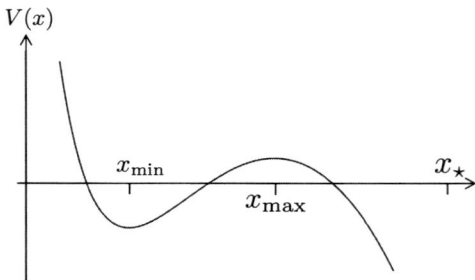

Fig. 7.3 Escape over a barrier.

The Fokker–Planck equation (7.119) tells us how the probability distribution of a system evolves. For some systems, such as those described by the diffusion equation, there is no end point to this evolution: the probability just spreads out more and more. However, for generic potentials V there are time-independent solutions to the Fokker–Planck equation obeying $\nabla \cdot \mathbf{J} = 0$. These are the equilibrium configurations. The solution is given by

$$P(\mathbf{x}) \sim e^{-V(\mathbf{x})/\gamma D} \ . \tag{7.123}$$

Using the Einstein relation (7.85), this becomes something very familiar. It is simply the Boltzmann distribution for a particle with energy $V(\mathbf{x})$ in thermal equilibrium

$$P(\mathbf{x}) \sim e^{-V(\mathbf{x})/k_B T} \ . \tag{7.124}$$

Isn't that nice! (Note that there's no kinetic energy in the exponent as we set $m = 0$ as our starting point.)

An Application: Escape Over a Barrier

Consider a bunch of particles sitting in the kind of 1d potential shown in Figure 7.3. We will assume that all the particles start off sitting close to the local minimum at $x_{\min}$. We model the potential close to this point as a harmonic oscillator

$$V(x) \approx \frac{1}{2}\omega_{\min}^2 (x - x_{\min})^2 \tag{7.125}$$

and we start our particles in a distribution that is effectively in local equilbrium (7.124), with

$$P(x, t = 0) = \sqrt{\frac{\omega_{\min}^2}{2\pi k_B T}} \, e^{-\omega_{\min}^2 (x - x_{\min})^2 / 2k_B T} \ . \tag{7.126}$$

But, globally, $x_{\min}$ is not the lowest energy configuration and this probability distribution is not the true equilibrium configuration. In fact, as drawn, the potential has no global minimum and there is no equilibrium distribution. So this isn't what we'll set out to find. Instead, we would like to calculate the rate at which particles leak out of the trap and over the barrier.

Although we're clearly interested in a time-dependent process, the way we proceed is to assume that the leakage is small and can be effectively treated as a steady-state process. This means that we think of the original probability distribution of particles (7.126) as a bath which, at least on the time scales of interest, is unchanging. The steady-state leakage is modelled by a constant probability current $J = J_0$, with J given by (7.121). Using the Einstein relation $D = k_B T/\gamma$, we can rewrite this as

$$J = -\frac{k_B T}{\gamma} e^{-V(x)/k_B T} \frac{\partial}{\partial x} \left(e^{+V(x)/k_B T} P \right) . \qquad (7.127)$$

The first step is to integrate $J_0 \, e^{+V(x)/k_B T}$ between the minimum $x_{\min}$ and some distance far from all the action, $x \gg x_{\max}$, which we call $x = x_\star$

$$\int_{x_{\min}}^{x_\star} dx \, J_0 \, e^{V(x)/k_B T} = -\frac{k_B T}{\gamma} \left[e^{V(x_\star)/k_B T} P(x_\star) - e^{V(x_{\min})/k_B T} P(x_{\min}) \right] .$$

We can take the probability $P(x_\star)$ to be vanishingly small compared to $P(x_{\min})$ given in (7.126), leaving us with

$$\int_{x_{\min}}^{x_\star} dx \, J_0 \, e^{V(x)/k_B T} \approx \frac{k_B T}{\gamma} \sqrt{\frac{\omega_{\min}^2}{2\pi k_B T}} . \qquad (7.128)$$

Meanwhile, the integral on the left-hand side is dominated by the maximum of the potential. Let's suppose that, close to the maximum, the potential looks like

$$V(x) \approx V_{\max} - \frac{1}{2}\omega_{\max}^2 (x - x_{\max})^2 . \qquad (7.129)$$

Then we'll write the integral as

$$J_0 \int_{x_{\min}}^{x_\star} dx \, e^{V(x)/k_B T} \approx J_0 \, e^{V_{\max}/k_B T} \sqrt{\frac{2\pi k_B T}{\omega_{\max}^2}} . \qquad (7.130)$$

Combining the two expressions (7.128) and (7.130), we get

$$J_0 \approx \frac{\omega_{\min}\omega_{\max}}{2\pi\gamma} e^{-V_{\max}/k_B T} . \qquad (7.131)$$

This is our final expression for the rate at which the particles escape over the barrier. We see that it depends on the height of the barrier, with the expected Boltzmann suppression factor $e^{-V_{\max}/k_B T}$. The exponential prefactor depends on both the curvatures of the potential at the minimum and maximum points, and the mobility.

This calculation is very similar to the tunnelling through a barrier that we met in the book on Quantum Mechanics. (See the section titled "WKB" in the chapter on "Approximation Methods".) The random microscopic bombardment in the present context acts, at least in this example, like uncertainty in the quantum world.

7.3.3 Velocity Diffusion

So far we've ignored the inertia term, setting $m = 0$ in the Langevin equation (7.108). Now we rectify this. We will, however, make things initially easier for ourselves by setting the potential to zero, so that the Langevin equation is

$$m\ddot{\mathbf{x}} = -\gamma\dot{\mathbf{x}} + \mathbf{f}(t) \ . \qquad (7.132)$$

We can trivially rewrite this as a first order equation involving $\mathbf{v} = \dot{\mathbf{x}}$

$$m\dot{\mathbf{v}} = -\gamma\mathbf{v} + \mathbf{f}(t) \ . \qquad (7.133)$$

This means that if we're only interested in the distribution over velocities, $P(\mathbf{v}, t)$, then the problem reduces to the one that we've just solved, simply by replacing $\mathbf{x} \to \mathbf{v}$ and $\gamma \to m$. Actually, you need to be a little more careful. The diffusion constant D that appears in (7.119) was really $D\gamma^2/\gamma^2$ where the numerator arose from the noise correlator and the denominator from the $\gamma\dot{\mathbf{x}}$ term in the Langevin equation. Only the latter changes, meaning that this combination gets replaced by $D\gamma^2/m^2$. The result is a Fokker–Planck equation of the form (7.119), but for a probability distribution over velocities rather than positions,

$$\frac{\partial P}{\partial t} = \frac{1}{m}\frac{\partial}{\partial \mathbf{v}} \cdot \left(\gamma P\mathbf{v} + \frac{D\gamma^2}{m}\frac{\partial P}{\partial \mathbf{v}}\right) \ . \qquad (7.134)$$

The equilibrium distribution that follows from this obeys $\partial P/\partial t = 0$, meaning

$$\frac{\partial P}{\partial \mathbf{v}} = -\frac{m}{D\gamma}P\mathbf{v} \quad \Longrightarrow \quad P = \left(\frac{m}{2\pi k_B T}\right)^{3/2} e^{-m\mathbf{v}^2/2k_B T} \qquad (7.135)$$

where we've again used the Einstein relation $D\gamma = k_B T$. This, of course, is the Maxwell–Boltzmann distribution.

In fact, we can do better than this. Suppose that we start all the particles off at $t = 0$ with some fixed velocity, $\mathbf{v} = \mathbf{v}_0$. This mean that the initial probability distribution is a delta function, $P(\mathbf{v}, t = 0) = \delta^3(\mathbf{v} - \mathbf{v}_0)$. We can write down a full time-dependent solution to the Fokker–Planck equation (7.134) with this initial condition.

$$
P(\mathbf{v}, t) = \left(\frac{m}{2\pi k_B T (1 - e^{-2\gamma t/m})} \right)^{3/2} \exp\left(-\frac{m(\mathbf{v} - \mathbf{v}_0 e^{-\gamma t/m})^2}{2 k_B T (1 - e^{-2\gamma t/m})} \right) . \quad (7.136)
$$

For $t \gg m/\gamma$, we return to the Maxwell–Boltzmann distribution. But now this tells us how we approach equilibrium.

Briefly, the Kramers–Chandrasekhar Fokker–Planck Equation

As our final example of a Fokker–Planck equation, we can consider the Langevin equation with both acceleration term and potential term,

$$
m\ddot{\mathbf{x}} = -\gamma \dot{\mathbf{x}} - \nabla V + \mathbf{f}(t) . \quad (7.137)
$$

Now we are looking for a probability distribution over phase space, $P(\mathbf{x}, \dot{\mathbf{x}}, t)$. The right way to proceed is to write this as two first order equations. The first of these is simply the definition of velocity $\mathbf{v} = \dot{x}$. The second is the Langevin equation

$$
m\dot{\mathbf{v}} = -\gamma \mathbf{v} - \nabla V + \mathbf{f}(t) . \quad (7.138)
$$

These can now be combined into a single Langevin equation for six variables. Once armed with this, we need only follow the method that we saw above to arrive at a Fokker–Planck equation for $P(\mathbf{x}, \mathbf{v}, t)$,

$$
\left(\frac{\partial}{\partial t} + v_i \frac{\partial}{\partial x_i} \right) P = \frac{1}{m} \frac{\partial}{\partial v_i} \left(\gamma v_i P + P \frac{\partial V}{\partial x_i} \right) + \frac{D\gamma^2}{m^2} \frac{\partial^2 P}{\partial v_i \partial v_i} . \quad (7.139)
$$

This form of the Fokker–Planck equation is sometimes called the Kramers equation and sometimes called the Chandrasekhar equation.

Note that this equation is now capturing the same physics that we saw in the Boltzmann equation: the probability distribution $P(\mathbf{x}, \mathbf{v}, t)$ is the same object that we called $f_1(\mathbf{x}, \mathbf{p}; t)$ in Chapter 6. Moreover, it is possible to derive this form of the Fokker–Planck equation starting from the Boltzmann equation describing a heavy particle in a surrounding bath of light particles. The key approximation is that, in small time intervals δt, the momentum of the heavy particle only changes by a small amount. Looking back, you can see that this was indeed an assumption in the derivation of the Fokker–Planck equation in Section 7.3.2, but not in the derivation of the Boltzmann equation.

7.3.4 Fokker, Planck, Schrödinger, and Feynman

Part of the purpose of writing a series of textbooks is to highlight the connections between different areas of physics that may, otherwise, seem unrelated. For this reason, rather unusually, we will finish this book on fluid mechanics by emphasising how the tools we're using are close to those that appear in quantum mechanics.

In quantum mechanics, probabilities are captured by the wavefunction $\psi(\mathbf{x})$, which obeys the Schrödinger equation

$$i\hbar\frac{\partial \psi}{\partial t} = -\frac{\hbar^2}{2m}\nabla^2\psi + U(\mathbf{x})\psi \ . \tag{7.140}$$

In classical mechanics, with diffusion, the evolution of probabilities is governed by the Fokker–Planck equation. As we now show, this takes a very similar form to the Schrödinger equation.

We'll return to the first order Langevin equation

$$\dot{\mathbf{x}} = \frac{1}{\gamma}\left(-\nabla V + \mathbf{f}\right) \tag{7.141}$$

and the corresponding Fokker–Planck equation (7.119). We can change variables to

$$P(x,t) = e^{-V(x)/2\gamma D}\tilde{P}(x,t) \ . \tag{7.142}$$

Substituting this into the Fokker–Planck equation (7.119), we see that the rescaled probability $\tilde{P}$ obeys

$$\frac{\partial \tilde{P}}{\partial t} = D\nabla^2\tilde{P} + \left(\frac{1}{2\gamma}\nabla^2 V - \frac{1}{4\gamma^2 D}(\nabla V)^2\right)\tilde{P} \ . \tag{7.143}$$

There are no first order gradients $\nabla\tilde{P}$, only $\nabla^2\tilde{P}$. Written in this way, the Fokker–Planck equation bears more than a passing resemblance to the Schrödinger equation (7.140). Some of the constants have different names, and different minus signs, and the potential $U(\mathbf{x})$ in the Schrödinger equation is a complicated function of the potential $V(\mathbf{x})$ in the Fokker–Planck equation

$$U = -\frac{1}{2\gamma}\nabla^2 V + \frac{1}{4D\gamma^2}(\nabla V)^2 \ . \tag{7.144}$$

But these are really just cosmetic differences. The only genuine difference between the Schrödinger equation (7.140) and the Fokker–Planck equation (7.143) is the factor of i on the left-hand side of the Schrödinger equation.

There is a formulation of quantum mechanics, due to Feynman, in which the quantum probabilities (strictly amplitudes) can be viewed as a sum over all paths. If the particle starts life at $\mathbf{x} = \mathbf{x}_A$ at time t_A, then the quantum amplitude for it to end up at $\mathbf{x} = \mathbf{x}_B$ at time t_B is

$$\langle \mathbf{x}_B, t_B | \mathbf{x}_A, t_A \rangle = \mathcal{N} \int \mathcal{D}x(t) \, \exp\left(\frac{i}{\hbar} \int dt \, \left(\frac{\dot{\mathbf{x}}^2}{2m} - U(\mathbf{x}) \right) \right) \quad (7.145)$$

where $\mathcal{N}$ is a normalisation factor. Here the integral is over all paths which start at $(\mathbf{x}_A, t_A)$ and end at $(\mathbf{x}_B, t_B)$. A chapter in Volume 3 on Quantum Mechanics is devoted to this result.

Given the similarity between the Schrödinger equation (7.140) and the Fokker–Planck equation (7.143), it is natural to wonder if there is also a path integral description of the evolution of the classical probability distribution. The purpose of this section is to show that indeed there is and that, moreover, it is rather natural. As an aside: of the first four volumes in this series of books, three of them have a section on path integrals. What can I say? All roads lead to Rome. . . .

A Path Integral for Fokker–Planck

The essence of the path integral can already be seen in the Chapman–Kolmogorov equation (7.115)

$$P(\mathbf{x}, t; \mathbf{x}_0, t_0) = \int_{-\infty}^{+\infty} d^3\mathbf{x}' \, P(\mathbf{x}, t; \mathbf{x}', t') \, P(\mathbf{x}', t'; \mathbf{x}_0, t_0) \, . \quad (7.146)$$

This simply says that to get from point A to point B, a particle has to pass through some position in between. And we sum up the probabilities for each position. Adding many more intervening time steps, as shown in Figure 7.4, naturally suggests that we should be summing over all possible paths.

Here we will sketch how these intermediate times are added, resulting in a path integral formulation for the Fokker–Planck equation. First note that this isn't the first time we've stumbled upon path integrals (also known as functional integrals) in this book: we also met them in Section 7.2.4 where we introduced the probability distribution for a given noise function $\mathbf{f}(t)$

$$\text{Prob}[f(t)] = \mathcal{N} \exp\left(-\int dt \, \frac{\mathbf{f}(t) \cdot \mathbf{f}(t)}{4D\gamma^2} \right) \quad (7.147)$$

subject to the normalisation condition

$$\int \mathcal{D}f(t) \, \text{Prob}[f(t)] = 1 \, . \quad (7.148)$$

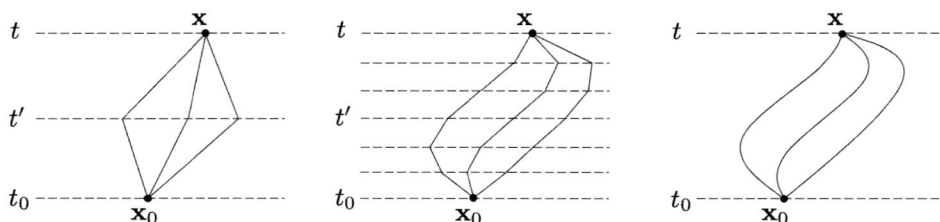

Fig. 7.4 From Chapman–Kolmogorov to Feynman.

Given a fixed noise profile $\mathbf{f}(t)$ and an initial condition, the path of the particle is fully determined by the Langevin equation (7.141). Let's call this solution $\mathbf{x}_f(t)$. Then the probability that the particle takes the path $\mathbf{x}_f$ is the same as the probability that the force is $\mathbf{f}$

$$\mathrm{Prob}[\mathbf{x}_f(t)] = \mathrm{Prob}[\mathbf{f}(t)]$$

$$= \mathcal{N} \exp\left(-\int dt \ \frac{\mathbf{f}(t)\cdot\mathbf{f}(t)}{4D\gamma^2}\right)$$

$$= \mathcal{N} \exp\left(-\frac{1}{4D\gamma^2}\int dt \ (\gamma\dot{\mathbf{x}}_f + \nabla V(\mathbf{x}_f))^2\right) \qquad (7.149)$$

where, in the last line, we've used the Langevin equation (7.141) to relate the force to the path taken. Because this equation holds for any path $\mathbf{x}_f$, we can simply drop the subscript $\mathbf{f}$. We have the probability that the particle takes a specific path $\mathbf{x}(t)$ given by

$$\mathrm{Prob}[\mathbf{x}(t)] = \mathcal{N} \exp\left(-\frac{1}{4D\gamma^2}\int dt \ (\gamma\dot{\mathbf{x}} + \nabla V)^2\right) \ . \qquad (7.150)$$

The total probability to go from $\mathbf{x}_A$ to $\mathbf{x}_B$ should therefore just be the sum over all these paths. With one, slightly fiddly, subtlety: the probability is normalised in (7.148) with respect to the integration measure over noise variable $\mathbf{f}$. And we want to integrate over paths. This means that we have to change integration variables and we pick up a Jacobian factor for our troubles. We have

$$\mathrm{Prob}[\mathbf{x}_B, t_B; \mathbf{x}_A, t_A] = \mathcal{N} \int \mathcal{D}f(t) \ \exp\left(-\frac{1}{4D\gamma^2}\int dt \ (\gamma\dot{\mathbf{x}}_f + \nabla V(\mathbf{x}_f))^2\right)$$

$$= \mathcal{N} \int \mathcal{D}x(t) \ \det \mathcal{M} \ \exp\left(-\frac{1}{4D\gamma^2}\int dt \ (\gamma\dot{\mathbf{x}} + \nabla V)^2\right). \qquad (7.151)$$

Here the operator $\mathcal{M}(t, t')$ that appears in the Jacobian can be thought of as $\delta f(t)/\delta x(t')$. It can be computed by returning to the Langevin equation

(7.141) which relates $\mathbf{f}$ and $\mathbf{x}$

$$\mathcal{M}(t,t') = \gamma \frac{\partial}{\partial t} \delta(t-t') + \nabla^2 V \, \delta(t-t') \,. \tag{7.152}$$

If we want to think in a simple minded way, we can consider this as a (very large) matrix $\mathcal{M}_{tt'}$, with columns labelled by the index t and rows labelled by t'. We'll write the two terms in this matrix as $\mathcal{M} = A+B$ so the determinant becomes

$$\det(A+B) = \det A \, \det(1+A^{-1}B) \,. \tag{7.153}$$

The first operator $A = \gamma \frac{\partial}{\partial t}\delta(t-t')$ doesn't depend on the path and its determinant just gives a constant factor which can be absorbed into the normalisation $\mathcal{N}$. The operator A^{-1} in the second term is defined as

$$\int dt' \, A(t,t')A^{-1}(t,'t'') = \delta(t-t'') \tag{7.154}$$

where the integral over $\int dt'$ is simply summing over the rows of A and the columns of A^{-1} as in usual matrix multiplication. It is simple to check that the inverse is simply the step function

$$A^{-1}(t',t'') = \frac{1}{\gamma}\theta(t'-t'') \,. \tag{7.155}$$

Now we write the second factor in (7.153) and expand

$$\det(1+A^{-1}B) = \exp\left(\operatorname{Tr}\log\left(1+A^{-1}B\right)\right)$$
$$= \exp\left(\sum_n \operatorname{Tr}(A^{-1}B)^n/n\right) \,. \tag{7.156}$$

Here we should look in more detail at what this compact notation means. The term $\operatorname{Tr} A^{-1}B$ is really short-hand for

$$\operatorname{Tr} A^{-1}B = \int dt\, dt' \, A^{-1}(t,t')B(t',t) \tag{7.157}$$

where the integral over $\int dt'$ is multiplying the "matrices" together while the integral over $\int dt$ comes from taking the trace. Using (7.155) we have

$$\operatorname{Tr} A^{-1}B = \frac{1}{\gamma}\int dt\, dt' \, \theta(t-t')\nabla^2 V \, \delta(t-t')$$
$$= \frac{\theta(0)}{\gamma}\int dt \, \nabla^2 V \,. \tag{7.158}$$

The appearance of $\theta(0)$ may look a little odd. This function is defined to be $\theta(x) = +1$ for $x > 0$ and $\theta(x) = 0$ for $x < 0$. The only really sensible value at the origin is $\theta(0) = 1/2$. Indeed, this follows from the standard regularisations of the step function, for example

$$\theta(x) = \lim_{\epsilon \to 0} \left(\frac{1}{2} + \frac{1}{\pi} \tan^{-1} \left(\frac{x}{\epsilon} \right) \right) \implies \theta(0) = \frac{1}{2} . \qquad (7.159)$$

(We met this same issue in Section 5.3 when deriving Kolmogorov's 4/5 law for turbulence.) What happens to the higher powers of $(A^{-1}B)^n$? Writing them out, we have

$$\mathrm{Tr}(A^{-1}B)^n = \int dt \int dt_1 \ldots dt_{2n-1} \; \theta(t - t_1)\delta(t_1 - t_2)\theta(t_2 - t_3)\delta(t_3 - t_4)$$

$$\cdots \theta(t_{2n-2} - t_{2n-1})\delta(t_{2n-1} - t) \frac{(\nabla^2 V)^n}{\gamma^n}$$

where we have been a little sloppy in writing $(\nabla^2 V)^n$ because each of these is actually computed at a different time. We can use the delta functions to do half of the integrals, say all the t_n for n odd. We get

$$\mathrm{Tr}(A^{-1}B)^n = \int dt \int dt_2 \, dt_4 \ldots \; \theta(t - t_2)\theta(t_2 - t_4)\theta(t_4 - t_6)$$

$$\cdots \theta(t_{2n-2} - t) \frac{(\nabla^2 V)^n}{\gamma^n}$$

But this integral is non-vanishing only if $t > t_2 > t_4 > t_6 > \ldots > t_n > t$. In other words, the integral vanishes. Note that you might think we could again get contributions from $\theta(0) = 1/2$, but the integrals now mean that the integrand has support on a set of zero measure. And with no more delta functions to rescue us, this gives zero. The upshot of this is that the determinant (7.156) can be expressed as a single exponential

$$\det(1 + A^{-1}B) = \exp \left(\frac{1}{2\gamma} \int dt \; \nabla^2 V \right) . \qquad (7.160)$$

We now have an expression for the measure factor in (7.151). Using this, the path integral for the probability becomes

$$\mathrm{Prob}[\mathbf{x}_B, t_B; \mathbf{x}_A, t_B]$$

$$= \mathcal{N}' \int \mathcal{D}x(t) \; \exp \left(-\frac{1}{4D\gamma^2} \int dt \; (\gamma \dot{\mathbf{x}} + \nabla V)^2 + \frac{1}{2\gamma} \int dt \; \nabla^2 V \right) .$$

Expanding this out, the terms linear in $\dot{\mathbf{x}}$ are total derivatives. Meanwhile, the potential terms have happily conspired to give the combination $U(\mathbf{x})$ defined in (7.144). The upshot is that the probability for a particle to move from $\mathbf{x}_A$ at time t_A to $\mathbf{x}_B$ at time t_B is

$$\text{Prob}[\mathbf{x}_B, t_B; \mathbf{x}_A, t_A]$$
$$= \mathcal{N}' e^{[V(x_B) - V(x_A)]/2\gamma D} \int \mathcal{D}x(t) \, \exp\left(-\int dt \, \left(\frac{\dot{\mathbf{x}}^2}{4D} + U\right)\right) \quad (7.161)$$

where U is given in (7.144). Notice that the prefactor $e^{[V(x_f) - V(x_i)]/2\gamma D}$ takes the same form as the map from probabilities P to the rescaled $\tilde{P}$ in (7.142). This completes our derivation of the path integral formulation of probabilities.

It is worth comparing our result (7.161) to the Feynman path integral (7.145). Apart from the overall rescaling, there are two key differences. First, the action $T - U$ appears in the measure of the quantum path integral, while the energy $T + U$ appears in the classical path integral. Second, the quantum path integral has a factor of i in the exponent, which means that we are summing over a bunch of phases. Meanwhile, the classical path integral is better behaved (more convergent) because it comes with a minus sign. Both of these differences persist in more sophisticated applications of the path integral and will reappear in the contexts of statistical mechanics and quantum field theory.

As a forward looking aside: one can formally change the Feynman path integral (7.145) into the classical path integral (7.161) by making time complex: $t \to -it$. This is known as *imaginary time*. It is a trick that was briefly mentioned in Volume 3 on Quantum Mechanics and one that we will employ to great effect in later books.

7.3.5 Briefly, Stochastic Calculus

There is one final generalisation of the Langevin equation that we will mention but will not pursue in detail. We'll return to the case $m = 0$, but generalise the noise term in the Langevin equation so that it is now spatially dependent. We write

$$\gamma \dot{\mathbf{x}} = -\nabla V + b(\mathbf{x}) \, \mathbf{f}(t) \, . \quad (7.162)$$

This is usually called the non-linear Langevin equation. The addition of the $b(\mathbf{x})$ multiplying the noise looks like a fairly innocuous change. But it's not. In fact, annoyingly, this equation is not even well defined!

The problem is that the system gets a random kick at time t, the strength of which depends on its position at time t. But if the system is getting a delta function impulse at time t, then its position is not well defined. Mathematically, this problem arises when we look at the position after some small time δt. Our equation (7.109) now becomes

$$\delta \mathbf{x} = \dot{\mathbf{x}} \, \delta t = -\frac{1}{\gamma} \nabla V \, \delta t + \frac{1}{\gamma} \int_t^{t+\delta t} dt' \, b(\mathbf{x}(t')) \, \mathbf{f}(t') \qquad (7.163)$$

and our trouble is in making sense of the last term. There are a couple of ways we could move forward:

- Ito: We could insist that the strength of the kick is related to the position of the particle immediately before the kick took place. Mathematically, we replace the integral with

$$\int_t^{t+\delta t} dt' \, b(\mathbf{x}(t')) \, \mathbf{f}(t') \; \longrightarrow \; b(\mathbf{x}(t)) \int_t^{t+\delta t} dt' \, \mathbf{f}(t') \; . \qquad (7.164)$$

  This choice is known as *Ito stochastic calculus*.

- Stratonovich: Alternatively, we might argue that the kick isn't really a delta function. It is really a process that takes place over a small, but finite, time. To model this, the strength of the kick should be determined by the average position over which this process takes place. Mathematically, we replace the integral with

$$\int_t^{t+\delta t} dt' \, b(\mathbf{x}(t')) \, \mathbf{f}(t') \; \longrightarrow \; \frac{1}{2} \left[b(\mathbf{x}(t + \delta t)) + b(\mathbf{x}(t)) \right] \int_t^{t+\delta t} dt' \, \mathbf{f}(t') \; .$$

  This choice is known as *Stratonovich stochastic calculus*.

Usually in physics, issues of this kind don't matter too much. Typically, any way of regulating microscopic infinitesimals leads to the same macroscopic answers. However, this is not the case here, and the Ito and Stratonovich methods give different answers in the continuum. In most applications of physics, including Brownian motion, the Stratonovich calculus is the right way to proceed because, as we argued when we first introduced noise, the delta function arising in the correlation function $\langle f(t)f(t') \rangle$ is just a convenient approximation to something more smooth. However, in other applications, such as financial modelling, Ito calculus is correct.

The subject of stochastic calculus is a long one and not something we're going to get into in these books. For the Stratonovich choice, the Fokker–Planck equation turns out to be

$$\frac{\partial P}{\partial t} = \frac{1}{\gamma} \nabla \cdot \left[P(\nabla V - D\gamma^2 b \nabla b) \right] + D\nabla^2(b^2 P) \; . \qquad (7.165)$$

This is also the form of the Fokker–Planck equation that you get by naively dividing (7.162) by $b(\mathbf{x})$ and then defining a new variable $\dot{\mathbf{y}} = \dot{\mathbf{x}}/b$ which reduces the problem to our previous Langevin equation (7.108). In contrast, if we use Ito stochastic calculus, the $b\nabla b$ term is absent in the resulting Fokker–Planck equation.

Our random diffusive walk has brought us a long way from the Navier–Stokes equation. The Ito vs Stratonovich issue of stochastic calculus will not be something we will need to embrace in later books. In contrast, the path integral will play an increasingly important role in our exploration of physics, starting in the book on Statistical Physics where we look at critical phenomena and the renormalisation group, and culminating in the books on Quantum Field Theory and The Standard Model where the path integral becomes an invaluable tool for understanding many aspects of the physics.

Further Reading

There are many books on fluid mechanics, ranging from the eminently accessible to the dauntingly comprehensive. Here are a collection that I found useful.

Fluid mechanics is a wonderfully visual subject. Of all the areas in physics, it's certainly the prettiest. If you're going to look at one book on fluid mechanics (other than this one!) then I would strongly recommend a book of pretty pictures. This isn't as lightweight as it sounds because, in this case, a picture really does paint twenty equations and helps build intuition for fluid flow. Traditionally, the gallery of choice was Milton Van Dyke's beautiful *An Album of Fluid Motion* [99]. It's out of press and therefore difficult to buy at a reasonable price (at the time of writing, Amazon offer a paperback version for £833.82). An updated version, with animations, can be found in the *Online Multimedia Fluid Mechanics* [40]. You can also find some gorgeous, if more random, pictures on the website

https://fuckyeahfluiddynamics.tumblr.com/

More traditional textbooks on fluid mechanics cater to physicists, applied mathematicians, or engineers. While the content is largely the same, the focus and presentation can be rather different. For physicists, the book by Acheson [1] covers the basics and is wonderfully done. A more modern, detailed rendering of the subject can be found in [37]. Other good books include [30, 52]. On the applied mathematics side, Grae Worster's book [102] gives a clear and elementary introduction, as does Childress' book [17].

Turning to the more exhaustive, and slightly exhausting, classics, Batchelor's famous book [9] is not particularly cuddly but the explanations are lucid. Meanwhile, there is aways a Landau and Lifshitz for every occasion. Their fluid mechanics book [51] has an astonishing amount of physics packed into it, but it's not the easiest read. Notably, both of these classics put thermodynamics front and centre, where it belongs, which is useful in making contact with other areas of physics which can otherwise feel hidden.

Finally, if you liked the beauty and rigour of Arnold's mathematical treatise on classical mechanics, then you're in luck. He has a book on fluid mechanics in much the same vein [4].

Chapter 1: Inviscid Flows

Much of the material in this chapter is covered in all the textbooks listed above. One further, classic resource is Feynman's chapter on "The Flow of Dry Water" in [31], the title taken from von Neumann's scathing remark that studying the equations of fluid dynamics without including viscosity is like studying "dry water" because you're missing an essential ingredient. The subsequent chapter, "The Flow of Wet Water", describes the effects of viscosity.

Some potted historical references: The Euler equations were first written down in [28]. A history of these equations can be found in [22, 23] and a summary of open mathematical questions in [18]. Kelvin introduced his idea of vortex atoms in [92], with the figure in Chapter 1 taken from his slightly later paper [93]. The ABC flow, in which the velocity is proportional to the vorticity, was named, and its chaotic streamlines studied, in [25]. Helicity was introduced by Keith Moffat in 1969 [58] (and somewhat earlier in [59]).

D'Alembert's paradox was first highlighted in [21]. A detailed review of the d'Alembert paradox can be found in [82]. The Rayleigh quote is taken from a review of Lamb's book, *Hydrodynamics*, written for Nature [70].

For a beautiful introduction to the variational principle in many areas of physics, including fluid mechanics, see [103]. The action principle for ideal fluids was revisited recently in [27]. For a review extending the action principle to include the effects of dissipation, see [56].

Chapter 2: The Navier–Stokes Equation

Again, this chapter contains mostly standard material that can be found in all fluid mechanics textbooks.

More potted history: Navier's pioneering work on fluid mechanics was published in [62] and the picture of the bridge shown in this chapter was first published in Navier's book [63]. More on Navier and his collapsing bridge can be found in [13]. Stokes' claim to the Navier–Stokes equation comes from his paper [83]. Poiseuille was a physicist and physiologist who first studied his eponymous flows in the context of blood. (He joins Julius von Mayer, one of the discovers of the first law of thermodynamics, in making a fundamental

Fig. 7.5   Sketches from Reynolds' original paper on the onset of turbulence.

discovery in physics motivated by blood.) Couette, by contrast, admitted to no such vampiric interests.

Osborne Reynolds introduced his dimensionless number in a classic 1883 paper with the catchy title *An Experimental Investigation of the Circumstances Which Determine Whether the Motion of Water Shall Be Direct or Sinuous, and of the Law of Resistance in Parallel Channels* [72]. Two pictures from this paper, showing the development of turbulence, are shown in Figure 7.5.

The equation for Stokes flow was introduced in [84], where Stokes' law for the drag on the sphere – possibly the most famous result in all of fluid mechanics – was also derived. More on Stokes flow can be found in the book [39]. Moffatt eddies were first predicted by Keith Moffatt in 1964 [57], and observed by Taneda in 1979 [89]. The story of swimming at low Reynolds number is described further in a famous and charming paper by Ed Purcell [69].

Prandtl's boundary layer was introduced in 1905 in the pioneering paper [68]. The numerical solution for a flat plate was found, two years later, by his student, Blasius [12]. Some historical context about Prandtl can be found in [3] and about Blasius in [38]. For more details on the boundary layer, see [20, 77]. Many of the photographs of flows exhibiting separation were taken by Taneda in a series of papers [86, 87, 88].

Chapter 3: Waves

"Waves" is a very comprehensive word.

William Thomson, in 1887 [94]

The material in this chapter is more eclectic and there are many books that scratch the surface waves to go deeper. For a very accessible book, covering all wavy topics in more detail, see Billingham and King [11]. For a classic, see Lighthill's monograph [54]. In addition, you might like to take a look at Thorne and Blandford's mammoth book *Modern Classical Physics* [95]. There's lots of good stuff about fluids and waves in there but, at 1200 pages, it gets tiring just holding it. Kip Thorne won the Nobel prize for gravitational waves (which are not the same thing as gravity waves!) and has a penchant for writing enormous books. His previous book, *Gravitation*, also needs a sturdy bookcase.

In this book, I cover internal gravity waves in just 3 pages. For another 350 pages or so on the same topic, see the book by Sutherland [85].

A comprehensive account of geophysical flows can be found in the weighty and completely excellent tome by Vallis [98]. For some historical colour, the wonderful 1922 book *Weather Prediction by Numerical Process* [34] describes Lewis Fry Richardson's complete failure to predict the weather six hours in advance. He was missing two of the most important and powerful ideas of the twentieth century. The first was computers, although he has a charming description of what he imagines a computing factory might look like. The second was low-energy effective field theory or, in the language of fluid mechanics, the need to filter out high frequency modes to focus on things that matter. The former problem was famously overcome by Mauchly and Eckert's ENIAC. The second by Charney's development of the quasi-geostrophic equations [16].

The connection between chiral waves and topological insulators deserves a special mention only because it is the most recent topic covered in this book. The connection was first pointed out in [24].

The story of sound waves is where thermodynamics and hydrodynamics meet. Landau and Lifshitz do an excellent job of explaining the physics [51], as do some of the general books mentioned above. The classic treatise for non-linear waves is Whitham [101], although Billingham and King [11] is something of an easier read. The Reverend Challis, whose quote from [14] kicks off Section 3.6, is better known for failing to discover Neptune, despite staring directly at it. One, important topic in non-linear waves that I do not discuss is the subject of solitons. This too can be found in [11]. We will remedy the omission when we get to the book on Quantum Field Theory.

One of the famous Millennium Prize problems is the existence and uniqueness of solutions to the Navier–Stokes equation. For a general mathematical discussion of fluid equations and their blow-up, you could do worse than look at Terence Tao's blog articles on the subject [90].

More details on traffic jams can be found in the original paper by Lighthill and Whitham [55]. For more on waves in biological systems, see the two volume series by Murray [60] or the shorter book by Goldsmith [35]. The original paper by Fisher, *The Wave of Advance of Advantageous Genes*, is published in the queasily named Annals of Eugenics [32].

Chapter 4: Instabilities

The go-to resource for all your instability needs is Drazin and Reid's comprehensive textbook [26].

The simulation of the Kelvin-Helmholtz instability is from [80], and the simulation of the Rayleigh-Taylor instability is from [19].

In addition, here is a (slightly haphazard) collection of key papers on the subject. Taylor's famous paper on the centrifugal instability, from which the picture in Section 4.2 is taken, is [91]. A biography of Taylor was written by his student, Batchelor [10].

The quote from Orr is from [64]. The rigorous justification of the Boussinesq approximation is due to Spiegel and Veronis and was given in [81]. Orzag's tour-de-force numerical work, showing that Poiseuille flow is linearly unstable, is [65]. The proof that Couette flow is linearly stable was given in [73]. The importance of non-normal operators for the instability of hydrodynamic flow was pointed out in [97].

Chapter 5: Turbulence

For a physicist, the best introduction to turbulence is Frisch's beautiful book [33], with the focus firmly on symmetries and scaling. For a more phenomenological perspective, see [67]. A review on numerical simulations can be found in [42]. The relevant chapter in Landau and Lifshitz [51] is one of the more readable parts of the book.

There are some good lecture notes online, including those of Eyink [29] and Schekochihin [75].

Kolmogorov's triumvirate of 1941 papers are [47], [48] and [49]. They are not the easiest read. They were advertised more widely by Batchelor, both in his paper [7] and his later book [8]. For a discussion of the constant that sits in front of the famous $E(k) \sim k^{-5/3}$ law, see [79]. For recent results on the scaling exponents in velocity correlations, see [43]. The connection between turbulence and chaos was made in [74].

In the turbulence community, much is made of a famous conference held in Marseille in 1961. Kolmogorov and Obukhov were there, as were von Kármán and Taylor, a rare meeting of the greats of the subject from East and West. The tidal channel data from [36] (reproduced in Figure 5.2) was shown at this meeting, the first time that unequivocal proof for Kolmogorov's $E(k) \sim k^{-5/3}$ scaling was found. Somewhat ironically, this was also where Kolmogorov started to distance himself from his claims, worried about the effects of intermittency on the scaling exponents [50].

In this book, I have focussed on just homogeneous, fully developed turbulence. There is much (much!) more to be said. For a review on the development of turbulence, see [46]. More recently, connections to phase transitions and critical phenomena were pointed out in [78]. A recent book on quantum turbulence is [6].

Chapter 6: Kinetic Theory

You won't find much kinetic theory in traditional books on fluid mechanics. It's a subject that is more likely to be found in books on statistical mechanics, and those by Kardar [45] and Huang [41] are particularly good in this regard. Both kick off with the Boltzmann equation. The classical book by Reif [71] ends with a wide ranging discussion of kinetic theory, transport, and (relevant for Chapter 7) stochastic processes. You can also find some excellent lecture notes online: I particularly like those of Alex Schekochihin [76].

More specialised books include the classic by Chapman and Cowling [15] and Lifshitz and Pitaevskii [53]. This latter book is volume 10 of the famous series by Landau and Lifshitz, albeit written several decades after Landau's death.

The data for the viscosity of oxygen was extracted from the National Institute of Standards and Technology website [61]. More details on determining the properties of liquids from first principles can be found in the book [96].

Chapter 7: Diffusion

The standard reference for the material in this chapter is Van Kampen's book [100]. However, it is not uncommon to find a discussion of the Langevin and Fokker-Planck equations in books on statistical mechanics or condensed matter physics. An example is the excellent all-round condensed matter textbook by Altland and Simons [2]. Finally, Daniel Arovas at UC San Diego has fantastically detailed lecture notes on many subjects, and those on statistical physics [5] include a discussion of the Langevin and Fokker-Planck equations.

The discussion of turkey cooking is taken from [44]. I'm grateful to Ray Goldstein for pointing me towards this paper.

References

[1] D. J. Acheson, *Elementary Fluid Dynamics*, Oxford University Press (1990)

[2] A. Altland and B. D. Simons, *Condensed Matter Field Theory*, Cambridge University Press (2010)

[3] J. D. Anderson, *Ludwig Prandtl's Boundary Layer*, Phys. Today 58 (12), 42 (2005)

[4] V. I. Arnold and B. A. Khesin, *Topological Methods in Hydrodynamics*, Springer (1998)

[5] D. Arovas, *Lecture Notes on Thermodynamics and Statistical Mechanics* (2019), available at: https://courses.physics.ucsd.edu/2010/Spring/physics 210a/LECTURES /210_COURSE.pdf

[6] C. F. Barenghi, L. Skrbek, and K. R. Sreenivasan, *Quantum Turbulence*, Cambridge University Press (2023)

[7] G. K. Batchelor, *Kolmogoroff's Theory of Locally Isotropic Turbulence*, Proc. Camb. Phil. Soc. 43, 533 (1947)

[8] G. K. Batchelor, *Theory of Homogeneous Turbulence*, Cambridge University Press (1958)

[9] G. K. Batchelor, *An Introduction to Fluid Mechanics*, Cambridge University Press (1967)

[10] G. K. Batchelor, *The Life and Legacy of G.I. Taylor*, Cambridge University Press (1996)

[11] J. Billingham and A. C. King, *Wave Motion*, Cambridge University Press (2012)

[12] H. Blasius, *Grenzschichten in Flüssigkeiten mit kleiner Reibung*, Z. Math. Phys. 56, 1 (1908)

[13] M. Cannone and S. Friedlander, *Navier: Blow-Up and Collapse*, Notices Amer. Math. Soc. 50 (1), 7 (2003)

[14] J. Challis, *On the Velocity of Sound, in Reply to the Remarks of the Astronomer Royal*, Lond., Edinb., Dublin Philos. Mag. J. Sci. 32 (218), 494 (1848)

[15] S. Chapman and T. G. Cowling, *The Mathematical Theory of Non-Uniform Gases: An Account of the Kinetic Theory of Viscosity, Thermal Conduction and Diffusion in Gases*, 3rd ed., Cambridge University Press (1991) [first published 1939]

[16] J. G. Charney, *On the Scale of Atmospheric Motions*, Geofys. Publ. Oslo 17, 1 (1948).

[17] S. Childress, *An Introduction to Theoretical Fluid Mechanics*, American Mathematical Society (2009)

[18] D. Christodoulou, *The Euler Equations of Compressible Fluid Flow*, Bulletin Amer. Math. Soc. 44 (4), 581 (2007)

[19] A. Cook, W. Cabot, and P. Miller, *The Mixing Transition in Rayleigh–Taylor Instability*, J. Fluid Mech. 511, 333 (2004)

[20] S. J. Cowley, *Laminar Boundary-Layer Theory: A 20th Century Paradox?*, in *Mechanics for a New Millennium*, ed. H. Aref and J. W. Phillips, Springer (2000)

[21] J. le Rond d'Alembert, *Memoir XXXIV*, Opuscules Mathématiques, vol. 5, p. 138 (1768)

[22] O. Darrigola, *Worlds of Flows: A History of Hydrodynamics from the Bernoullis to Prandtl*, Oxford University Press (2008)

[23] O. Darrigola and U. Frisch, *From Newton's Mechanics to Euler's Equations*, Physica D: Nonlinear Phenom. 237 (14), 1855 (2008)

[24] P. Delplace, J. B. Marston, and A. Venaille, *Topological Origin of Equatorial Waves*, Science 358, 1075 (2017) [arXiv:1702.07583 [cond-mat]]

[25] T. Dombre, U. Frisch, J. M. Greene, M. Hénon, A. Mehr, and A.M. Soward, *Chaotic Streamlines in the ABC Flows*, J. Fluid Mech. 167, 353 (1986)

[26] P. G. Drazin and W. H. Reid, *Hydrodynamic Stability*, Cambridge University Press (1981)

[27] S. Dubovsky, T. Gregoire, A. Nicolis, and R. Rattazzi, *Null Energy Condition and Superluminal Propagation*, J. High Energy Phys. 3, 25 (2005) [arXiv:hep-th/0512260 [hep-th]]

[28] L. Euler, *Principes Generaux du Mouvement des Fluides*, Mém. Acad. Sci. Berlin 11, 274 (1757)

[29] G. L. Eyink, *Turbulence Theory*, course notes, available at: www.ams.jhu.edu/~eyink/Turbulence/notes.html

[30] G. Falkovich, *Fluid Mechanics: A Short Course for Physicists*, Cambridge University Press (2011)

[31] R. P. Feynman, R. B. Leighton, and M. Sands, *The Feynman Lectures on Physics*, vol. II, Pearson (1971)

[32] R. A. Fisher, *The Wave of Advance of Advantageous Genes*, Annals of Eugenics, 7 355 (1937)

[33] U. Frisch, *Turbulence: The Legacy of A.N. Kolmogorov*, Cambridge University Press (2010)

[34] L. Fry Richardson, *Weather Prediction by Numerical Process*, Cambridge University Press (1922)

[35] M. Goldsmith, *Waves: A Short Introduction*, Oxford University Press (2018)

[36] H. L. Grant, R. W. Stewart, and A. Moilliet *Turbulence Spectra From a Tidal Channel*, J. Fluid Mech. 12 (2), 241–268 (1962)

[37] E. Guyon, J.-P. Hulin, L. Petit, and C. D. Mitescu, *Physical Hydrodynamics*, Oxford University Press (2015)

[38] W. H. Hager, *Blasius: A Life in Research and Education*, Experiments in Fluids 34, 566 (2003)

[39] J. Happel and H. Brenner, *Low Reynolds Number Hydrodynamics*, Springer (1983)

[40] G. Homsy et al., *Multimedia Fluid Mechanics Online*, ed. G. Homsy, Cambridge University Press (2008)

[41] K. Huang, *Statistical Mechanics*, Wiley (2008)

[42] T. Ishihara, T. Gotoh, and Y. Kaneda, *Study of High-Reynolds Number Isotropic Turbulence by Direct Numerical Simulation*, Annu. Rev. Fluid Mech. 41, 165 (2009)

[43] K. P. Iyer, K. R. Sreenivasan, and P. K. Yeung, *Scaling Exponents Saturate in Three-Dimensional Isotropic Turbulence*, Phys. Rev. Fluids 5, 054605 (2020) [arXiv:2002.11900 [physics.flu-dyn]]

[44] Y. Jin, L. R. Wang, and J. J. Wang, *Physics in Turkey Cooking: Revisit the Panofsky formula*, AIP Advances 11, 115316 (2021)

[45] M. Kardar *Statistical Physics of Particles*, Cambridge University Press (2007)

[46] R. R. Kerswell, *Recent Progress in Understanding the Transition to Turbulence in a Pipe*, Nonlinearity 18 (6), R17 (2005)

[47] A. N. Kolmogorov, *The Local Structure of Turbulence in Incompressible Viscous Fluid for Very Large Reynolds Number*, Dokl. Akad. Nauk SSSR 30, 299 (1941); reprinted in Proc. Roy. Soc. Lond. A 434, 9 (1991)

[48] A. N. Kolmogorov, *On Degeneration (Decay) of Isotropic Turbulence in an Incompressible Viscous Fluid*, Dokl. Akad. Nauk SSSR 31, 538 (1941)

[49] A. N. Kolmogorov, *Dissipation of Energy in Locally Isotropic Turbulence in an Incompressible Viscous Fluid*, Dokl. Akad. Nauk SSSR 32, 16–18 (1941); reprinted in Proc. Roy. Soc. Lond. A 434, 15–17 (1991)

[50] A. N. Kolmogorov, *A Refinement of Previous Hypothesis Concerning the Local Structure of Turbulence in a Viscous Incompressible Fluid at High Reynolds Number*, J. Fluid Mech. 13, 82–85 (1962)

[51] L. D. Landau and E. M. Lifshitz, *Fluid Mechanics: A Course in Theoretical Physics*, vol. 6, Pergamon (2013)

[52] E. Lauga, *Fluid Mechanics: A Very Short Introduction*, Oxford University Press (2022)

[53] E. M. Lifshitz and L. P. Pitaevskii, *Physical Kinetics: A Course in Theoretical Physics*, vol 10, Butterworth-Heinemann (1981)

[54] J. Lighthill, *Waves in Fluids*, Cambridge University Press (1978)

[55] M. J. Lighthill and G. B. Whitham, *On Kinematic Waves. II. A Theory of Traffic Flow on Long Crowded Roads*, Proc. R. Soc. Lond. A 229, 317 (1955)

[56] H. Liu and P. Glorioso, *Lectures on Non-Equilibrium Effective Field Theories and fluctuating hydrodynamics*, Theoretical Advanced Study Institute, Summer School 2017 (TASI 2017), Proceedings of Science 008, 49pp. (2018) [arXiv:1805.09331 [hep-th]]

[57] H. K. Moffatt, *Viscous and Resistive Eddies Near a Sharp Corner*, J. Fluid Mech. 18 (1), 1 (1964)

[58] H. K. Moffat, *The Degree of Knottedness of Tangled Vortex Lines*, J. Fluid Mech. 35, 117 (1969)

[59] J. J. Moreau, *Constantes d'un Îlot Tourbillonnaire en Fluide Parfait Barotrope*, C. R. Hebdom. Séances Acad. Sci. 252 (19), 2810 (1961)

[60] J. D. Murray, *Mathematical Biology Vol I and II*, Springer (1993)

[61] National Institute for Standards and Technology, *Thermophysical Properties of Fluid Systems*, available at: https://webbook.nist.gov/chemistry/fluid/

[62] C.-L. M. H. Navier, *Mémoire Sur les Lois du Mouvement des Fluides*, Mém. Acad. Sci. Inst. France 6, 389 (1822)

[63] C.-L. M. H. Navier, *Rapport à Monsieur Becquey et Mémoire sur les Ponts Suspendus*, Imprimerie Royale, Paris (1823)

[64] W. McF. Orr, *The Stability or Instability of the Steady Motions of a Perfect Liquid and of a Viscous Liquid. Part II: A Viscous Liquid*, Proc. Roy. Irish Acad. A 27, 69 (1907)

[65] S. Orzag, *Accurate Solution of the Orr–Sommerfeld Stability Equation*, J. Fluid Mech. 50 (4), 689 (1971)

[66] Q. Ouyang and H. L. Swinney, *Transition From a Uniform State to Hexagonal and Striped Turing Patterns*, Nature 352, 610 (1991)

[67] S. B. Pope, *Turbulent Flows*, Cambridge University Press (2000)

[68] L. Prandtl, *Über Flüssigkeitsbewegung bei sehr kleiner Reibung*, in *Verhandlungen des dritten internationalen Mathematiker-Kongresses, Heidelberg, 1904*, pp. 485, B.G. Teubner (1905)

[69] E. Purcell, *Life at Low Reynolds Number*, Amer. J. Phys. 45, 3 (1977)

[70] Rayleigh, Lord, *Review of "Hydrodynamics" by H. Lamb*, Nature 97, 318 (1916)

[71] F. Reif, *Fundamentals of Statistical and Thermal Physics*, Waveland Press (2008)

[72] O. Reynolds, *An Experimental Investigation of the Circumstances Which Determine Whether the Motion of Water Shall Be Direct or Sinuous, and of the Law of Resistance in Parallel Channels*, Phil. Trans. Roy. Soc. 174, 935 (1883)

[73] V. A. Romanov, *Stability of Planar-Parallel Couette Flow*, Funct. Anal. Its Applic. 7, 137 (1973)

[74] D. Ruelle and F. Takens, *On the Nature of Turbulence*, Commun. Math. Phys. 20, 167 (1970)

[75] A. A. Schekochihin, *5 Lectures on Turbulence*, available at: www-thphys.physics.ox.ac.uk/people/AlexanderSchekochihin/notes/SummerSchool07/

[76] A. A. Schekochihin, *Lectures on Kinetic Theory and Magnetohydrodynamics of Plasmas*, available at: www-thphys.physics.ox.ac.uk/people/AlexanderSchekochihin/KT/

[77] H. Schlichting, *Boundary Layer Theory*, McGraw-Hill (1979)

[78] H. Y. Shih, T. L. Hsieh, and N. Goldenfeld, *Ecological Collapse and the Emergence of Travelling Waves at the Onset of Shear Turbulence*, Nature Phys. 12, 245–248 (2016)

[79] K. R. Sreenivasan, *On the Universality of the Kolmogorov Constant*, Phys. Fluids 7, 11 (1995)

[80] W. D. Smyth and J. N. Moum, *Ocean Mixing by Kelvin–Helmholtz Instability*, Oceanography 25 (2), 140 (2012)

[81] E. A. Spiegel and G. Veronis, *On the Boussinesq Approximation for a Compressible Fluid*, Astrophys. J. 131, 442 (1960)

[82] K. Stewartson, *d'Alembert's Paradox*, SIAM Rev. 23 (3), 308 (1981)

[83] G. G. Stokes, *On the Theories of the Internal Friction of Fluids in Motion, and of the Equilibrium and Motion of Elastic Solids*, Trans. Camb. Philos. Soc. 8, 286 (1849) [read April 1845]

[84] G. G. Stokes, *On the Effect of the Internal Friction on the Motion of Pendulums*, Trans. Camb. Philos. Soc. 9, 8 (1851) [read December 1850]

[85] B. Sutherland, *Internal Gravity Waves*, Cambridge University Press (2014)

[86] S. Taneda, *Experimental Investigation of the Wakes Behind Cylinders and Plates at Low Reynolds Number*, J. Phys. Soc. Japan 11, 302 (1956)

[87] S. Taneda, *Experimental Investigation of the Wake Behind a Sphere at Low Reynolds Number*, J. Phys. Soc. Japan 11, 1104 (1956)

[88] S. Taneda, *Experimental Investigation of Vortex Streets*, J. Phys. Soc. Japan 20, 1714 (1965)

[89] S. Taneda, *Visualization of Separating Stokes Flows*, J. Phys. Soc. Japan 46, 1935 (1979)

[90] T. Tao, *Blog posts on the Navier–Stokes Equation*, available at: https://terrytao.wordpress.com/tag/navier-stokes-equations/

[91] G. I. Taylor, *Stability of a Viscous Liquid Contained Between Two Rotating Cylinders*, Phil. Trans. Roy. Soc. Lond. 223, 289 (1923)

[92] W. Thomson, *On Vortex Atoms*, Proc. Roy. Soc. Edinb. VI, 94 (1867)

[93] W. Thomson, *On Vortex Motion*, Trans. Roy. Soc. Edinb. 25, 217 (1869), in *Mathematical and Physical Papers by Sir William Thomson*, ed. J. Larmor, vol. 4, p.13, Cambridge University Press (2012)

[94] W. Thomson, *On Ship Waves*, Proc. Inst. Mech. Eng. 38 (1), 409 (1887)

[95] K. S. Thorne and R. D. Blandford, *Modern Classical Physics: Optics, Fluids, Plasmas, Elasticity, Relativity, and Statistical Physics*, Princeton University Press (2017)

[96] K. Trachenko, *Theory of Liquids, from Excitations to Thermodynamics*, Cambridge University Press (2023)

[97] L. N. Trefethen, A. E. Trefethen, S. C. Reddy, and T. A . Driscoll, *Hydrodynamics Without Eigenvalues*, Science 261 (5121), 578 (1993)

[98] G. K. Vallis, *Atmospheric and Oceanic Fluid Dynamics: Fundamentals and Large-Scale Circulation*, Cambridge University Press (2017)

[99] M. Van Dyke, *An Album of Fluid Motion*, Parabolic Press (1982)

[100] N. G. Van Kampen, *Stochastic Processes in Physics and Chemistry*, 3rd ed., North Holland (2007)

[101] G. B. Whitham, *Linear and Nonlinear Waves*, Wiley (1974)

[102] G. Worster, *Understanding Fluid Flow*, Cambridge University Press (2010)

[103] W. Yourgrau and S. Mandelstam, *Variational Principles in Dynamics and Quantum Theory*, Dover (1968)

Index